Student Solutions Manual for Smith's
Calculus with Applications

TERRY L. SHELL
Santa Rosa Junior College

Brooks/Cole Publishing Company
Pacific Grove, California

Brooks/Cole Publishing Company
A Division of Wadsworth, Inc.

© 1988 by Wadsworth, Inc., Belmont, California 94002.
All rights reserved. No part of this book may be reproduced,
stored in a retrieval system, or transcribed, in any form or
by any means—electronic, mechanical, photocopying, recording,
or otherwise—without the prior written permission of the
publisher, Brooks/Cole Publishing Company, Pacific Grove,
California 93950, a division of Wadsworth, Inc.

Printed in the United States of America

10 9 8 7 6 5 4 3 2 1

QA303.S63   1988   87-19402
515—dc19   CIP

ISBN 0-534-08899-6

Sponsoring Editor: Jeremy Hayhurst
Cover Design: Katherine Minerva
Cover Photo: Lee Hocker

CONTENTS

Chapter 1   Functions and Graphs   1

Chapter 2   The Derivative   35

Chapter 3   Additional Derivative Topics   63

Chapter 4   Applications and Differentiation   86

Chapter 5   Exponential Functions, Logarithmic Functions   114

Chapter 6   Integration   138

Chapter 7   Additional Integration Topics   161

Chapter 8   Applications and Integration   183

Chapter 9   Functions of Several Variables   200

Chapter 10   Differential Equations   243

Appendix A   Review of Algebra   259

CHAPTER 1
FUNCTIONS AND GRAPHS

1.1 WHAT IS CALCULUS?, PAGES 9 - 11

1. (Answers vary.) A mathematical model is a body of mathematics based on certain assumptions and data about the real world which is used to predict future occurrences.
3. (Answers vary.) A function of a variable is a rule that assigns to each value of x exactly one value for f(x).   **5. a.** y is a function of x, since for each year, x, there will be one and only one closing price, y.   **b.** y is not a function of x, since for each closing price, x, we may not be able to uniquely determine the year, y.   **7. a.** g(1954) = $.29 **b.** g(1974) = $.53
9. a. f(8) = 2(8) - 5 = 11   **b.** f(-3) = 2(-3) - 5 = -11   **11. a.** f(2) = 5(2) - 3 = 7
b. f(10) = 5(10) - 3 = 47   **c.** f(-25) = 5(-25) - 3 = -78   **d.** f(100) - 5(100) - 3 = 497
13. a. h(0) = 6 - 4(0) = 6   **b.** h(8) = 6 - 4(8) = -26   **c.** h(-7) = 6 - 4(-7) = 34
d. h(100) = 6 - 4(100) = -394   **15. a.** m(0) = 0^2 - 3(0) + 1 = 1
b. m(1) = 1^2 - 3(1) + 1 = -1   **c.** m(2) = 2^2 - 3(2) + 1 = -1   **d.** m(3) = 3^2 - 3(3) + 1 = 1
17. a. Since any number can be substituted for x, the domain is all real numbers.   **b.** Since any number can be substituted for x, the domain is all real numbers.   **19. a.** Since the denominator can not be zero, x + 5 ≠ 0. The domain is all real numbers except 5.   **b.** Since the denominator can not be zero, 2x + 1 ≠ 0 => x ≠ - 1/2. Since the square root of a negative number is not a real number, 2x - 1 ≥ 0 => x ≥ 1/2. The domain is all real numbers x ≥ 1/2.
21. a. f(t) = 5t - 2   **b.** f(w) = 5w - 2   **23. a.** f(t + h) = 5(t + h) - 2
b. f(s + t) = 5(s + t) - 2   **25.** f(t + h + 8) = 5(t + h + 8) - 2 = 5t + 5h + 38
27. g(3 + h) = 2$(3 + h)^2$ - 4(3 + h) - 5 = 2(9 + 6h + h^2) - 4(3 + h) - 5
= 18 + 12h + 2h^2 - 12 - 4h - 5 = 2h^2 + 8h + 1   **29.** f(2x^2) = 5(2x^2) - 2 = 10x^2 - 2
31. g(2x^2 - 4x) = 2$(2x^2 - 4x)^2$ - 4(2x^2 - 4x) - 5 = 2(4x^4 - 16x^3 + 16x^2) - 4(2x^2 - 4x) - 5
= 8x^4 - 32x^3 + 32x^2 - 8x^2 + 16x - 5 = 8x^4 - 32x^3 + 24x^2 + 16x - 5
33. g(5x) = 2$(5x)^2$ - 4(5x) - 5 = 50x^2 - 20x - 5
35. g[f(x)] = g[5x - 2] = 2$(5x - 2)^2$ - 4(5x - 2) - 5 = 2(25x^2 - 20x + 4) - 4(5x - 2) 5
= 50x^2 - 40x + 8 - 20x + 8 - 5 = 50x^2 - 60x + 11
37. $\dfrac{f(x+h) - f(x)}{h} = \dfrac{2(x+h) - 2x}{h} = \dfrac{2x + 2h - 2x}{h} = \dfrac{2h}{h} = 2$

Chapter 1

39. $\dfrac{f(x+h) - f(x)}{h} = \dfrac{2(x+h)^2 - 2x^2}{h} = \dfrac{2(x^2 + 2xh + h^2) - 2x^2}{h} = \dfrac{4xh + 2h^2}{h} = 4x + 2h$

41. $\dfrac{f(x+h) - f(x)}{h} = \dfrac{[2(x+h)^2 - 3] - [2x^2 - 3]}{h} = \dfrac{2(x^2 + 2xh + h^2) - 3 - 2x^2 + 3}{h}$

$= \dfrac{4xh + 2h^2}{h} = 4x + 2h$

43. $\dfrac{f(x+h) - f(x)}{h} = \dfrac{[(x+h)^2 - 2(x+h) + 1] - [x^2 - 2x + 1]}{h}$

$= \dfrac{[x^2 + 2xh + h^2 - 2x - 2h + 1] - [x^2 - 2x + 1]}{h} = \dfrac{2xh + h^2 - 2h}{h} = 2x + h - 2$

45. **a.** Locating 1954 in the left column, the price for round steak was r(1954) = \$.92.

b. Locating 1954 in the left column, the price for 1/2 gallon of milk was m(1954)
= \$.45.   47. s(1984) - s(1944) = 1.48 - .34 = \$1.14.   49. **a.** 1.15 - .64 = \$.51
b. e(1984) - e(1944)

51. **a.** $\dfrac{g(1944 + 40) - g(1944)}{40} = \dfrac{g(1984) - g(1944)}{40} = \dfrac{1.52 - .21}{40} = .03275 \approx \$.03$

b. The change in price (over the last 40 years) is g(1984) - g(1944). The average change in price per year is: $\dfrac{g(1984) - g(1944)}{40} = \$.03$

53. **a.** $\dfrac{s(1954) - s(1944)}{1954 - 1944} = \dfrac{\$.52 - \$.34}{10} = .018 \approx \$.02$

b. $\dfrac{s(1964) - s(1944)}{1964 - 1944} = \dfrac{\$.59 - \$.34}{20} = .0125 \approx \$.01$

c. $\dfrac{s(1974) - s(1944)}{1974 - 1944} = \dfrac{\$1.08 - \$.34}{30} = .02\overline{46} \approx \$.02$

d. $\dfrac{s(1984) - s(1944)}{1984 - 1944} = \dfrac{\$1.48 - \$.34}{40} = .0285 \approx \$.03$ **e.** $\dfrac{s(1944 + h) - s(1944)}{h}$

55. $A(2) = .106(2) - .0015(2)^3 = .224$. About .2%

57. $P(20) = 230(1.02)^{20} = 230(1.4859) \approx 342$. The population will be about 342,000,000.

59. **a.** $\dfrac{M(1982) - M(1977)}{5} = \dfrac{2,495,000 - 2,176,000}{5} = 63,800$ **b.** The average change in number of marriages from 1977 to year 1977 + h.

1.2 FUNCTIONS AND GRAPHS, PAGES 14 - 16

1. A picture of the ordered pairs for the function.   3. The x-intercepts are the points where the graph crosses (or touches) the x-axis. To find the x-intercept, set y = 0. The y-intercept is

Chapter 1

the point where the graph crosses the y-axis. To find the y-intercept, set x = 0. **5.** (2, f(2))
7. $(x_0, f(x_0))$ **9.** $(x_0, g(x_0))$ **11.** (3, h(3)) **13.** $(x_0 + t, h(x_0 + t))$ **15.** Since all points on the vertical line have an x-coordinate of x = -3, the domain is x = -3. Since the y–coordinates of all points on the line range through all values from -∞ to ∞, the range is all real numbers. The x-intercept is (-3, 0). There is no y-intercept. Since there is a vertical line that contains more than one point on the graph, this relation is not a function. **17.** For all points on the graph, the x-coordinates vary from -2 to 5, so the domain is $-2 \le x \le 5$. For all points on the graph, the y-coordinates vary from -5 to 3, so the range is $-5 \le y \le 3$. The x–intercepts are $(-\frac{3}{2}, 0)$ and $(\frac{15}{4}, 0)$. The y-intercept is (0, 3). Since there is no vertical line that contains more than one point on the graph, this relation is a function. **19.** For all points on the graph, the x-coordinates and the y-coordinates vary from -∞ to ∞ . This means the domain and the range is the set of all real numbers . The x-intercepts are (-3, 0), (-1, 0) and (2, 0). The y-intercept is $(0, \frac{5}{2})$. Since there is no vertical line that contains more than one point on the graph, the relation is a function.

21. A few points are:
x = -3, f(-3) = 2(-3) + 3 = 3
x = -1, f(-1) = 2(-1) + 3 = 1
x = 0, f(0) = 2(0) + 3 = 3
x = 2, f(2) = 2(2) + 3 = 7
The graph is shown on the right.

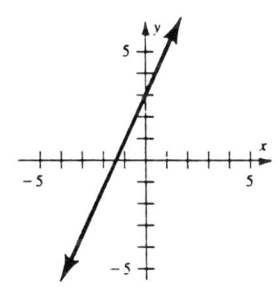

23. A few points are:
x = -2, f(-2) = 6 - 2(-2) = 10
x = 0, f(0) = 6 - 2(0) = 6
x = 1, f(1) = 6 - 2(-) = 4
The graph is shown on the right.

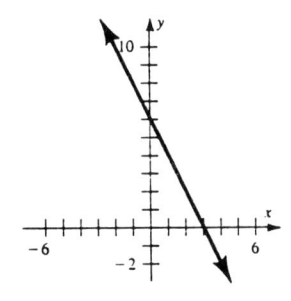

25. A few points are:
x = -1, f(-1) = -3(-1) - 1 = 2
x = 0, f(0) = -3(0) - 1 = -1
x = 2, f(2) = -3(2) - 1 = -7
The graph is shown on the right.

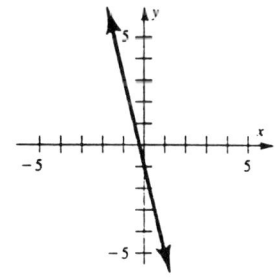

Chapter 1

27. A few points are:
$x = -4, g(-4) = (1/2)(-4)^2 = 8$
$x = -2, g(-2) = (1/2)(-2)^2 = 2$
$x = -1, g(-1) = (1/2)(-1)^2 = 1/2$
$x = 0, g(0) = (1/2)(0^2) = 0$
$x = 2, g(2) = (1/2)(2^2) = 2$
$x = 4, g(4) = (1/2)(4^2) = 8$
The graph is shown on the right.

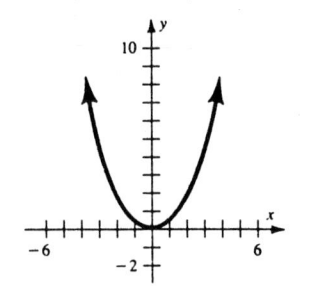

29. A few points are:
$x = -5, g(-5) = (-5)^2 + 4(-5) + 4 = 9$
$x = -3, g(-3) = (-3)^2 + 4(-3) + 4 = 1$
$x = -2, g(-2) = (-2)^2 + 4(-2) + 4 = 0$
$x = -1, g(-1) = (-1)^2 + 4(-1) + 4 = 1$
$x = 0, g(0) = (0)^2 + 4(0) + 4 = 4$
The graph is shown on the right.

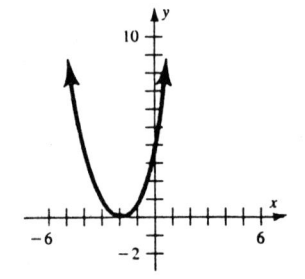

31. A few points are:
$x = 0, g(0) = 0^2 - 6(0) + 9 = 9$
$x = 2, g(2) = 2^2 - 6(2) + 9 = 1$
$x = 3, g(3) = 3^2 - 6(3) + 9 = 0$
$x = 4, g(4) = 4^2 - 6(4) + 9 = 1$
$x = 5, g(5) = 5^2 - 6(5) + 9 = 4$
The graph is shown on the right.

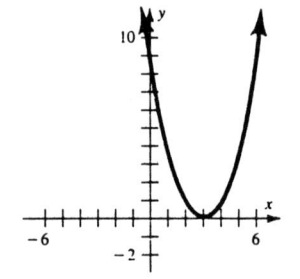

33. A few points are:
$x = -2, g(-2) = 2(-2)^2 - 4(-2) + 5 = 21$
$x = 0, g(0) = 2(0)^2 - 4(0) + 5 = 5$
$x = 1, g(1) = 2(1)^2 - 4(1) + 5 = 3$
$x = 2, g(2) = 2(2)^2 - 4(2) + 5 = 5$
$x = 3, g(3) = 2(3)^2 - 4(3) + 5 = 11$
The graph is shown on the right.

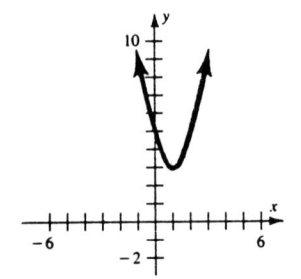

35. A few points are:
$x = -2, h(-2) = \dfrac{-1}{-2} = \dfrac{1}{2}$
$x = -1, h(-1) = \dfrac{-1}{-1} = 1$

Chapter 1

$x = -\frac{1}{2}$, $h(-\frac{1}{2}) = \frac{-1}{-\frac{1}{2}} = 2$

$x = 0$, $h(0)$ is undefined.

$x = \frac{1}{2}$, $h(\frac{1}{2}) = -\frac{1}{\frac{1}{2}} = -2$

$x = 1$, $h(1) = -\frac{1}{1} = -1$

The graph is shown on the right.

37. A few points are:

$x = -2$, $h(-2) = \frac{-2}{-2} = 1$

$x = -1$, $h(-1) = \frac{-2}{-1} = 2$

$x = -\frac{1}{2}$, $h(-\frac{1}{2}) = \frac{-2}{-\frac{1}{2}} = 4$

$x = 0$, $h(0)$ is undefined.

$x = \frac{1}{2}$, $h(\frac{1}{2}) = \frac{-2}{\frac{1}{2}} = -4$

$x = 1$, $h(1) = \frac{-2}{1} = -2$

The graph is shown on the right.

39. A few points are:

$x = -1$, $h(-1) = \frac{2}{-1-2} = -\frac{2}{3}$

$x = 0$, $h(0) = \frac{2}{0-2} = -1$

$x = 1$, $h(1) = \frac{2}{1-2} = -2$

$x = 2$, $h(2)$ is undefined.

$x = 3$, $h(3) = \frac{2}{3-2} = 2$

$x = 4$, $h(4) = \frac{2}{4-2} = 1$

The graph is shown on the right

41. A few points are:

Let $a = 0$, then $0 + c = 1500$ => $c = 1500$

Let $a = 100$, then $100 + c = 1500$ => $c = 1400$

Let $a = 700$, then $700 + c = 1500$ => $c = 800$

Let $a = 1500$, then $1500 + c = 1500$ => $c = 0$

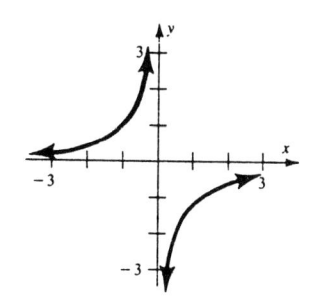

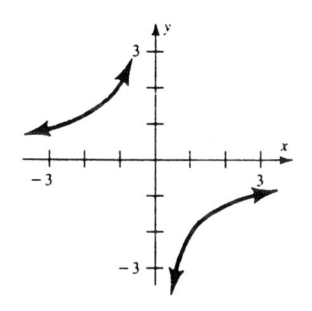

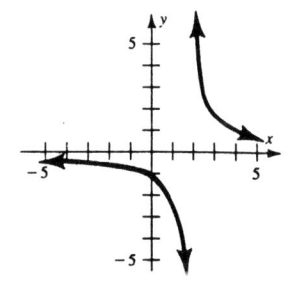

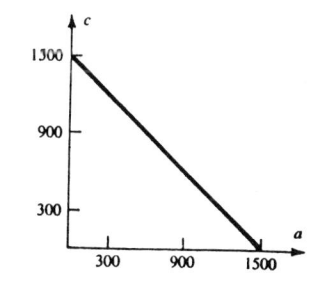

Chapter 1

Since a and c cannot be negative, the graph does not extend out of the first quadrant.

43. A few points are:
d = 0, P = .4(0) + 7.6 = 7.6
d = 5, p = .4(5) + 7.6 = 9.6
d = 10, p = .4(10) + 7.6 = 11.6
The graph is shown on the right.

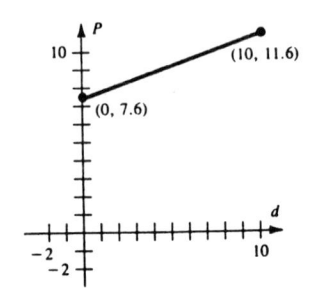

45. A few points are:
t = 0, n = 150 - (0 - 10)^2 = 50
t = 4, n = 150 - (4 - 10)^2 = 114
t = 8, n = 150 - (8 - 10)^2 = 146
t = 10, n = 150 - (10 - 10)^2 = 150
t = 12, n = 150 - (12 - 10)^2 = 146
t = 22, n = 150 - (22 - 10)^2 = 6
The graph is shown on the right.

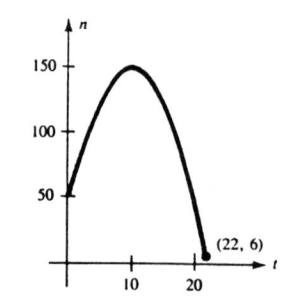

1.3 LINEAR FUNCTIONS, PAGES 24 - 25

1. To find the x-intercept, set y = 0: 0 = 2x + 4 => x = -2. To find the y-intercept, set x = 0: y = 2(0) + 4 => y = 4. The x-intercept is (-2, 0); the y-intercept is (0, 4). **3.** To find the x-intercept, set y = 0: 4x + 3(0) + 4 = 0 => x = -1. To find the y-intercept, set x = 0: 4(0) + 3y + 4 = 0 => y = - 4/3. The x-intercept is (-1, 0); the y-intercept is (0, - 4/3).
5. First divide both sides of the equation by 50; 2x - 5y + 10 = 0. To find the x-intercept, set y = 0: 2x - 5(0) + 10 = 0 => x = -5. To find the y-intercept, set x = 0: 2(0) - 5y + 10 = 0 => y = 2. The x-intercept is (-5, 0); the y-intercept is (0, 2). **7.** Since we can not set y = 0, there is no x-intercept. To find the y-intercept, set x = 0: y + 2 = 0 => y = -2. (If this bothers you, write the equation as 0x + y + 2 = 0. Now set x = 0.) The y-intercept is (0, -2).
9. The slope is $\frac{4-3}{5-2} = \frac{1}{3}$. **11.** The slope is $\frac{-3-2}{-2-5} = \frac{-5}{-7} = \frac{5}{7}$.
13. The slope is $\frac{-2-(-3)}{-1-(-2)} = \frac{1}{1} = 1$.
15. This is in the form y = mx + b, where m is the slope and b is the y-intercept. The slope is m = 2; the y-intercept is b = 4 or (0, 4). **17.** This is in the form y = mx + b, where m is the

Chapter 1   7

slope and b is the y-intercept. The slope is m = 9; the y-intercept is b = 1 or (0, 1).

19. First isolate y to get the form y = mx + b:

   2x - 3y + 5 = 0   Subtract 2x and 5 from both sides.
   -3y = -2x - 5   Divide both sides by -3.
   $y = \frac{2}{3}x + \frac{5}{3}$

   The slope is m = 2/3; the y-intercept is b = 5/3 or (0, 5/3).

21. Write the equation in the form y = mx + b:

   y - 5 = 0
   y = 5
   y = 0x + 5

   The slope is m = 0; the y-intercept is b = 5 or (0, 5).

23. This is a vertical line; hence it has no slope nor a y-intercept.

25. Start with the y-intercept, (0, -5).

   Since the slope is m = 2 = 2/1,
   from the y-intercept move 2 units
   up and 1 unit right. (The coordinates
   of this point are (1, -3).) Draw a
   line through these two points.

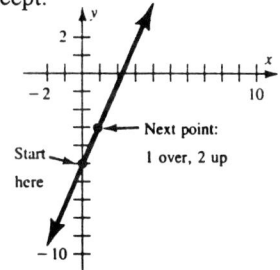

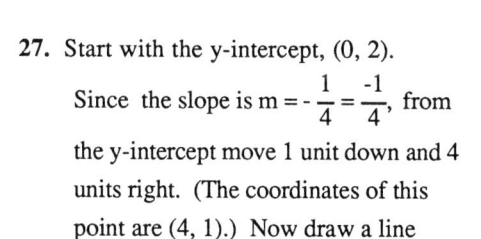

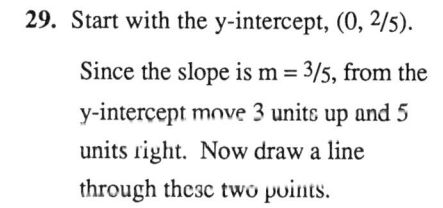

27. Start with the y-intercept, (0, 2).

   Since the slope is m = $-\frac{1}{4} = \frac{-1}{4}$, from
   the y-intercept move 1 unit down and 4
   units right. (The coordinates of this
   point are (4, 1).) Now draw a line
   through these two points.

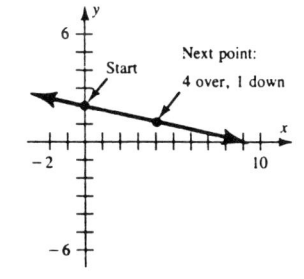

29. Start with the y-intercept, (0, 2/5).

   Since the slope is m = 3/5, from the
   y-intercept move 3 units up and 5
   units right. Now draw a line
   through these two points.

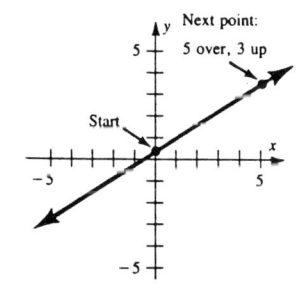

Chapter 1

31. First write x + 3y - 9 = 0 in the form

$y = mx + b \Rightarrow y = -\frac{1}{3}x + 3$. Start

with the y-intercept, (0, 3). Since the

slope is $m = -\frac{1}{3} = \frac{-1}{3}$, from the

y-intercept move 1 unit down and 3 units right. Now draw a line through these two points.

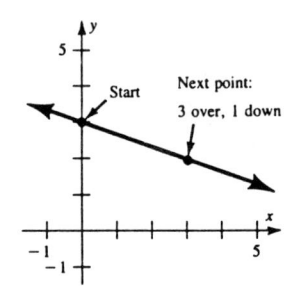

33. First write x = 2/3 y in the form

$y = mx + b \Rightarrow y = \frac{3}{2}x + 0$. Starting

with the y-intercept, (0, 0) move 3 units up and 2 units right. Draw a line through these two points.

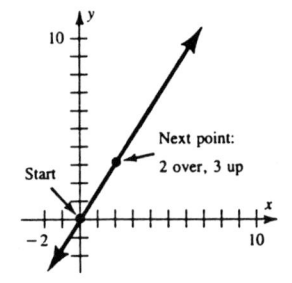

35. This equation has no slope or y-intercept. It is the vertical line x = -5/2.

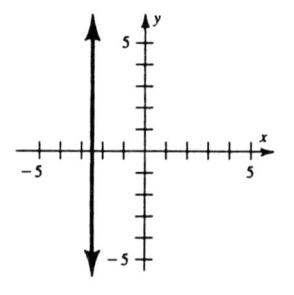

37. When x = -4, y = -3(-4) + 2 = 14. When x = 3, y = -3(3) + 2 - =7. Graph the line segment between these two points.

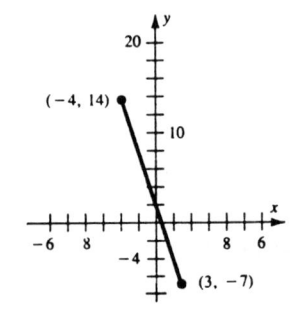

Chapter 1   9

39. Let x = -3, then 3(-3) - 2y + 8 = 0.

Solving for y gives y = -$\frac{1}{2}$. Let x = 4, then 3(4) - 2y + 8 = 0. Solving for y gives y = 10. Graph the line segment between these two points.

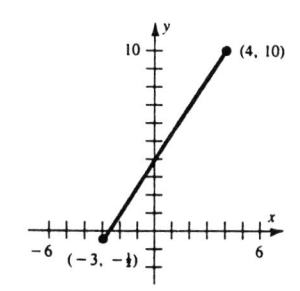

41. Let x = 0, then y = 3(0) + 2 = 2. Let x = 2, then y = 3(2) + 2 = 8. Connect (0, 2) and (2, 8) with a line segment. Graph the horizontal line y = 8 between x = 2 and x = 5 excluding the endpoints. Let x = 5, then 5 - 2y + 11 = 0. Solving for y gives y = 8. Let y = 7, then 7 - 2y + 11 = 0 => y = 9. Graph the half-line starting at (5, 8) through (7, 9).

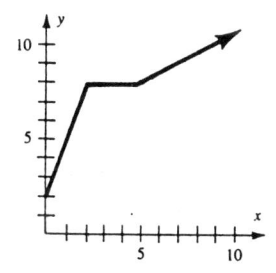

43. This is $\begin{cases} y = 3x - 6 \text{ if } 3x - 6 \geq 0 \Rightarrow y = 3x - 6 \text{ if } x \geq 2. \\ y = -(3x - 6) \text{ if } 3x - 6 < 0 \Rightarrow y = -3x + 6 \text{ if } x < 0. \end{cases}$

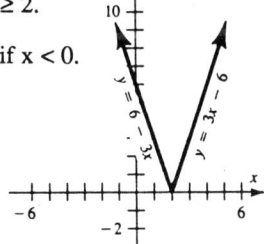

45. This is $\begin{cases} y = -3x \text{ if } x \geq 0 \Rightarrow y = -3x \text{ if } x \geq 0. \\ y = -3(-x) \text{ if } x < 0 \Rightarrow y = 3x \text{ if } x < 0. \end{cases}$

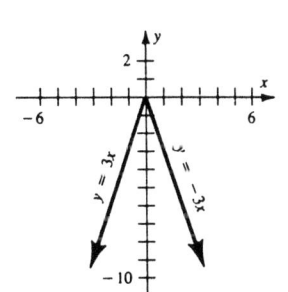

47. Using the form y = mx + b where m is the slope and b is the y-intercept, y = -2x - 3. In

Chapter 1

standard form this is $2x + y + 3 = 0$. **49.** Using the form $y = mx + b$, $y = -x + 4$. In standard form this is $y - 4 = 0$. **51.** Using the point-slope form, $y - y_1 = m(x - x_1)$ we get $y - 5 = -1(x - -3)$ => $y - 5 = -x - 3$. In standard form this is $x + y - 2 = 0$. **53.** Using the point-slope form, $y - y_1 = m(x - x_1)$ we get $y - -3 = \frac{3}{5}(x - 4)$ => $y + 3 = \frac{3}{5}x - \frac{12}{5}$. Multiplying both sides by 5 gives $5y + 15 = 3x - 12$. In standard form this is $3x - 5y - 27 = 0$. **55.** First, the slope is $\frac{-2 - 6}{1 - 5} = \frac{8}{4} = 2$. Using the point-slope form, we have $y - 6 = 2(x - 5)$. In standard form this is $2x - y - 4 = 0$. **57.** First, the slope is $\frac{6 - 6}{7 - 5} = 0$. Using the point-slope form, we have $y - 6 = 0(x - 5)$ => $y - 6 = 0$. **59.** We want the equation of the line through (5, 10) and (10, 20). The slope is $\frac{20 - 10}{10 - 5} = 2$. Using the point-slope form $y - y_1 = m(x - x_1)$, $y - 10 = 2(x - 5)$. In standard form this is $2x - y = 0$.

61. We want the equation of the line through (10, 11.2) and (20, 14.2). The slope is $\frac{14.2 - 11.2}{20 - 10} = \frac{3}{10}$. The equation is $y - 11.2 = \frac{3}{10}(x - 10)$. In standard form this is $3x - 10y + 82 = 0$. To predict the population in 1990 let $x = 30$: $3(30) - 10y + 82 = 0$ => $10y + 172 = 0$ => $y = 17.2$ (million). **63.** We want the equation of the line through (50, 60) and (260, 60). The slope is 0. The equation is $y - 60 = 0(x - 50)$ => $y - 60 = 0$.

65.

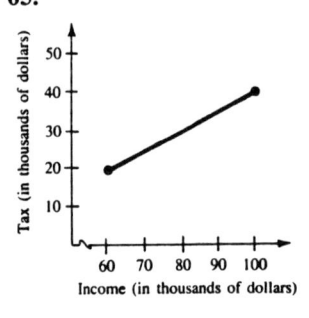

67. The equation for the line through (15000, 2097) and (18200, 2865) is:

slope = $\frac{2865 - 2097}{18200 - 15000} = .24$, so $T - 2097 = .24(I - 15000)$.

69. The domain splits between 25,000 and 30,000 at 28,800. The equation of the line through (23500, 4349) and (28800, 6045) is $T - 4349 = .32(I - 23500)$. The equation of the line through (28800, 6045) and (34100, 7953) is $T - 6045 = .36(I - 28800)$. The equations for income between 25,000 and 30,000 are: $\begin{cases} T = .32(I - 23500) + 4349 \text{ if } I \leq 28{,}800 \\ T = .36(I - 28{,}800) + 6045 \text{ if } 28{,}800 < I \end{cases}$

Chapter 1

1.4 QUADRATIC AND POLYNOMIAL FUNCTIONS, PAGES 30 - 31

1. See the discussion following Example 2.
This has the form $y = ax^2$ where $a = 1$.
The parabola opens up from the vertex
(0, 0) since $a > 0$. Some points are:
Let $x = 1$, then $y = (1)^2 = 1$
Let $x = 2$, then $y = (2)^2 = 4$
Let $x = 3$, then $y = (3)^2 = 9$
Using the symmetry with respect to the
vertical axis, the graph is shown on the right.

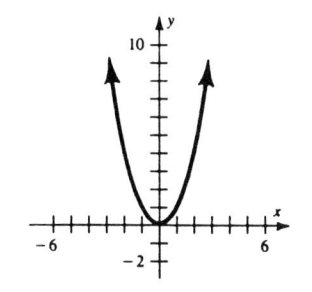

3. See the discussion following Example 2.
This has the form $y = ax^2$ where $a = -2$.
The parabola opens down from the vertex
(0, 0) since $a < 0$. Some points on the graph are:
Let $x = 1$, then $y = -2(1)^2 = -2$
Let $x = 2$, then $y = -2(2)^2 = -8$
Let $x = 3$, then $y = -2(3)^2 = -18$
Using the symmetry with respect to the
vertical axis, the graph is shown on the right.

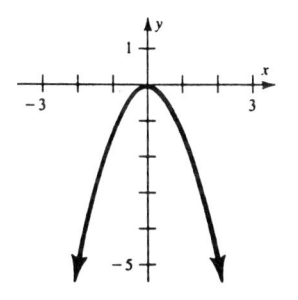

5. See the discussion following Example 2.
This has the form $y = ax^2$ where $a = -5$.
The parabola opens down from the vertex
(0, 0) since $a < 0$.
Some points on the graph are:
Let $x = 1$, then $y = -5(1)^2 = -5$
Let $x = 2$, then $y = -5(2)^2 = -20$
Let $x = \frac{1}{2}$, then $y = -5(\frac{1}{2})^2 = -\frac{5}{4}$
Using the symmetry with respect to the
vertical axis, the graph is shown on the right.

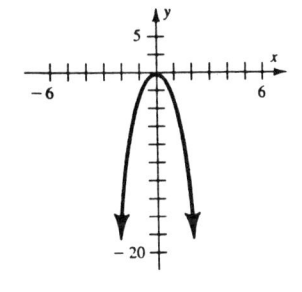

7. See the discussion following Example 2.
This has the form $y = ax^2$ where $a = \frac{1}{3}$.
The parabola opens up from the vertex
(0, 0) since $a > 0$. Some points are:
Let $x = 1$, then $y = (\frac{1}{3})(1)^2 = \frac{1}{3}$
Let $x = 3$, then $y = (\frac{1}{3})(3)^2 = 3$

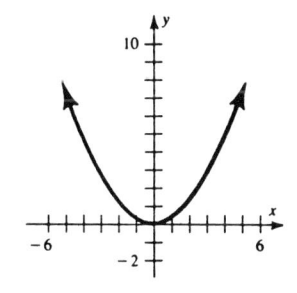

Chapter 1

Let x = 6, then $y = (\frac{1}{3})(6)^2 = 12$

Using the symmetry with respect to the vertical axis, the graph is on the previous page.

9. See the discussion following Example 2.

This has the form $y = ax^2$ where $a = \frac{1}{10}$.

The parabola opens up from the vertex (0, 0) since a > 0. Some points on the graph are:

Let x = 1, then $y = \frac{1}{10}(1)^2 = \frac{1}{10}$

Let x = 3, then $y = \frac{1}{10}(3)^2 = \frac{9}{10}$

Let x = 5, then $y = \frac{1}{10}(5)^2 = \frac{5}{2}$

Using the symmetry with respect to the vertical axis, the graph is shown on the right.

11. See the discussion following Example 2.

This has the form $y = ax^2$ where $a = \frac{2}{3}$.

The parabola opens up from the vertex (0, 0) since a > 0. Some points on the graph are:

Let x = 1, then $y = \frac{2}{3}(1)^2 = \frac{2}{3}$

Let x = 2, then $y = \frac{2}{3}(2)^2 = \frac{8}{3}$

Let x = 3, then $y = \frac{2}{3}(3)^2 = 6$

Using the symmetry with respect to the vertical axis, the graph is shown on the right.

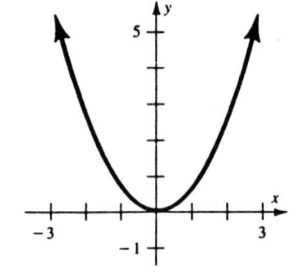

13. See Example 3 and the preceeding discussion. This has the form $y - k = a(x - h)^2$ with (h, k) = (1, 0) and a = 1. The parabola opens up (since a > 0) from the vertex (1, 0).

15. See Example 3 and the preceeding discussion. This has the form $y - k = a(x - h)^2$ with (h, k) = (-3, 0) and a = 1. The parabola opens up (since a > 0) from the vertex (-3, 0).

17. See Example 3 and the preceeding discussion. This has the form $y - k = a(x - h)^2$ with (h, k) = (1, 0) and a = 1/4. The parabola opens up (since a > 0) from the vertex (1, 0).

19. See Example 3 and the preceeding discussion. This has the form $y - k = a(x - h)^2$ with

Chapter 1

(h, k) = (1, 2) and a = 1. The parabola opens up (since a > 0) from the vertex (1, 2).

21. This has the form $y - k = a(x - h)^2$ with (h, k) = (1, 2) and a = -3/5. The parabola opens down (since a < 0) from the vertex (1, 2). **23.** This has the form $y - k = a(x - h)^2$ with (h, k) = $(\frac{3}{4}, -\frac{1}{2})$ and a = 1. The parabola opens up (since a > 0) from the vertex $(\frac{3}{4}, -\frac{1}{2})$.

25. See Problem 13. Some points are:
Let x = 2, then $y = (2 - 1)^2 = 1$
Let x = 0, then $y = (0 - 1)^2 = 1$
Let x = 3, then $y = (3 - 1)^2 = 4$
Let x = -1, then $y = (-1 - 1)^2 = 4$

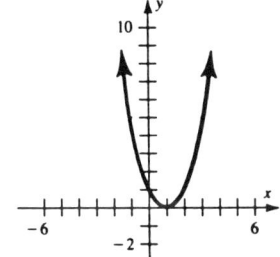

27. See Problem 15. Some points are:
Let x = -2, then $y = (-2 + 3)^2 = 1$
Let x = -4, then $y = (-4 + 3)^2 = 1$
Let x = -1, then $y = (-1 + 3)^2 = 4$
Let x = -5, then $y = (-5 + 3)^2 = 4$
Let x = 0, then $y = (0 + 3)^2 = 9$

29. See Problem 17. Some points are:
Let x = 0, then $y = \frac{1}{4}(0 - 1)^2 = \frac{1}{4}$
Let x = 2, then $y = \frac{1}{4}(2 - 1)^2 = \frac{1}{4}$
Let x = -1, then $y = \frac{1}{4}(-1 - 1)^2 = 1$
Let x = 3, then $y = \frac{1}{4}(3 - 1)^2 = 1$

31. See Problem 19. Some points are:
Let x = 0, then $y - 2 = (0 - 1)^2 = 1 \Rightarrow y = 3$
Let x = 2, then $y - 2 = (2 - 1)^2 = 1 \Rightarrow y = 3$
Let x = 3, then $y - 2 = (3 - 1)^2 = 4 \Rightarrow y = 6$

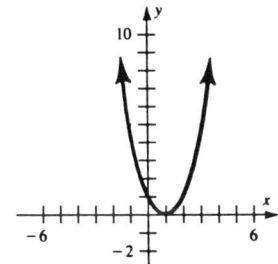

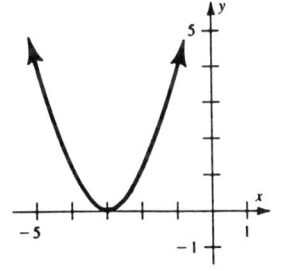

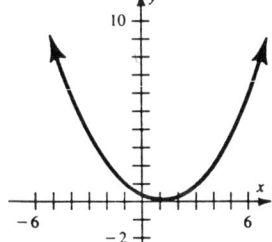

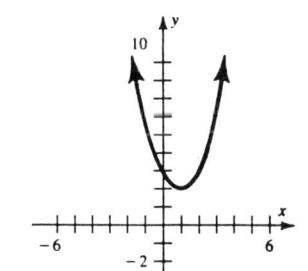

Chapter 1

33. See Problem 21. Some points are:

Let x = 0, then y - 2 = -$\frac{3}{5}$ (0 - 1)^2 = -$\frac{3}{5}$ => y = $\frac{7}{5}$

Let x = 2, then y - 2 = -$\frac{3}{5}$ (2 - 1)^2 = -$\frac{3}{5}$ => y = $\frac{7}{5}$

Let x = 3, then y - 2 = -$\frac{3}{5}$ (3 - 1)^2 = -$\frac{12}{5}$ => y = -$\frac{2}{5}$

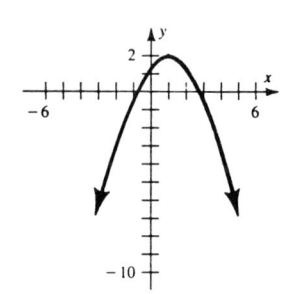

35. This has the form y - k = a(x - h)^2 where (h, k) = (-$\frac{1}{3}$, -$\frac{2}{3}$) and a = 1. The parabola opens up (since a > 0) from the vertex (-$\frac{1}{3}$, -$\frac{2}{3}$). Some points are:

Let x = -1, then y + $\frac{2}{3}$ = (-1 + $\frac{1}{3}$)^2 = $\frac{4}{9}$ => y = -$\frac{2}{9}$

Let x = 1, then y + $\frac{2}{3}$ = (1 + $\frac{1}{3}$)^2 = $\frac{16}{9}$ => y = $\frac{10}{9}$

Let x = 2, then y + $\frac{2}{3}$ = (2 + $\frac{1}{3}$)^2 = $\frac{49}{9}$ => y = $\frac{43}{9}$

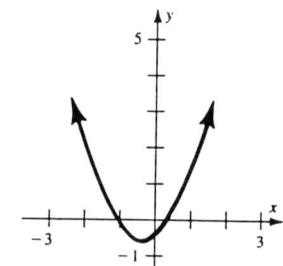

37. Subtract 3 from both sides to get y - 3 = -4(x + 1)^2. This has the form y - k = a(x - h)^2 where a = -4 and (h, k) = (-1, 3). The graph is a parabola opening down (since a < 0) with vertex (-1, 3). Hence, the maximum value of y is 3, occurring when x = -1.   **39.** Subtract 1250 from both sides to get y - 1250 = -10(x - 450)^2. This has the form y - k = a(x - h)^2 where a = -10 and (h, k) = (450, 1250). The graph is a parabola opening down (since a < 0) with vertex (450, 1250). Hence, the maximum value of y is 1250, occurring when x = 450.
41. Add 1400 to both sides to get y + 1400 = 25(x - 560)^2. This has the form y - k = a(x - h)^2 where a = 25 and (h, k) = (560, -1400). The graph is a parabola opening up (since a > 0) with vertex (560, -1400). Hence, the minimum value of y is –1400, occurring when x = 560.   **43.** This has the form y - k = a(x - h)^2 where a = -3 and (h, k) = (-3, -14). The graph is a parabola opening down (since a < 0) with vertex (-3, -14). Hence, the maximum value of y is -14, occurring when x = -3.   **45.** Subtract 2(x + 3)^2 from both sides. Then multiply both sides by -1 to get y - 13 = 2(x + 3)^2. This has the form y - k = a(x - h)^2 where a = 2 and (h, k) = (-3, 13). The graph is a parabola opening up (since a > 0) with vertex (-3, 13). Hence, the minimum value for y is 13, occurring at x = -3.

47. Subtract 6 (x - 1)^2 from both sides, then multiply both sides by 1/3 to get y + 4 = -2 (x - 1)^2. The graph is a parabola opening down, since a = -2, with vertex (1, -4). Hence, the maximum is y = -4, which occurs when x = 1.   **49. a.** The revenue is (demand)(number of items) = (1000 - 10x)x. The revenue function is

R(x) = (1000 - 10x)x = 1000x - 10x². **b.** The break-even points occur when R(x) = C(x):
1000x - 10x² = 50,000 - 500x
$$0 = 10x^2 - 1500x + 50,000$$
$$0 = x^2 - 150x + 5000$$
$$0 = (x - 50)(x - 100)$$
$$x = 50 \text{ or } x = 100$$

The break-even points are (50, R(50)) = (50, 25000) and (100, R(100)) = (100, 0).

51. a. First write the profit function in the form P(x) - k = a(x - h)². The graph of P(x) - 1156250 = -10(x - 375)² is a parabola which opens downward from the vertex (375,1156250). The maximum profit is 1156250 which occurs when 375 boats are produced; that is 375 should be produced. **b.** The loss is $250,000, since when x = 0, P(0) = -10(0 - 375)² + 1156250 = - $250,000. **c.** The maximum profit is $1,156,150. (See Part a.) **53.** Subtract 650 from both sides to get P(x) - 650 = -2(x - 25)². This is a parabola that opens down, since a = -2, with vertex at (25, 650). Hence, the maximum value of P(x) is 650, which occurs when x = 25.

55. The graphs of R(x) = 50x - x² and C(x) = 1200 - 20x are shown on the right. They intersect at (30, 600) and (40, 400).

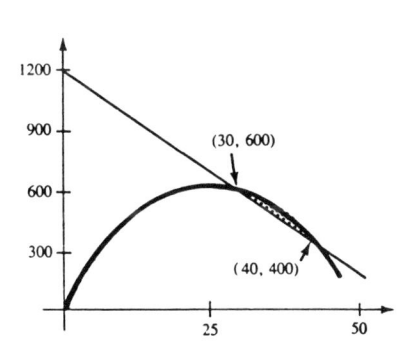

57. The graph of d(x) = -.0005(x - 2390)² + 3456 is a parabola with vertex at (2390, 3456). Thus, his maximum height was 3,456 units.

1.5 RATIONAL FUNCTIONS, PAGES 37 - 38

1. A rational function is a function of the form $\frac{P(x)}{Q(x)}$ where P(x) and Q(x) are polynomials and Q(x) ≠ 0. **3** A horizontal asymptote is a horizontal line that the function approaches as |x| gets larger. A horizontal asymptote will be present if: a) the degrees of the numerator and denominator are equal, in which case the ratio of leading coefficients gives the value of the asymptote or b) The degree of the numerator is less than that of the denominator, in which case the x-axis will be the asymptote. **5.** Set the denominator equal to zero and solve:

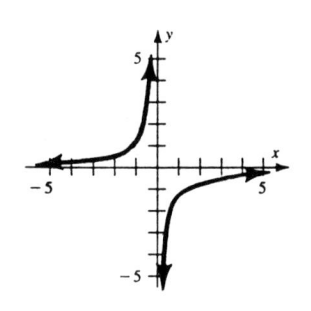

Chapter 1 16

$2x - 3 = 0 \Rightarrow x = 3/2$. There is a vertical asymptote at $x = 3/2$. Since the denominator has degree 1 which is greater than the degree of the numerator (which is 0), the horizontal asymptote is $y = 0$ (the x-axis). **7.** Set the denominator equal to zero: $x^2 + 2x - 3 = 0 \Rightarrow (x - 1)(x + 3) = 0 \Rightarrow x = 1$ or $x = -3$. The vertical asymptotes are $x = 1$ and $x = -3$. Since the degree of the denominator (2) is greater than that of the numerator (0), the horizontal asymptote is $y = 0$ (the x-axis). **9.** Set the denominator equal to zero: $6x^2 + 5x - 6 = 0 \Rightarrow (2x + 3)(3x - 2) = 0 \Rightarrow x = -3/2$ or $x = 2/3$. The vertical asymptotes are $x = -3/2$ and $x = 2/3$. Since the degree of the denominator (2) is greater than the degree of the numerator (0), the horizontal asymptote is $y = 0$ (the x-axis).

11. Set the denominator equal to zero: $2x^2 + 4x + 1 = 0 \Rightarrow x = \frac{-2 \pm \sqrt{2}}{2}$. There is a vertical asymptote at $x = \frac{-2 \pm \sqrt{2}}{2}$. Since the degree of the denominator equals that of the numerator $(2 = 2)$, the horizontal asymptote is at the ratio of leading coefficients: $y = 5/2$. **13.** Set the denominator equal to zero: $x^2 + x + 2 = 0$. Since this quadratic equation has no solution, there is no vertical asymptote. Since the degree of the denominator (2) is greater than the degree of the numerator (1), the horizontal asymptote is $y = 0$ (the x-axis). **15.** Set the denominator equal to zero: $x^2 + x - 2 = 0 \Rightarrow (x - 2)(x - 1) = 0 \Rightarrow x = -2$ or $x = 1$. The vertical asymptotes are $x = -2$ and $x = 1$. Since the degree of the denominator (2) is greater than that of the numerator (1) the horizontal asymptote is $y = 0$ (the x-axis). **17.** Set the denominator equal to zero: $x = 0$ is the vertical asymptote. Since the degree of the denominator (1) is greater than that of the numerator (0), the horizontal asymptote is $y = 0$. A few points are:

Let $x = -2$, then $y = -\left(\frac{1}{2}\right) = \frac{1}{2}$

Let $x = -1$, then $y = -\left(\frac{1}{-1}\right) = 1$

Let $x = -\frac{1}{2}$, then $y = -\begin{pmatrix} 1 \\ -\frac{1}{2} \end{pmatrix} = 2$

Let $x = \frac{1}{2}$, then $y = -\begin{pmatrix} 1 \\ \frac{1}{2} \end{pmatrix} = -2$

Let $x = 1$, then $y = -\left(\frac{1}{1}\right) = -1$

Let $x = 2$, then $y = -\left(\frac{1}{2}\right) = -\frac{1}{2}$

19. Set the denominator equal to zero: $x - 4 = 0 \Rightarrow x = 4$ is the vertical asymptote. Since

Chapter 1

the degree of the denominator (1) is greater than that of the numerator (0), the horizontal asymptote is y = 0. A few points are:

Let x = 0, then $y = \dfrac{1}{0-4} = -\dfrac{1}{4}$

Let x = 2, then $y = \dfrac{1}{2-4} = -\dfrac{1}{2}$

Let x = 3, then $y = \dfrac{1}{3-4} = -1$

Let x = 3.5, then $y = \dfrac{1}{3.5-4} = -2$

Let x = 4.5, then $y = \dfrac{1}{4.5-4} = 2$

Let x = 6, then $y = \dfrac{1}{6-4} = \dfrac{1}{2}$

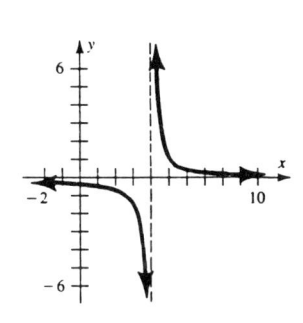

21. Set the denominator equal to zero and solve: 3x - 2 = 0 => x = 2/3, so there is a vertical asymptote at x = 2/3. Since the degree of the numerator is less than the degree of the denominator (0 < 1), the horizontal asymptote is y = 0 (the x=axis). A few points are:

Let x = 1, then $y = \dfrac{1}{3(1)-2} = 1$

Let x = 2, then $y = \dfrac{1}{3(2)-2} = \dfrac{1}{4}$

Let x = -1, then $y = \dfrac{1}{3(-1)-2} = -\dfrac{1}{5}$

Let x = 0, then $y = \dfrac{1}{3(0)-2} = -\dfrac{1}{2}$

You should generate more points of your own.

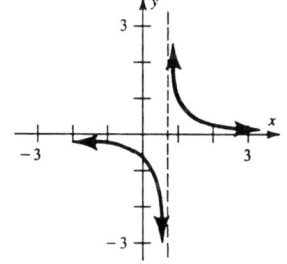

23. Add $\dfrac{1}{x} + 3 = \dfrac{1}{x} + \dfrac{3x}{x} = \dfrac{3x+1}{x}$. Setting the denominator equal to zero, the vertical asymptote is x = 0. Since the degree of the numerator equals that of the denominator, the horizontal asymptote is the ratio of leading coefficients: y = 3/1. A few points are:

Let x = -1, then $y = \dfrac{3(-1)+1}{-1} = 2$

Let x = $-\dfrac{1}{3}$, then $y = \dfrac{3(-\tfrac{1}{3})+1}{-\tfrac{1}{3}} = 0$

Let x = $\dfrac{1}{3}$, then y - $\dfrac{3(\tfrac{1}{3})+1}{\tfrac{1}{3}} = 6$

Let x = 1, then $y = \dfrac{3(1)+1}{1} = 4$

You should generate more ordered pairs.

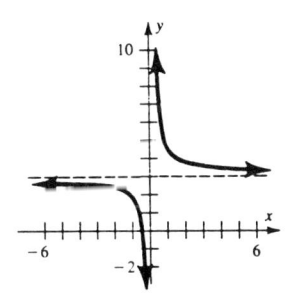

Chapter 1

25. First factor and reduce: $y = \dfrac{2x^2 - 3x + 1}{x^2 + 2x - 3} = \dfrac{(2x - 1)(x - 1)}{(x + 3)(x - 1)} = \dfrac{2x - 1}{x + 3}$. Note that $x \neq 1$.

Now set the denominator equal to zero: $x + 3 = 0 \Rightarrow x = -3$ is the vertical asymptote. Since the degrees of numerator and denominator are equal, the horizontal asymptote is the ratio of leading coefficients: $y = \dfrac{2}{1} = 2$. A few points are:

Let $x = -5$, then $y = \dfrac{2(-5) - 1}{-5 + 3} = \dfrac{11}{2}$

Let $x = -4$, then $y = \dfrac{2(-4) - 1}{-4 + 3} = 9$

Let $x = -2$, then $y = \dfrac{2(-2) - 1}{-2 + 3} = -5$

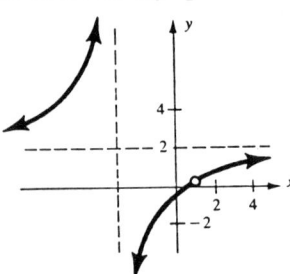

27. First factor and reduce: $y = \dfrac{2x^2 - x - 1}{4x^2 + 4x + 1} = \dfrac{(2x + 1)(x - 1)}{(2x + 1)(2x + 1)} = \dfrac{x - 1}{2x + 1}$. Note that $x \neq -\dfrac{1}{2}$.

The denominator is zero when $x = -1/2$; hence there is a vertical asymptote there. Since the degrees of both numerator and denominator are equal, there is a horizontal asymptote at $y = 1/2$. A few points are:

Let $x = 0$, then $y = \dfrac{0 - 1}{2(0) + 1} = -1$

Let $x = 1$, then $y = \dfrac{1 - 1}{2(1) + 1} = 0$

Let $x = -1$, then $y = \dfrac{-1 - 1}{2(-1) + 1} = 2$

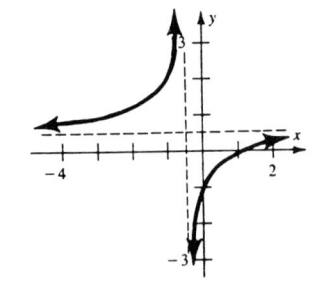

You should generate more points.

29. Let $p = 80$: $C = \dfrac{40{,}000(80)}{110 - 80} = \$106{,}667$.

31. Setting the denominator equal to zero, we get a vertical asymptote at $p = 110$. However, this is not in the restricted domain $0 \leq p \leq 100$. A few points are:

Let $p = 0$, $C = \dfrac{40{,}000(0)}{110 - 0} = 0$

Let $p = 20$, $C = \dfrac{40{,}000(20)}{110 - 20} \approx 8{,}889$

Let $p = 40$, $C = \dfrac{40{,}000(40)}{110 - 40} \approx 22{,}857$

Chapter 1

Let $p = 60$, $C = \dfrac{40{,}000(60)}{110 - 60} = 48{,}000$

You should generate more points.

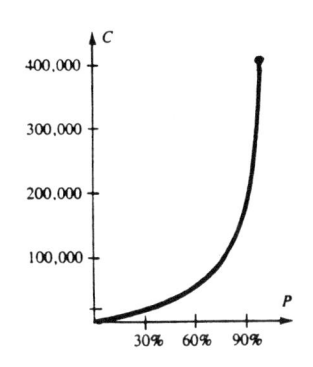

33. Let $p = 80$: $C = \dfrac{18{,}000(80)}{100 - 80} = \$72{,}000$

35. This is a rational function with vertical asymptote at $p = 100$. A few points are:

Let $p = 0$, $C = \dfrac{18{,}000(0)}{100 - 0} = 0$

Let $p = 20$, $C = \dfrac{18{,}000(20)}{100 - 20} = 4500$

Let $p = 40$, $C = \dfrac{18{,}000(40)}{100 - 40} = 12{,}000$

Let $p = 60$, $C = \dfrac{18{,}000(60)}{100 - 60} = 27{,}000$

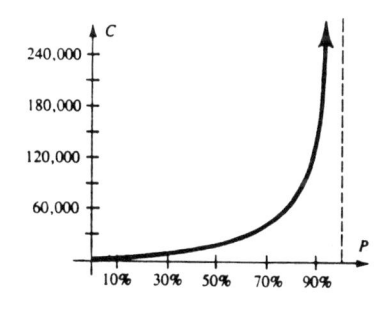

37. This is a rational function with vertical asymptote at FFD $= 0$. A few points are:

Let $x = 1$, MA $= \dfrac{256}{1^2} = 256$

Let $x = 2$, MA $= \dfrac{256}{2^2} = 64$

Let $x = 4$, MA $= \dfrac{256}{4^2} = 16$

Let $x = 6$, MA $= \dfrac{256}{6^2} \approx 7$

Let $x = 10$, MA $= \dfrac{256}{10^2} \approx 2.6$

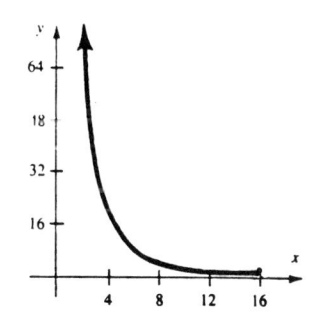

Chapter 1

Let $x = 16$, $MA = \dfrac{256}{16^2} = 1$

39. Using $\dfrac{\text{original pressure}}{\text{new pressure}} = \dfrac{\text{new volume}}{\text{original volume}}$,

replace original pressure with 2800 and original volume with 400: $\dfrac{2800}{\text{new pressure}} = \dfrac{\text{new volume}}{400}$ =>

new pressure = $\dfrac{1{,}120{,}000}{\text{new volume}}$.

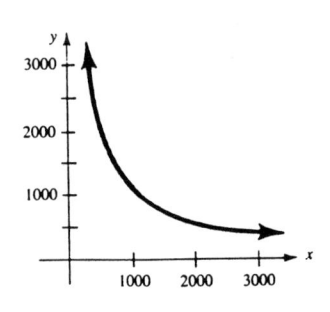

1.6 CHAPTER REVIEW, PAGE 39

1. $f(3) = 2(3)^2 - 3(3) + 5 = 14$ **3.** $g(t + 1) = 1 - 4(t + 1) = -4t - 3$

5. $\dfrac{g(x + h) - g(x)}{h} = \dfrac{[1 - 4(x + h)] - [1 - 4x]}{h} = \dfrac{1 - 4x - 4h - 1 + 4x}{h} = \dfrac{-4h}{h} = -4$

7. This is a linear equation. A few points are:

Let $x = -1$, then $g(-1) = 1 - 4(-1) = 5$

Let $x = 0$, then $g(0) = 1 - 4(0) = 1$

Let $x = 1$, then $g(1) = 1 - 4(1) = -3$

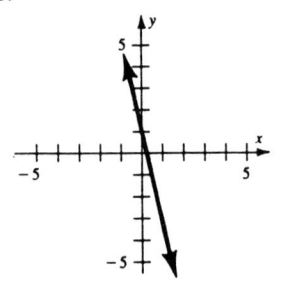

9. slope = $\dfrac{y_2 - y_1}{x_2 - x_1} = \dfrac{(-7) - (9)}{(-3) - (-5)} = \dfrac{-16}{2} = -8$

11. A few points are:

Let $x = -4$, then $y = |-4 + 3| = 1$

Let $x = -3$, then $y = |-3 + 3| = 0$

Let $x = -2$, then $y = |-2 + 3| = 1$

Also, remember that $|a| = \begin{cases} a & \text{if } a \geq 0 \\ -a & \text{if } a < 0 \end{cases}$.

Therefore, $y = \begin{cases} x + 3, & \text{if } x + 3 \geq 0 \\ -(x + 3), & \text{if } x + 3 < 0 \end{cases}$.

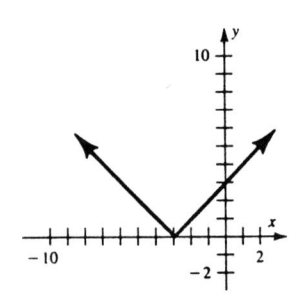

This simplifies to

Chapter 1

$$y = \begin{cases} x + 3, & \text{if } x \geq -3 \\ -x - 3, & \text{if } x < -3 \end{cases}.$$

13. Using the point-slope form; $y - y_1 = m(x - x_1)$: $y - 20 = -5(x - 30)$
=> $y - 20 = -5x + 150$ => $5x + y - 170 = 0$.   **15.** Vertical lines have no slope. The vertical line passing through (-3, -2) is $x = -3$ or $x + 3 = 0$.   **17.** This has the form
$y - k = a(x - h)^2$, where $a = -\frac{2}{3}$ and $(h, k) = (40, 550)$. The graph is a parabola opening down (because $a < 0$) from the vertex (40, 550). Hence, the maximum value of y is 550, which occurs when $x = 40$.   **19.** Substitute $p = 90$: $C = \frac{500(90)}{100 - 90} = \$4,500$.

1.7 STUDENT'S TEST REVIEW AND ADDITIONAL PRACTICE
OBJECTIVES

The material of this chapter is reviewed in the following list of objectives. After each objective there are some practice questions. Answers to these problems immediately follow. Detailed solutions are given for every third problem. For a sample test select the first question of each set and check your answers. Additional practice is given by the other questions in each set. If you are having trouble with a particular type of problem, or if you want additional practice, look back at the indicated section in the text.

[1.1] Objective 1: *Evaluate a function.* Let $f(x) = 5x - 3$ and $g(x) = 2x^2 - 3x + 1$ and find the requested values.

    1. f(10)   **2.** f(-3)   **3.** g(-2)   **4.** g(4)

[1.2] Objective 2: *Evaluate a function and simplify.* Let $f(x) = 3x^2$ and $g(x) = x^2 - 5$.

    5. f(s + 3)   **6.** g(t - h)   **7.** f(x + h) - f(x)   **8.** $\dfrac{g(x + h) - g(x)}{h}$

Objective 3: *Given a graph, find coordinates of indicated points, the domain, range, and intercepts.*

9.

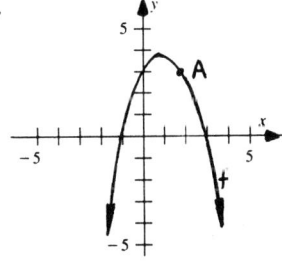

10.

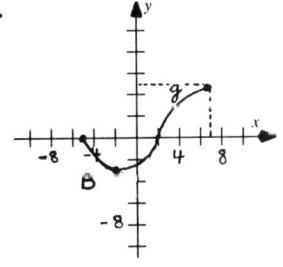

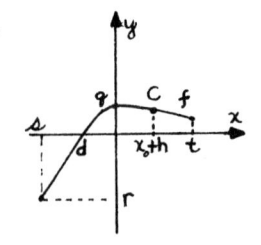

Chapter 1

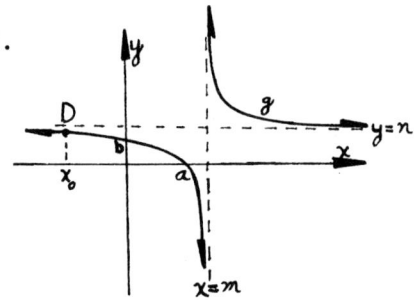

11.
12.

Objective 4: *Graph a given function by plotting points.*

13. $f(x) = 9 - 5x$ **14.** $g(x) = 25 - 5x - x^2$

15. $h(x) = \dfrac{1}{x - 3}$ **16.** $t(x) = \dfrac{x^2 - 3}{x}$

[1.3] Objective 5: *Find the x- and y-intercepts for a line whose equation is given.*

17. $y = 5x - 4$ **18.** $3x + 2y - 6 = 0$
19. $3x - 4y + 12 = 0$ **20.** $50x - 250y = 1000$

Objective 6: *Find the slope of the line passing through two given points.*

21. $(4, 1), (-3, -2)$ **22.** $(-3, -1), (2, 6)$
23. $(-3, 5), (-1, -5)$ **24.** $(-1, -2), (-6, -8)$

Objective 7: *Graph a line by finding the slope and y-intercept.*

25. $y = -\dfrac{2}{5}x + 2$ **26.** $y = \dfrac{2}{3}x - \dfrac{1}{3}$

27. $x + 2y - 8 = 0$ **28.** $3x - 5y - 10 = 0$

Objective 8: *Graph a line segment or piecewise linear function.*

29. $2x - 3y - 6 = 0; -6 \le x \le 3$ **30.** $y = -5x + 3; -4 \le x \le 3$
31. $\begin{cases} x - y + 5 = 0 \text{ if } -5 \le x \le 0 \\ x + y + 5 = 0 \text{ if } 0 < x < 5 \end{cases}$ **32.** $y = 2|x - 1|$

Objective 9: *Find the standard form equation of a line satisfying given conditions.*

33. Passing through $(2, 10)$ and $(5, 25)$

34. Slope $-2/3$; passing through $(20, -100)$

35. y-intercept -4; slope $1/5$

36. no y-intercept and no slope passing through $(1, 5)$

[1.4] Objective 10: *Sketch the graph of a parabola.*

37. $y = 2(x + 3)^2$ **38.** $y + 1 = (\dfrac{1}{2})(x - 2)^2$

39. $y - 4 = (-\dfrac{2}{3})(x + 2)^2$ **40.** $3(x - 1)^2 + y - 2 = 0$

Objective 11: *Find the maximum or minimum value of y for quadratic function.*

Chapter 1

41. $y + 250 = (-\frac{1}{3})(x - 1300)^2$ 42. $y = -5(x - 300)^2 + 1100$

43. $2(x - 3)^2 - y + 250 = 0$ 44. $2y - 3(x + 40)^2 - 100 = 0$

[1.5] Objective 12: *Name the asymptotes for a rational function.*

45. $f(x) = \dfrac{1}{x - 5}$ 46. $f(x) = \dfrac{3x + 2}{2x - 5}$

47. $g(x) = \dfrac{4}{x^2 - x - 6}$ 48. $g(x) = \dfrac{3x^2 + 2x + 1}{2x^2 - 7x - 4}$

Objective 13: *Graph rational functions.*

49. $f(x) = \dfrac{4}{x}$ 50. $f(x) = \dfrac{2}{(x + 2)}$

51. $g(x) = \dfrac{4}{x} + 3$ 52. $g(x) = \dfrac{2x + 2}{x^2 + 3x + 2}$

Objective 14: *Solve applied problems based on the preceeding objectives. For specific examples of the types of applications look at the list of applications in this chapter on page 1.*

53. **Rate of Change** If $Z(x)$ = the price of Xerox on the first trading day in year x, write an expression for the average rate of change in price of Xerox stock from 1980 to 1985.

54. **Supply** If a distributor can supply 2000 items when the cost is $200 can supply 8000 if the cost is $400. Write the supply equation if you know that supply is a linear function.

55. **Revenue** After market research it was found that the demand equation for a certain product is 4000 - 50x dollars, where x is the number of items produced. The cost function for this product is $C(x) = 25{,}000 + 1000x$. Find the revenue function.

56. **Equilibrium Point** The demand for the item described in Problem 54 is 7000 items if the price is $100 and only 1000 items if the price is $700. What is the equilibrium point for which the supply and demand are the same?

57. **Break Even** What is the break even-point for the information described in Problem 55?

58. **Maximum Profit** A manufacturer can produce no more than 200 items and it is determined that the profit (in dollars) is given by the following function:

$$P(x) = -10(x - 170)^2 + 14{,}320$$

What is the maximum profit, and how many items should be manufactured in order to achieve this maximum profit?

Chapter 1

59. Cost-benefit Model Use the cost-benefit model

$$C = \frac{20{,}000p}{100 - p}$$

to find the cost of removing 95% of the pollutants.

60. Radiology Technology Graph the relationship

$$mA = \frac{200}{t}$$

which relates the time that an x-ray machine is on and the current used in the x–ray. You need only graph this function for $0 < t \leq 2$.

ANSWERS TO STUDENT'S TEST REVIEW AND PRACTICE QUESTIONS
1. $f(10) = 47$ **2.** $f(-3) = -18$ **3.** $g(-2) = 2(-2)^2 - 3(-2) + 1 = 15$ **4.** $g(4) = 21$
5. $3s^2 + 18s + 27$ **6.** $g(t - h) = (t - h)^2 - 5 = t^2 - 2h + h^2 - 5$ **7.** $6xh + 3h^2$
8. $2x + h$ **9.** The coordinates of A are (2, 3). Since for any value on the x-axis we choose there is a cooresponding point on the graph, the domain is all real numbers. Since for any value on the y-axis less than or equal to 4 there is a corresponding point (in fact, 2 points) on the graph, the range is $y \leq 4$. The x-intercepts are -1 and 3, and the y-intercept is 3.
10. Coordinates of B are (-2, -3). The domain is $-5 \leq x \leq 7$; the range is $-4 \leq y \leq 5$. The x–intercepts are at -5 and 2; the y-intercept is at -4. **11.** Coordinates of C are $(x_0 + h, f(x_0 + h))$. The domain is $s \leq x \leq t$; the range is $r \leq y \leq q$. The x–intercept is at d; the y–intercept is at q. **12.** The coordinates for D are $(x_0, g(x_0))$. Since there is a point on the graph for any value on the x–axis we choose, except $x = m$, the domain is all real numbers except m, abreviated as $x \neq m$; since there is a point on the graph for any value on the y–axis we choose, except $y = n$, the range is $y \neq n$. The x–intercept is at a or (a, 0); the y–intercept is at b or (0, b).

13.

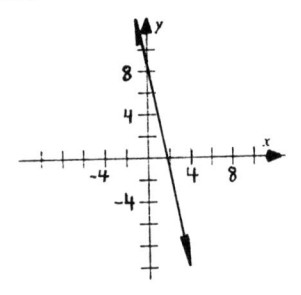

14.

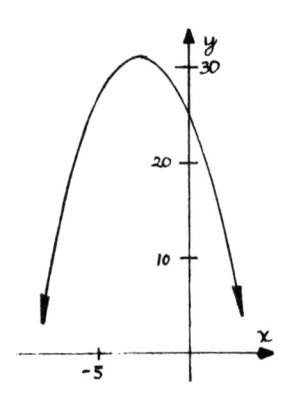

Chapter 1

15. Some points are:

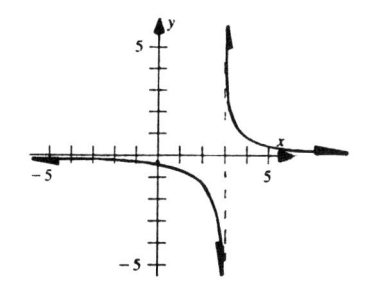

Let $x = 0$, then $y = \dfrac{1}{(0-3)} = -\dfrac{1}{3}$

Let $x = 1$, then $y = \dfrac{1}{(1-3)} = -\dfrac{1}{2}$

Let $x = 2$, then $y = \dfrac{1}{(2-3)} = -1$

Let $x = 4$, then $y = \dfrac{1}{(4-3)} = 1$

Remembering that $x \neq 3$, you should generate more points to get the graph shown on the right.

16.

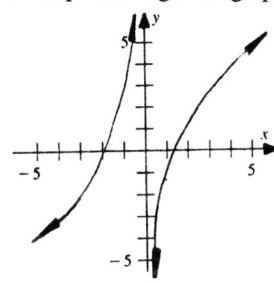

17. x-intercept $(4/5, 0)$; y-intercept $(0, -4)$ **18.** For the x-intercept, set $y = 0$ and solve: $3x + 2(0) - 6 = 0 \Rightarrow 3x = 6 \Rightarrow x = 2$. So the x-intercept is $(2, 0)$. For the y-intercept, set $x = 0$ and solve: $3(0) + 2y - 6 = 0 \Rightarrow 2y = 6 \Rightarrow y = 3$. So the y-intercept is $(0, 3)$.

19. x-intercept $(-4, 0)$; y-intercept $(0, 3)$ **20.** x-intercept $(20, 0)$; y-intercept $(0, -4)$

21. $m = \dfrac{y_2 - y_1}{x_2 - x_1} = \dfrac{-2 - 1}{-3 - 4} = \dfrac{-3}{-7} = \dfrac{3}{7}$ **22.** $m = \dfrac{7}{5}$ **23.** $m = -5$

24. $m = \dfrac{y_2 - y_1}{x_2 - x_1} = \dfrac{-8 - (-2)}{-6 - (-1)} = \dfrac{-6}{-5} = \dfrac{6}{5}$

25.

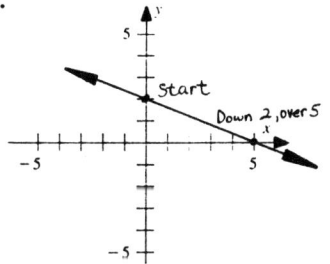

start

Down 2, over 5

26.

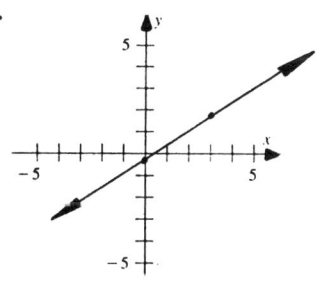

27. First, isolate y:

$x + 2y = 8$

$2y = -x + 8$

$y = \left(-\dfrac{1}{2}\right)x + 4$

Chapter 1

So we start at 4 on the y–axis and count down 1, over 2:

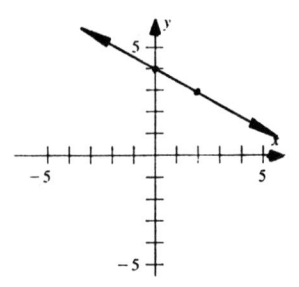

28.

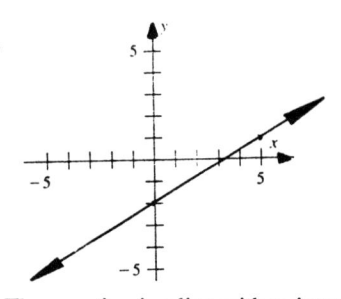

29.

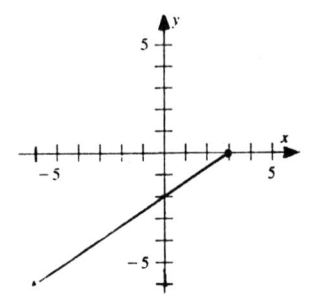

30. The equation is a line with y–intercept at 3 and a slope of $-5/1$; but the graph is restricted between $x = -4$ and $x = 3$. The endpoints are:
Let $x = -4$, then $y = -5(-4) + 3 = 23$
Let $x = 3$, then $y = -5(3) + 3 = -12$

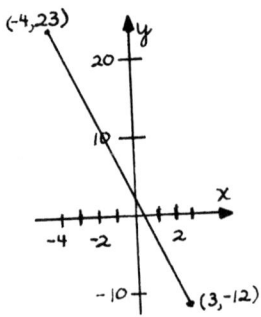

31.

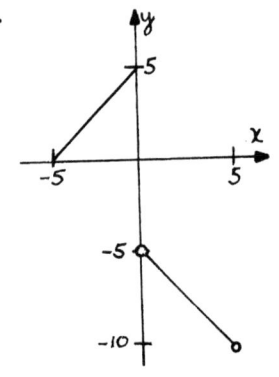

32.

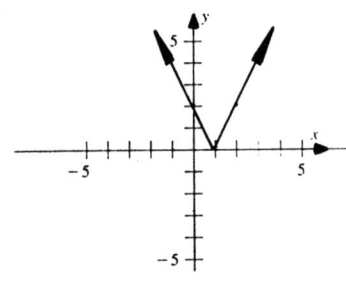

Chapter 1

33. First calculate the slope, $m = \dfrac{25 - 10}{5 - 2} = 5$. Next use either ordered pair and the point-slope form:

$y - y_1 = m(x - x_1)$
$y - 10 = 5(x - 2)$ using (2, 10).

Finally, put this into standard form:

$y - 10 = 5x - 10$
$y = 5x$
$5x - y = 0$

34. $2x + 3y + 260 = 0$ **35.** $x - 5y - 20 = 0$ **36.** Both conditions indicate the line is vertical (parallel to the y-axis). Since the x-intercept is 1, the equation has the form $x = 1$, so the standard form is $x - 1 = 0$.

37.

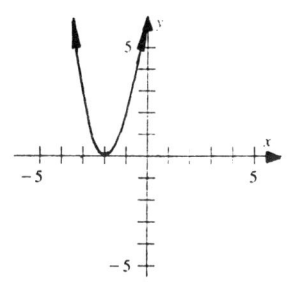

38.

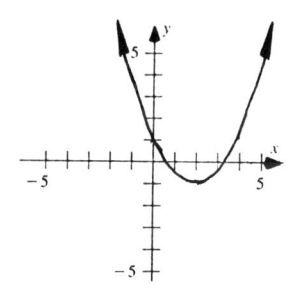

39. This has the form $y - k = a(x - h)^2$ where (h, k) is the vertex, so the vertex is at (-2, 4) and the related equation is $y = -\dfrac{2}{3}x^2$. Using $y = -\dfrac{2}{3}x^2$ some points <u>relative to the vertex</u> are:

Let $x = 0$, then $y = -\dfrac{2}{3}(0)^2 = 0$; this is the vertex.

Let $x = 3$, then $y = -\dfrac{2}{3}(3)^2 = -6$; 3 right and 6 down from the vertex.

Let $x = -1$, then $y = -\dfrac{2}{3}(-1)^2 = -\dfrac{2}{3}$; 1 left and $\dfrac{2}{3}$ down from the vertex.

You should continue the above to generate more points. Also, use the symmetry of the parabola.

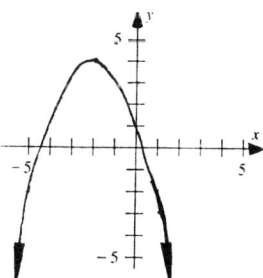

Chapter 1

40.

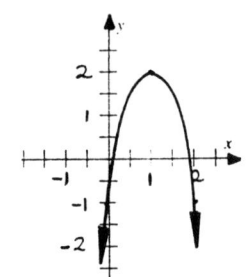

41. Maximum: y = -250; occurs when x = 1300   **42.** The graph of the equation is a parabola with vertex at (300, 1100) which opens down (a = -5 < 0). Therefore, the maximum value is y = 1100 which occurs when x = 300.   **43.** Minimum: y = 250; occurs when x = 3   **44.** Minimum: y = 50; occurs when x = -40   **45.** The denominator is zero when x = 5, so there is a vertical asymptote at x = 5. Since the degree of the numerator is less than the degree of the denominator (0 < 1), there is a horizontal asymptote at y = 0.   **46.** Vertical asymptote at x = 5/2; horizontal asymptote at y = 3/2.   **47.** Vertical asymptotes at x = 3 and x = -2; horizontal asymptote at y = 0.

48. Set the denominator equal to zero and solve:
$$2x^2 - 7x - 4 = 0$$
$$(2x + 1)(x - 4) = 0$$
$$x = -\frac{1}{2}, 4$$

There are vertical asymptotes at x = -1/2 and x = 4. Since the degrees of numerator and denominator are equal, the horizontal asymptote is the ratio of leading coefficients, y = 3/2.

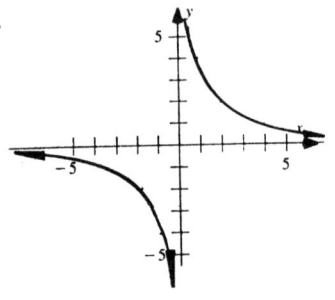

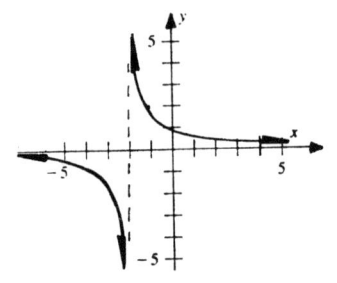

49.   **50.**

51. There are at least two ways to graph this. One way is to realize it is the same graph as in Problem 49 but shifted 3 units up. Another way is to add:
$$g(x) = \frac{4}{x} + 3 = \frac{4}{x} + \frac{3x}{x} = \frac{3x + 4}{x}$$

The denominator is zero at x = 0, which is the equation of the vertical asymptote. Since the degree of 3x + 4 is the same as the degree of x, there is a horizontal asymptote at

$y = 3/1 = 3$. Some points are:

Let $x = 1$, then $y = \frac{4}{1} + 3 = 7$

Let $x = -1$, then $y = \frac{4}{(-1)} + 3 = -1$

Let $x = 2$, then $y = \frac{4}{2} + 3 = 5$

You should generate more points as above.

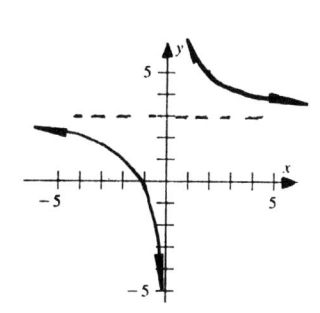

52.

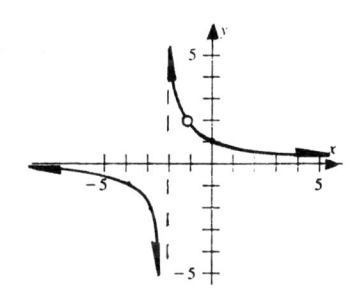

53. $\dfrac{z(1985) - z(1980)}{5}$

54. We have two points: (200, 2000) and (400, 8000). To find a linear equation relating supply (y) with cost (x), first compute the slope:
$$m = \frac{y_2 - y_1}{x_2 - x_1} = \frac{8000 - 2000}{400 - 200} = 30$$
Now use either ordered pair in the point-slope form; we will use (200, 2000):
$y - y_1 = m(x - x_1) \Rightarrow y - 2000 = 30(x - 200)$
Now solve for y: $y = 30x - 4000$. In function notation, call S(x) the supply function:
$S(x) = 30x - 4000$.

55. $R(x) = 4000x - 50x^2$

56. Demand equation: $y = -10x + 8000$
Supply equation (see Problem 54): $y = 30x - 4000$
Supply equals Demand when $x = 300$. Equilibrium at ($300, 5000).

57. Break-even occurs when Cost equals Revenue:
$25,000 + 1000x = C(x) = R(x) = 4000x - 50x^2$
$50x^2 - 3000x + 25000 = 0$
$x^2 - 60x + 500 = 0$ (divided by 50)
$(x - 10)(x - 50) = 0$ (factor)
$x = 10$ or $x = 50$
Break-even points occur at (10, 35000) and (50, 75000).

58. Maximum profit is p = $14,320; occurs when x = 170 items are sold.

59. $380,000

60. Some points for $0 < t \le 2$ are:

Let $t = \dfrac{1}{4}$, then $mA = \dfrac{200}{\left(\dfrac{1}{4}\right)} = 800$

Let $t = \dfrac{1}{2}$, then $mA = \dfrac{200}{\left(\dfrac{1}{2}\right)} = 400$

Let $t = 1$, then $mA = \dfrac{200}{1} = 200$

Let $t = \dfrac{3}{2}$, then $mA = \dfrac{200}{\left(\dfrac{3}{2}\right)} \approx 133$

You should generate more points.

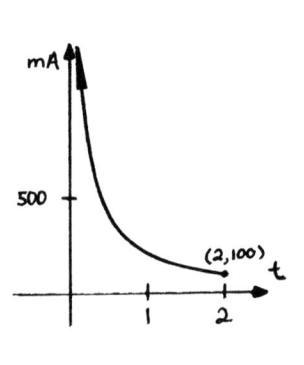

1.8 MODELING APPLICATION 1: SAMPLE ESSAY

An Analysis of Two Car-Buying Strategies

Suppose you are considering buying a car and that you have $4,000 to spend and have narrowed your choices down to two cars. The first car is a two-year-old car in excellent condition costing $5,500 with a combined city-highway EPA rating of 16 miles per gallon. The second car is a small new car costing $8,600 with a combined city-highway rating of 31 miles per gallon. Suppose, also, that you like both of these cars equally well and, therefore, decide to make a decision about which car to buy based only on which car will be the least expensive.*

Since you have only $4,000 you will need to finance at a local bank the difference between this and the price of whichever car you decided to purchase at an interest rate of 16.5% APR (Annual Percentage Rate). Also, suppose that gasoline currently costs $1.25 per gallon and that the price will increase at an average rate of $0.15 per gallon per year.

Whenever you build a mathematical model you will need to make certain assumptions in order to simplify the analysis. For this extended application we will not take into consideration the cost of maintaining the cars. This information would be difficult to consider, but since one car is new and the other is relatively new and in excellent condition, we will assume that the total maintenance costs will be equal over the life of the cars, and so these costs will not be considered when total operating costs are computed.

The solution of the problem depends on finding a cost function for each car.

---

*This essay is based on the article "An analysis of Two Car-Buying Stragegies" by Paul Bland and Betty Givan in The Mathematics Teacher, February 1982, pp. 124-127

Chapter 1

Suppose $C(x) = 4000 + g(x) + f(x)$
where $C(x)$ = cost of operating an automobile for x years;
 $g(x)$ = cost of gasoline for x years;
 $f(x)$ = cost of financing for x years;
We need to find g and f for each of the cars.

First, find g.

Suppose that you know that you drive 12,000 miles per year. Then, the number of gallons for each choice can easily be found:

Car A: $\dfrac{12,000}{16} = 750$ gallons per year

Car B: $\dfrac{12,000}{31} = 387$ gallons per year

Next, find the cost of gasoline for x years. Consider the entries in Table 1.

Table 1 Cost of Gasoline by Year

Year	Cost Car A	Car B
1	750(1.25)	387 (1.25)
2	750([1.25 + .15]	387[1.25 + .15]
3	750[1.25 + .15(2)]	387[1.25 + .15(2)]
4	750[1.25 + .15(3)]	387[1.25 + .15(3)]
⋮		
x	750[1.25 + .15(x - 1)]	387[1.25 + .15(x - 1)]

In order to find the total operating cost add the columns in Table 1. In order to obtain a closed form, you will need to know

$$1 + 2 + 3 + \cdots + x = \dfrac{x(x+1)}{2}$$

We can now calculate the operating cost for Car A, denoted by g_A:

$g_A = 750[1.25] + 750[1.25 + .15(1)] + 750[1.25 + .15(2)]$
$\qquad + \cdots + 750[1.25 + .15(x - 1)]$
$= 937.5 + [937.5 + 112.5(1)] + [937.5 + 112.5(2))]$
$\qquad + \cdots + [937.5 + 112.5(x - 1)]$
$= 937.5x + 112.5[1 + 2 + 3 + \cdots + x - 1]$
$= 937.5x + 112.5\left[\dfrac{(x-1)x}{2}\right]$
$= 937.5x + 56.25x(x - 1)$
$= 937.5x + 56.25x^2 - 56.25x$
$= 56.25x^2 + 881.25x$

The operating cost for Car B (g_B) can be found in the same way (the details are left as an exercise):

$$g_B = 29.03x^2 + 454.85x$$

Next, find f.

For the cost due to financing, you will need to know the formula for the monthly payment on a loan. This formula is

$$m = \frac{Pi}{1 - (1+i)^{-N}}$$

where m is the monthly payment, P is the amount borrowed, i is the interest rate divided by 12 and N is the number of months the loan will be paid. Since you have $4,000 to buy the car and since the difference between $4,000 and the amount you will have to pay must be financed at 16.5%, we see that

Car A: P = $1,500; obtain a one-year loan
Car B: P = $4,600; obtain a three-year loan

You can use a calculator to find m using $i = \frac{.165}{12} = .01375$ and N = 12 for car A and N = 36 for car B:

$$m_A = \$136.45 \qquad m_B = \$162.86$$

Therefore,

$$f_A = \$136.45 \times 12 = \$1{,}627.4 \text{ and } f_B = \$162.86 \times 36 = \$5{,}862.96$$

We now calculate the cost function for each car:

$$C_A = 4000 + g_A + f_A \; 5637.4 + 881.25x + 56.25x^2$$

$$C_B \; 4000 + g_B + f_B = 9862.96 + 454.85x + 29.03x^2$$

The graph of these functions is shown in Figure 1.

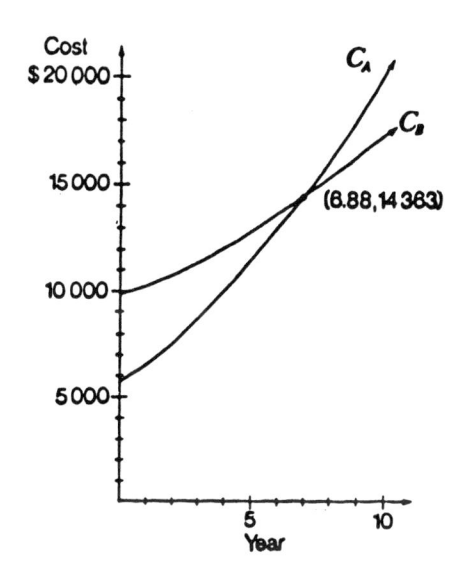

Figure 1 Total Cost of Options A and B

From Figure 1 it looks like the break-even point is about 7 years. This means that if you plan on keeping the car less than 7 years then Car A is the less expensive choice; whereas, if you plan on keeping it longer than 7 years, then choice B is the better choice. If you want a closer approximation than you can get from Figure 1, you can set C_A and C_B equal and solve:

$$C_A = C_B$$

$$5637.4 + 881.25x + 56.25x^2 = 9862.96 + 454.85x + 29.03x^2$$

$$27.22x^2 + 426.4x - 4225.56 = 0$$

$$x = \frac{-426.4 \pm \sqrt{426.4^2 - 4(27.22)(-4225.56)}}{2(27.22)}$$

$$= 6.88, -22.55$$
$$\uparrow$$
$$\text{reject}$$

The break-even point is $x = 6.88$ or after 6 years 11 months. Table 2 shows the cumulative costs of Options A and B.

Table 1 Cumulative Costs for Options A and B

	Cost			Mileage	
Year	Option A	Option B	Year	Option A	Option B
1	6,575	10,347	1	36,000	12,000
2	7,625	10,889	2	48,000	24,000
3	8,787	11,489	3	60,000	26,000
4	10,062	12,147	4	72,000	48,000
5	11,450	12,863	5	84,000	60,000
6	12,950	13,637	6	96,000	72,000
6.88	14,297	14,367	6.88	106,560	82,560
7	14,562	14,469	7	108,000	84,000
8	16,287	15,360	8	120,000	96,000
9	18,125	16,308	9	132,000	108,000
10	20,075	17,315	10	144,000	120,000

CHAPTER 2
THE DERIVATIVE

2.1 LIMITS, PAGES 50 - 52

1. On the graph, as x gets closer and closer to x = 4, the y-coordinates get closer and closer to y = 0. The limit is 0.   **3.** As x gets closer and closer to 0, the graph approaches 8, so the limit is 8.   **5.** On the graph, as x gets closer to -1 (approaching from the left), the y values are always y = 7, so the limit is 7.   **7.** On the graph, as x gets closer to 3 (approaching from the right), the y-values approach (but never reach) 6. The limit is 6.
9. As x approaches 2 from the right, the graph remains a constant value of 2, so the limit is 2.   **11.** On the graph, as x gets closer and closer to x = 4, the y-coordinates approach y = 2, so the limit is 2.

13.

x	2	3	4	4.5	4.9	4.99
f(x)	3	7	11	13	14.6	14.96

, limit is 15.

15.

x	0	-1	-1.5	-1.9	-1.99	-1.999
f(x)	250	302	329.5	352.22	357.4202	357.942002

, limit is 358.

17.

x	0	.5	.9	.99	.999	1.5	1.1	1.01	1.001
f(x)	-1	-2.$\overline{3}$	-4.$\overline{27}$	-4.$\overline{9207}$	-4.992	-13	-5.$\overline{8}$	-5.$\overline{08}$	-5.$\overline{008}$

, limit is -5.

19.

x	1	1.5	1.9	1.99	1.999	2.5	2.01	2.001
f(x)	-1	-2	-10	-100	-1000	2	100	1000

, as x approaches 2, the f(x) values appear to be increasing or decreasing without bound. The limit does not exist.

21.

x	-1	-10	-100	-1000	-10000	-100000	-1000000
f(x)	11.5	-52.143	314.588	-3014.06	-30014.006	-30014.0	-3000014

as x decreases without bound, the f(x) values seem to be decreasing without bound (about 3 times as fast as x). The limit does not exist.

23. Since $x^2 - 4$ is a polynomial, $\lim_{x \to 2} (x^2 - 4) = 2^2 - 4 = 0$

25. As x approaches 1, $\dfrac{1}{x-3}$ approaches $\dfrac{1}{1-3} = -\dfrac{1}{2}$.

27. A table might help:

x	2.9	2.99	2.999	3.1	3.01	3.001
1/(x-3)	-10	-100	-1000	10	100	1000

As x approaches 3 from the left, 1/(x-3) gets smaller and smaller (negatively) without bound; likewise as x approaches 3 from the right 1/(x-3) gets larger and larger without bound. A limit does not exist. It would be incorrect to say the limit is infinite.

Chapter 2   36

29. A table might help:

x	10	100	1000	10,000
f(x)	.0099	.0001	.000001	.00000001

$\dfrac{1}{x^2+1}$ approaches 0 as x increases without bound, so the limit is 0.

31. We could look at a table of values, but clearly as x grows without bound, 2x also grows without bound (twice as fast). The limit does not exist.

33. If we first factor and then reduce: $\dfrac{x^2-4}{x-2} = \dfrac{(x+2)(x-2)}{x-2} = x+2, \ x \neq 2$.

The restriction $x \neq 2$ is important! Fortunately, when evaluating the limit as x approaches 2 we do not include $x = 2$, but rather only values close to 2. So for our discussion,

$$\lim_{x \to 2} \dfrac{x^2-4}{x-2} = \lim_{x \to 2} (x+2) = 4$$

35. $\lim_{x \to 3} \dfrac{x^2+3x-10}{x-2} = \lim_{x \to 3} \dfrac{(x+5)(x-2)}{x-2} = \lim_{x \to 3} (x+5) = 8$

37. $\lim_{x \to 4} \dfrac{\sqrt{x}-4}{x-16} = \lim_{x \to 4} \dfrac{\sqrt{x}-4}{x-16} \cdot \dfrac{(\sqrt{x}+4)}{(\sqrt{x}+4)} = \lim_{x \to 4} \dfrac{x-16}{(x-16)(\sqrt{x}+4)} = \lim_{x \to 4} \dfrac{1}{\sqrt{x}+4} = \dfrac{1}{6}$

39. We could substitute $x = 9$ to find the limit: $\lim_{x \to 9} \dfrac{\sqrt{x}-3}{x-3} = \dfrac{\sqrt{9}-3}{9-3} = 0$

41. A table might help:

x	1.9	1.99	1.999	2.01	2.001
f(x)	-36.1	-296.01	-2996.001	304.01	3004.001

f(x) seems to be increasing or decreasing without bound. As x approaches 2 the numerator approaches 3, but the denominator approaches 0, hence the limit does not exist.

43. $\lim_{x \to 2} \dfrac{x^3-8}{x^2+2x+4} = \lim_{x \to 2} \dfrac{(x-2)(x^2+2x+4)}{x^2+2x+4} = \lim_{x \to 2} (x-2) = 0$. (The factoring was not really necessary; you could have substituted $x = 2$ to get $\dfrac{0}{12} = 0$.)

45. $\lim_{x \to 2} \dfrac{x+2}{x^3+8} = \lim_{x \to 2} \dfrac{x+2}{(x+2)(x^2-2x+4)} = \lim_{x \to 2} \dfrac{1}{x^2-2x+4} = \dfrac{1}{4}$. (The factoring was not really necessary; you could have tried substituting $x = 2$ to get $\dfrac{4}{16} = \dfrac{1}{4}$).

47. $\lim_{|x| \to \infty} \dfrac{2x^2-5x-3}{x^2-9} \cdot \dfrac{\frac{1}{x^2}}{\frac{1}{x^2}} = \lim_{|x| \to \infty} \dfrac{2-\frac{5}{x}-\frac{3}{x^2}}{1-\frac{9}{x^2}} = \dfrac{2-0-0}{1-0} = 2$

49. $\lim\limits_{|x|\to\infty} \dfrac{3x-1}{2x+3} \cdot \dfrac{\frac{1}{x}}{\frac{1}{x}} = \lim\limits_{|x|\to\infty} \dfrac{3-\frac{1}{x}}{2+\frac{3}{x}} = \dfrac{3-0}{2+0} = \dfrac{3}{2}$

51. $\lim\limits_{x\to-\infty} \dfrac{5x+10{,}000}{x-1} \cdot \dfrac{\frac{1}{x}}{\frac{1}{x}} = \lim\limits_{x\to-\infty} \dfrac{5+\frac{10{,}000}{x}}{1-\frac{1}{x}} = \dfrac{5+0}{1-0} = 5$

53. As $x \to \infty$, $x+2$ will grow without bound and $\dfrac{3}{x-1}$ will approach 0:

$\lim\limits_{x\to\infty}(x+2+\dfrac{3}{x-1}) = \lim\limits_{x\to\infty}(x+2)$ which grows without bound; the limit does not exist.

55. $\lim\limits_{x\to-\infty} \dfrac{4x^4-3x^3+2x+1}{3x^4-9} \cdot \dfrac{\frac{1}{x^4}}{\frac{1}{x^4}} = \lim\limits_{x\to-\infty} \dfrac{4-\frac{3}{x}+\frac{2}{x^2}+\frac{1}{x^4}}{3-\frac{9}{x^4}} = \dfrac{4-0+0+0}{3-0} = \dfrac{4}{3}$

57. a. As n approaches 10 from the left, the graph drops closer and closer to $7.00 on the cost axis. So the limit is $7.00.   **b.** As n approaches 10 from the right, the graph rises closer and closer to $12.00 on the cost axis. So the limit is $12.00.   **c.** As n approaches 13 from either side, the graph approaches $7.00 on the cost axis, so the limit is $7.00   **d.** Since in parts (a) and (b) we get different limiting values as we approach from both sides, the limit does not exist.

59. a. $\lim\limits_{t\to t_1^-} P(t) = 40\%$ and $\lim\limits_{t\to t_1^+} P(t) = 60\%$; since these are not equal, the limit does not exist.   **b.** $\lim\limits_{t\to t_2^-} P(t) = 80\%$, since the closer t gets to t_2 (from the left) the closer P(t) gets to 80% (even though it never reaches 80%).   **c.** $\lim\limits_{t\to t_3} P(t) = 70\%$ **d.** $\lim\limits_{t\to t_4} P(t) = 50\%$

61. $\lim\limits_{h\to 0} \dfrac{f(1+h)-f(1)}{h} = \lim\limits_{h\to 0} \dfrac{\sqrt{1+h}-\sqrt{1}}{h} = \lim\limits_{h\to 0} \dfrac{\sqrt{1+h}-1}{h} \cdot \dfrac{\sqrt{1+h}+1}{\sqrt{1+h}+1}$

$= \lim\limits_{h\to 0} \dfrac{h}{h(\sqrt{1+h}+1)} = \lim\limits_{h\to 0} \dfrac{1}{\sqrt{1+h}+1} = \dfrac{1}{\sqrt{1+0}+1} = \dfrac{1}{2}$

2.2 CONTINUITY, PAGES 57 - 60

1. Since the graph has a hole as $x = 4$, this is a point of discontinuity. (The function is not defined at $x = 4$.)   **3.** Since the function is not defined at $x = 2$, this is a point of discontinuity.   **5.** Suspicious points at $x = 1$ and $x = 4$; however, $f(1) = 6$ which matches

Chapter 2

the limiting value as x approaches 1 so the function is continuous at x = 1. The function does not exist, so x = 4 is a point of discontinuity.   **7.** Since the function is not defined at x = 1 or x = 6, both of these are points of discontinuity.   **9.** Since the function is not defined at x = -2, it is a point of discontinuity. Even though the function is defined at x = 2, f(2) = 4, the left- and right-hand limits do not match, hence the $\lim_{x \to 2} f(x)$ doesn't exist; x = 2 is also a point of discontinuity.   **11.** There are no suspicious points; hence the function is continuous for x > 0.   **13.** $\lim_{x \to 1} f(x) = f(1)$, yes.   **15.** Since $\lim_{x \to .001} f(x) = f(.001)$, yes.
17. Since there are no abrupt jumps or undefined values, it is continuous. The domain is $0 \le x < 24$, where x is the hour of the day.   **19.** If the selling price of IBM stock changes, even by only 1 cent, this would be a jump in the graph; hence, this function is not continuous (unless the price remains constant). The domain is $0 \le x < 24$ where x is the hour of the day.
21. As with the solution to Problem 19, even a 1 cent change is a sudden jump in the graph; hence, this function is not continuous. The domain is t > 0 where t is the number of minutes.
23. Read the Continuity Theorem, especially 4. The numerator, 1, is a continuous function and the denominator, $x^2 + 5$, is continuous. All we need to do is check when the denominator is zero, but $x^2 + 5 \ne 0$ so $f(x) = \dfrac{1}{x^2 + 5}$ is continuous through $-5 \le x \le 5$.   **25.** Read the Continuity Theorem, especially 4. The numerator, 1, is a continuous function and the denominator, $x^2 - 9$, is continuous. All we need to do is check when the denominator is zero:
$$x^2 - 9 = 0$$
$$x = \pm 3$$
But since neither of these values fall in the domain, the f(x) is continuous throughout $5 \le x \le 10$.   **27.** The function is not defined at x = 1 or -1, so the function has discontinuities at x = 1 and x = -1.

29. $f(x) = \dfrac{x + 2}{x^2 - 6x - 16}, -5 \le x \le 5 = \dfrac{x + 2}{(x - 8)(x + 2)}, -5 \le x \le 5$ Although both x = 8 and x = -2 will cause division by zero, only x = -2 is in the domain $-5 \le x \le 5$; hence, x = -2 is the only discontinuity.   **31.** x = 3 is a suspicious point. Now, $\lim_{x \to 3} f(x) = \lim_{x \to 3} \dfrac{1}{x - 3}$ which does not exist, hence x = 3 is a discontinuity.   **33.** x = -2 is the only suspicious point. We have two things to check:
1. f(-2) exists, and 2. $\lim_{x \to -2} f(x) = f(-2)$. The first condition is easy: f(-2) = -4.

The second condition is harder:

$$\lim_{x \to -2} f(x) = \lim_{x \to -2} \dfrac{x^2 - x - 6}{x + 2} = \lim_{x \to -2} \dfrac{(x - 3)(x + 2)}{x + 2} = \lim_{x \to -2} (x - 3) = -5$$

38

Since this limit does not match the value for f(-2), the function is discontinuous at x = -2.

35. $\dfrac{x^2 - 3x - 10}{x + 2} = \dfrac{(x - 5)(x + 2)}{x + 2}$ which is undefined at x = -2, but equal to x - 5 for all other x. Since the domain is $0 \le x \le 5$, f(x) = x - 5 which is a polynomial and, therefore, continuous.   **37.** x = -2 is the only suspicious point. We have two things to check: 1. f(-2) exists, and 2. $\lim_{x \to -2} f(x) = f(-2)$. The first is easy, f(-2) = -3. The second can be done as follows, $\lim_{x \to -2} f(x) = \lim_{x \to -2} \dfrac{(x - 5)(x + 2)}{x + 2} = \lim_{x \to -2} (x - 5) = -7 \ne f(-2)$.
So x = -2 is a discontinuity.   **39.** There are two things to check if f(x) is continuous at some number c: 1. f(c) exists, and 2. $\lim_{x \to c} f(x) = f(c)$. The first condition is true for all values of c. The second contition is not met for any value of c because as x approaches c, f(x) will jump back and forth between 1 and -1. That is, f(x) will not approach a specific number, so $\lim_{x \to c} f(x) \ne f(c)$. Discontinuous at all points.   **41.** See Example 11.

$f(x) = |x - 2| = \begin{cases} 2 - x & \text{if } -5 \le x < 2 \\ x - 2 & \text{if } 2 \le x \le 5 \end{cases}$ Continuous throughout.   **43. a.** This is not a continuous function because, for example, at w = 1 there is a jump from c = .25 to c = .45, formally: $\lim_{w \to 1^+} c(w) = .45$ and $\lim_{w \to 1^-} c(w) = .25$; hence, $\lim_{w \to 1} c(w)$ does not exist. The domain is w > 0.   **b.** c(1.9) = .25 + .20(1) = $.45; c(2.01) = .25 + .20(2) = $.65; c(2.89) = .25 + .20(2) = $.65

c.

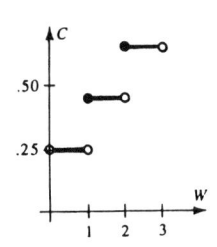

45. Answers vary. A few examples are: $f(x) = \dfrac{1}{x - 2}$ or $g(x) = \dfrac{x^2 - 2x}{x - 2}$ or $h(x) = \dfrac{|x - 2|}{x - 2}$

2.3 RATES OF CHANGE, PAGES 68 - 71

1. $\dfrac{\text{change in height}}{\text{change in time}} = \dfrac{80 - 40}{7 - 1} = \dfrac{20}{3}$ ft/sec   **3.** $\dfrac{\text{change in height}}{\text{change in time}} = \dfrac{80 - 40}{2 - 1} = 40$ ft/sec

5. $\dfrac{\text{change in output}}{\text{change in number of workers}} = \dfrac{2000 - 1000}{800 - 100} = \dfrac{10}{7}$

Chapter 2

7. $\dfrac{\text{change in output}}{\text{change in number of workers}} = \dfrac{2000 - 1750}{800 - 500} = \dfrac{5}{6}$

9. $\dfrac{\text{change in scores}}{\text{change in time}} = \dfrac{530 - 580}{1984 - 1979} = -10\,\dfrac{\text{points}}{\text{year}}$

11. $\dfrac{\text{change in scores}}{\text{change in time}} = \dfrac{530 - 548}{1984 - 1982} = -9\,\dfrac{\text{points}}{\text{year}}$

13. $\dfrac{\text{change in GNP}}{\text{change in time}} = \dfrac{3.66 - 2.16}{1984 - 1978} = \dfrac{1}{4}$

15. $\dfrac{\text{change in GNP}}{\text{change in time}} = \dfrac{3.66 - 2.63}{1984 - 1980} = \dfrac{1.03}{4} \approx .2575$

17. Average speed $= \dfrac{26.7}{6{:}36 - 6{:}09} \approx .99\,\dfrac{\text{miles}}{\text{minute}} \cdot \dfrac{60\ \text{minutes}}{\text{hour}} \approx 59.3\,\dfrac{\text{miles}}{\text{hour}}$

19. Average speed $= \dfrac{13.2}{7{:}28 - 7{:}06} = .6\,\dfrac{\text{miles}}{\text{minute}} \cdot \dfrac{60\ \text{minutes}}{\text{hour}} = 36\,\dfrac{\text{miles}}{\text{hour}}$

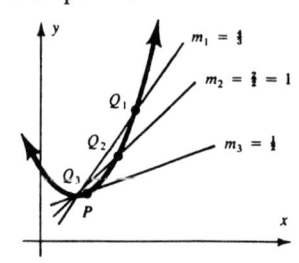

21. slope $\to 0$ **23.** slope $= 2$ for <u>each</u> value of Q.

25. slope $\to 0$

27. $\dfrac{f(2) - f(-3)}{2 - (-3)} = \dfrac{[4 - 3(2)] - [4 - 3(-3)]}{5} = \dfrac{-15}{5} = -3$

29. $\dfrac{f(3) - f(-3)}{3 - (-3)} = \dfrac{5 - 5}{6} = 0$ **31.** $\dfrac{f(3) - f(1)}{3 - 1} = \dfrac{3(3)^2 - 3(1)^2}{2} = 12$

33. $\dfrac{[-2(4)^2 + (4) + 4] - [-2(1)^2 + (1) + 4]}{4 - 1} = \dfrac{-24 - 3}{3} = -9$

35. $\dfrac{\left[\dfrac{-2}{(5)+1}\right] - \left[\dfrac{-2}{(1)+1}\right]}{5 - 1} = \dfrac{-\dfrac{1}{3} + 1}{4} = \dfrac{1}{6}$

37. $\lim\limits_{h \to 0} \dfrac{f(-5 + h) - f(-5)}{h} = \lim\limits_{h \to 0} \dfrac{[5(-5 + h) - 1] - [5(-5) - 1]}{h} = \lim\limits_{h \to 0} \dfrac{5h}{h} = 5$

39. $\lim\limits_{h \to 0} \dfrac{f(0 + h) - f(0)}{h} = \lim\limits_{h \to 0} \dfrac{(-2) - (-2)}{h} = \lim\limits_{h \to 0} 0 = 0$

Chapter 2

41. $\lim\limits_{h \to 0} \dfrac{f(0+h) - f(0)}{h} = \lim\limits_{h \to 0} \dfrac{[(0+h)^2 - 3(0+h)] - [0^2 - 3(0)]}{h} = \lim\limits_{h \to 0} \dfrac{h^2 - 3h}{h}$

$= \lim\limits_{h \to 0} (h - 3) = -3$

43. $\lim\limits_{h \to 0} \dfrac{\sqrt{(4+h)} - \sqrt{4}}{h} = \lim\limits_{h \to 0} \dfrac{\sqrt{4+h} - 2}{h} \cdot \dfrac{\sqrt{4+h} + 2}{\sqrt{4+h} + 2} = \lim\limits_{h \to 0} \dfrac{(4+h) - 4}{h(\sqrt{4+h} + 2)}$

$= \lim\limits_{h \to 0} \dfrac{1}{\sqrt{4+h} + 2} = \dfrac{1}{\sqrt{4} + 2} = \dfrac{1}{4}$

45. $\dfrac{P(30) - P(20)}{30 - 20} = \dfrac{[50(30) - 30^2] - [50(20) - 20^2]}{10} = 0$

47. $\dfrac{P(21) - P(20)}{21 - 20} = \dfrac{[50(21) - 21^2] - [50(20) - 20^2]}{1} = 9$

49. $\dfrac{C(200) - C(100)}{200 - 100} = \dfrac{[30(200)^2 - 100(200)] - [30(100)^2 - 100(100)]}{100} = 8900$

51. $\dfrac{C(101) - C(100)}{101 - 100} = \dfrac{[30(101)^2 - 100(101)] - [30(100)^2 - 100(100)]}{1} = 5930$

53. $\dfrac{C(x+h) - C(x)}{h} = \dfrac{[30(x+h)^2 - 100(x+h)] - [30(x)^2 - 100(x)]}{h}$

$= \dfrac{[30(x^2 + 2xh + h^2) - 100(x+h)] - [30x^2 - 100x]}{h}$

$= \dfrac{30x^2 + 60xh + 30h^2 - 100x - 100h - 30x^2 + 100x}{h} = \dfrac{60xh + 30h^2 - 100h}{h}$

$= 60x + 30h - 100$

55. See Problem 53 for details.

$\lim\limits_{h \to 0} \dfrac{C(x+h) - C(x)}{h} = \lim\limits_{h \to 0} (60x + 30h - 100) = 60x + 0 - 100 = 60x - 100$

57. The instantaneous rate of change of cost at a given x.

2.4 DEFINITION OF DERIVATIVE, PAGES 75 - 76

1. The "corner" point at (1,0) has no derivative because

$\lim\limits_{h \to 0^+} \dfrac{f(1+h) - f(1)}{h} = -2$ and $\lim\limits_{h \to 0^-} \dfrac{f(1+h) - f(1)}{h} = 2$,

Chapter 2

so the $\lim_{h \to 0} \dfrac{f(1+h) - f(1)}{h}$ does not exist.   **3.** Since the function is discontinuous at $x = 3$, the derivative does not exist at $(3,1)$. The function is differentiable everywhere else.

5. Since the function is discontinuous at $x = 4$, the derivative does not exist at $x = 4$.

7. $f'(x) = \lim\limits_{h \to 0} \dfrac{f(x+h) - f(x)}{h} = \lim\limits_{h \to 0} \dfrac{2(x+h)^2 - 2x^2}{h} = \lim\limits_{h \to 0} \dfrac{2(x^2 + 2xh + h^2) - 2x^2}{h}$

$= \lim\limits_{h \to 0} \dfrac{4xh + 2h^2}{h} = \lim\limits_{h \to 0} (4x + 2h) = 4x$

9. $f'(x) = \lim\limits_{h \to 0} \dfrac{f(x+h) - f(x)}{h} = \lim\limits_{h \to 0} \dfrac{[-3(x+h)^2] - [-3x^2]}{h} = \lim\limits_{h \to 0} \dfrac{-3(x^2 + 2xh + h^2) + 3x^2}{h}$

$= \lim\limits_{h \to 0} \dfrac{-6xh - 3h^2}{h} = \lim\limits_{h \to 0} (-6x - 3h) = -6x$

11. $f'(x) = \lim\limits_{h \to 0} \dfrac{f(x+h) - f(x)}{h} = \lim\limits_{h \to 0} \dfrac{[4 - 5(x+h)] - [4 - 5x]}{h} = \lim\limits_{h \to 0} \dfrac{4 - 5x - 5h - 4 + 5x}{h}$

$= \lim\limits_{h \to 0} (-5) = -5$

13. $f'(x) = \lim\limits_{h \to 0} \dfrac{f(x+h) - f(x)}{h} = \lim\limits_{h \to 0} \dfrac{[3(x+h)^2 + 4(x+h)] - [3x^2 + 4x]}{h}$

$= \lim\limits_{h \to 0} \dfrac{3(x^2 + 2xh + h^2) + 4x + 4h - 3x^2 - 4x}{h} = \lim\limits_{h \to 0} \dfrac{6xh + 3h^2 + 4h}{h}$

$= \lim\limits_{h \to 0} 6x + 3h + 4 = 6x + 4$

15. $f'(x) = \lim\limits_{h \to 0} \dfrac{f(x+h) - f(x)}{h} = \lim\limits_{h \to 0} \dfrac{[-3(x+h)^2 - 50(x+h) + 125] - [-3x^2 - 50x + 125]}{h}$

$= \lim\limits_{h \to 0} \dfrac{-3(x^2 + 2xh + h^2) - 50x - 50h + 125 + 3x^2 + 50x - 125}{h} = \lim\limits_{h \to 0} \dfrac{-6xh - 3h^2 - 50h}{h}$

$= \lim\limits_{h \to 0} (-6x - 3h - 50) = -6x - 50$

17. The point of tangency is at $x = -3$, $y = 5x^2 = 5(-3)^2 = 45$. The slope of the tangent line at any x is

Chapter 2

$$f'(x) = \lim_{h \to 0} \frac{f(x+h) - f(x)}{h} = \lim_{h \to 0} \frac{5(x+h)^2 - 5x^2}{h} = \lim_{h \to 0} \frac{5(x^2 + 2xh + h^2) - 5x^2}{h}$$

$= \lim_{h \to 0} 10x + 5h = 10x$. In particular, the slope at $x = -3$ is $f'(-3) = 10(-3) = -30$.

Using the point-slope form with $m = -30$ and $(x_1, y_1) = (-3, 45)$:
$y - 45 = -30(x + 3)$ or $30x + y + 45 = 0$

19. The point of tangency is at $x = -2$, $y = 4 - 5x = 4 - 5(-2) = 14$. The slope of the tangent line is $f'(x) = -5$. (See Problem 11.) Using the point slope form with $m = -5$ and $(x_1, y_1) = (-2, 14)$: $y - 14 = -5(x + 2)$ or $5x + y - 4 = 0$.

21. The point of tangency is at $x = -1$, $y = 3 + 2x - 3x^2 = -2$. The slope of the tangent line is

$$f'(x) = \lim_{h \to 0} \frac{f(x+h) - f(x)}{h} = \lim_{h \to 0} \frac{[3 + 2(x+h) - 3(x+h)^2] - [3 + 2x - 3x^2]}{h}$$

$$= \lim_{h \to 0} \frac{3 + 2x + 2h - 3(x^2 + 2xh + h^2) - 3 - 2x + 3x^2}{h} = \lim_{h \to 0} \frac{2h - 6xh - 3h^2}{h}$$

$= \lim_{h \to 0} (2 - 6x - 3h) = 2 - 6x$

So, at $x = -1$, the slope is $f'(-1) = 2 - 6(-1) = 8$. Using the point-slope form with $m = 8$ and $(x_1, y_1) = (-1, -2)$: $y + 2 = 8(x + 1)$ or $8x - y + 6 = 0$

23. $C'(20) = \lim_{h \to 0} \frac{C(20 + h) - C(20)}{h}$

$$= \lim_{h \to 0} \frac{[(20+h)^2 - 60(20+h) + 2500] - [20^2 - 60(20) + 2500]}{h}$$

$$= \lim_{h \to 0} \frac{400 + 40h + h^2 - 1200 - 60h + 2500 - 400 + 1200 - 2500}{h} = \lim_{h \to 0} (40 + h - 60) = -20$$

25. $N'(t) = \lim_{h \to 0} \frac{N(t+h) - N(t)}{h} = \lim_{h \to 0} \frac{[2(t+h)^2 - 200(t+h) + 1000] - [2t^2 - 200t + 1000]}{h}$

$$= \lim_{h \to 0} \frac{2(t^2 + 2th + h^2) - 200t - 200h + 1000 - 2t^2 + 200t - 1000}{h} = \lim_{h \to 0} \frac{4th + 2h^2 - 200h}{h}$$

$= \lim_{h \to 0} (4t + 2h - 200) = 4t - 200$

27. (See Problem 25.) The rate of change of number of bacteria with respect to time is

Chapter 2

$N'(t) = 4t - 200$. At the beginning, $t = 0$, so $N'(0) = 4(0) - 200 = -200$ bacteria per unit time.

29. $g'(x) = \lim_{h \to 0} \dfrac{g(x+h) - g(x)}{h} = \lim_{h \to 0} \dfrac{(x+h)^3 - x^3}{h} = \lim_{h \to 0} \dfrac{x^3 + 3x^2h + 3xh^2 + h^3 - x^3}{h}$

$= \lim_{h \to 0} (3x^2 + 3xh + h^2) = 3x^2$

31. $f'(x) = \lim_{h \to 0} \dfrac{f(x+h) - f(x)}{h} = \lim_{h \to 0} \dfrac{\left(-\dfrac{3}{x+h}\right) - \left(-\dfrac{3}{x}\right)}{h}$

$= \lim_{h \to 0} \dfrac{-\dfrac{3}{x+h} \cdot \dfrac{x}{x} + \dfrac{3}{x} \cdot \dfrac{(x+h)}{(x+h)}}{h} = \lim_{h \to 0} \dfrac{1}{h}\left(\dfrac{-3x + 3(x+h)}{x(x+h)}\right) = \lim_{h \to 0} \dfrac{1}{h}\left(\dfrac{3h}{x(x+h)}\right)$

$= \lim_{h \to 0} \dfrac{3}{x(x+h)} = \dfrac{3}{x^2}$

33. $y' = \lim_{h \to 0} \dfrac{2\sqrt{x+h} - 2\sqrt{x}}{h} = \lim_{h \to 0} \dfrac{2\sqrt{x+h} - 2\sqrt{x}}{h} \cdot \dfrac{2\sqrt{x+h} + 2\sqrt{x}}{2\sqrt{x+h} + 2\sqrt{x}}$

$= \lim_{h \to 0} \dfrac{4(x+h) - 4x}{h(2\sqrt{x+h} + 2\sqrt{x})} = \lim_{h \to 0} \dfrac{4}{2\sqrt{x+h} + 2\sqrt{x}} = \dfrac{4}{2\sqrt{x} + 2\sqrt{x}} = \dfrac{1}{\sqrt{x}}$

2.5 DIFFERENTIATION TECHNIQUES, PART I, PAGES 80 - 81

1. $y' = 7x^6$ **3.** $y' = 12x^{11}$ **5.** $y = -5x^{-6}$ **7.** $y' = 0$ **9.** $y' = -4(-8x^{-9}) = 32x^{-9}$

11. First, $y = -\dfrac{1}{2}x^{\frac{1}{2}}$; so $y' = -\dfrac{1}{2}\left(\dfrac{1}{2}x^{-\frac{1}{2}}\right) = -\dfrac{1}{4}x^{-\frac{1}{2}} = -\dfrac{1}{4\sqrt{x}} = -\dfrac{\sqrt{x}}{4x}$

13. $y' = 5(-8x^{-9}) = -40x^{-9}$ **15.** $y' = 12\left(\dfrac{5}{4}x^{\frac{1}{4}}\right) = 15 x^{\frac{1}{4}}$

17. Use the sum rule: $y' = (3x^2)' + (x)' = 6x + 1$

19. Use the sum and difference rules: $y' = (2x^2)' - (5x)' - (6)' = 4x - 5 - 0 = 4x - 5$

21. Use the sum and difference rules: $y' = (5x^3)' - (5x^2)' + (4x)' - (5)'$

$= 15x^2 - 10x + 4 - 0 = 15x^2 - 10x + 4$

23. Use the sum rule: $y' = (x^{-3})' + (x^2)' + (x^{-1})' = -3x^{-4} + 2x + (-1)x^{-2} = -\dfrac{3}{x^4} + 2x - \dfrac{1}{x^2}$

$= \dfrac{-3 + 2x^5 - x^2}{x^4} = \dfrac{2x^5 - x^2 - 3}{x^4}$

Chapter 2

25. First, $y = (x^4 + 2)^2 = x^8 + 4x^4 + 4$; so $y' = 8x^7 + 16x^3 + 0 = 8x^7 + 16x^3$

27. First, $y = \dfrac{2}{x} + \dfrac{5}{x^2} = 2x^{-1} + 5x^{-2}$; so $y' = -2x^{-2} + (-10)x^{-3} = \dfrac{-2}{x^2} - \dfrac{10}{x^3} = \dfrac{-2x - 10}{x^3}$

29. First, $y = -5x^7 + 2x^{\frac{1}{2}} - 3x^{-1}$; so $y' = -35x^6 + 1x^{-\frac{1}{2}} - (-3)x^{-2} = -35x^6 + x^{-\frac{1}{2}} + 3x^{-2}$

$= \dfrac{-35x^8 + x\sqrt{x} + 3}{x^2}$

31. Marginal cost is $C'(x) = (20x^2)' + (500x)' + (250,000)' = 40x + 500 + 0 = 40x + 500$

33. $m = \dfrac{400}{t} = 400t^{-1}$, so $\dfrac{dm}{dt} = 400(-1)t^{-2} = \dfrac{-400}{t^2}$

35. The rate of change of earnings at any t is $A'(t) = (.05t^2)' + (25t)' + (5)' = .10t + 25$. Since 1982 is the initial time (t = 0), 1988 corresponds to t = 6, $A'(6) = .10(6) + 25 = 25.6$; that is, the rate of change of earnings is 25.6 (thousand dollars per year).

37. The instantaneous rate of change of number of tasks (N) per hour (unit change in x) is

$N' = \dfrac{dN}{dt} = 25\left(\dfrac{1}{2}\right)x^{-\frac{1}{2}} = \dfrac{25}{2\sqrt{x}}$. At the end of the fifth hour (x = 5), the subjects are learning

at the rate of $\dfrac{25}{2\sqrt{5}} \approx 6$ tasks per hour.

39. The rate of change of P with respect to x is $P'(x) = .0005 + .00001(2x)$
$= .0005 + .00002x$, so the population is changing at the rate of $P'(1,000,000)$
$= .0005 + (.00002)(1,000,000) \approx 20$ foxes per rabbit.

41. $f'(x) = \lim\limits_{h \to 0} \dfrac{f(x+h) - f(x)}{h} = \lim\limits_{h \to 0} \dfrac{(x+h)^4 - x^4}{h}$

$= \lim\limits_{h \to 0} \dfrac{x^4 + 4x^3h + 6x^2h^2 + 4xh^3 + h^4 - x^4}{h} = \lim\limits_{h \to 0} \dfrac{4x^3h + 6x^2h^2 + 4xh^3 + h^4}{h}$

$= \lim\limits_{h \to 0} (4x^3 + 6x^2h + 4xh^2 + h^3) = 4x^3 + 0 + 0 + 0 = 4x^3$

43. Let $d(x) = f(x) - g(x)$

$d'(x) = \lim\limits_{h \to 0} \dfrac{d(x+h) - d(x)}{h} = \lim\limits_{h \to 0} \dfrac{[f(x+h) - g(x+h)] - [f(x) - g(x)]}{h}$

$= \lim\limits_{h \to 0} \dfrac{f(x+h) - f(x) - g(x+h) + g(x)}{h} = \lim\limits_{h \to 0} \left[\dfrac{f(x+h) - f(x)}{h} - \dfrac{g(x+h) - g(x)}{h}\right]$

Chapter 2

$$= \lim_{h \to 0} \frac{f(x + h) - f(x)}{h} - \lim_{h \to 0} \frac{g(x + h) - g(x)}{h} = f'(x) - g'(x)$$

2.6 DIFFERENTIATION TECHNIQUES, PART II, PAGE 85

1. Use the Product Rule: $f'(x) = (5x^2)'(x^2 - 6) + 5x^2(x^2 - 6)' = 10x(x^2 - 6) + 5x^2(2x - 0)$

$= 10x^3 - 60x + 10x^3 = 20x^3 - 60x$ or multiply first: $f(x) = 5x^4 - 30x^2$ so $f'(x) = 20x^3 - 60x$

3. Use the Product Rule: $f'(x) = (x + 1)'(x - 2) + (x + 1)(x - 2)'$

$= (1 + 0)(x - 2) + (x + 1)(1 - 0) = 2x - 1$ or multiply first: $f(x) = x^2 - x - 2$ so $f'(x) = 2x - 1$

5. Use the Product Rule: $g'(x) = (3x^2 + 5)'(2x^2 - 5) + (3x^2 + 5)(2x^2 - 5)'$

$= (6x + 0)(2x^2 - 5) + (3x^2 + 5)(4x - 0) = 12x^3 - 30x + 12x^3 + 20x = 24x^3 - 10x$ or multiply first: $g(x) = 6x^4 - 5x^2 - 25$ so $g'(x) = 24x^3 - 10x$

7. Use the Product Rule: $g'(x) = (5x^4)'(2x^2 - 5x + 1) + (5x^4)(2x^2 - 5x + 1)'$

$= 20x^3(2x^2 - 5x + 1) + 5x^4(4x - 5 + 0) = 40x^5 - 100x^4 + 20x^3 + 20x^5 - 25x^4$

$= 60x^5 - 125x^4 + 20x^3$ or multiply first: $g(x) = 10x^6 - 25x^5 + 5x^4$

so $g'(x) = 60x^5 - 125x^4 + 20x^3$

9. Use the Quotient Rule: $y' = \dfrac{(x - 3)x' - x(x - 3)'}{(x - 3)^2} = \dfrac{(x - 3)(1) - (x)(1)}{(x - 3)^2} = \dfrac{-3}{(x - 3)^2}$

11. Use the Quotient Rule: $y' = \dfrac{(x - 3)(x + 5)' - (x + 5)(x - 3)'}{(x - 3)^2} = \dfrac{(x - 3)(1) - (x + 5)(1)}{(x - 3)^2}$

$= \dfrac{x - 3 - x - 5}{(x - 3)^2} = \dfrac{-8}{(x - 3)^2}$

13. Use the Quotient Rule: $y' = \dfrac{(x^2 - 5)(x^2 + 3)' - (x^2 + 3)(x^2 - 5)'}{(x^2 - 5)^2}$

$= \dfrac{(x^2 - 5)(2x) - (x^2 + 3)(2x)}{(x^2 - 5)^2} = \dfrac{2x^3 - 10x - 2x^3 - 6x}{(x^2 - 5)^2} = \dfrac{-16x}{(x^2 - 5)^2}$

15. Use the Quotient Rule: $y' = \dfrac{(8x^2 - 3x + 1)(x^2 + 2x - 5)' - (x^2 + 2x - 5)(8x^2 - 3x + 1)'}{(8x^2 - 3x + 1)^2}$

$= \dfrac{(8x^2 - 3x + 1)(2x + 2) - (x^2 + 2x - 5)(16x - 3)}{(8x^2 - 3x + 1)^2}$

Chapter 2

$$= \frac{(16x^3 + 10x^2 - 4x + 2) - (16x^3 + 29x^2 - 86x + 15)}{(8x^2 - 3x + 1)^2} = \frac{-19x^2 + 82x - 13}{(8x^2 - 3x + 1)^2}$$

17. Use the Product Rule: $y' = (2x^2 - 1)'(x^3 + 2x^2 - 3) + (2x^2 - 1)(x^3 + 2x^2 - 3)'$

$= (4x)(x^3 + 2x^2 - 3) + (2x^2 - 1)(3x^2 + 4x) = 4x^4 + 8x^3 - 12x + 6x^4 + 8x^3 - 3x^2 - 4x$

$= 10x^4 + 16x^3 - 3x^2 - 16x$

19. Use the Product Rule: $y' = (4x^2 + x)'(x^3 - 3x^2 + 13) + (4x^2 + x)(x^3 - 3x^2 + 13)'$

$= (8x + 1)(x^3 - 3x + 13) + (4x^2 + x)(3x^2 - 6x)$

$= (8x^4 - 23x^3 - 3x^2 + 104x + 13) + (12x^4 - 21x^3 - 6x^2) = 20x^4 - 44x^3 - 9x^2 + 104x + 13$

21. Use the Product Rule: $y' = (3x^3 - 2x + 5)'(2x^4 + 5x - 9) + (3x^3 - 2x + 5)(2x^4 + 5x - 9)'$

$= (9x^2 - 2)(2x^4 + 5x - 9) + (3x^3 - 2x + 5)(8x^3 + 5)$

$= (18x^6 - 4x^4 + 45x^3 - 81x^2 - 10x + 18) + (24x^6 - 16x^4 + 55x^3 - 10x + 25)$

$= 42x^6 - 20x^4 + 100x^3 - 81x^2 - 20x + 43$

23. Use the Product Rule: $y' = \left(5x^{\frac{2}{3}}\right)'\left(5x^{-1} + 3x\right) + 5x^{\frac{2}{3}}\left(5x^{-1} + 3x\right)'$

$= \left(\frac{10}{3}x^{-\frac{1}{3}}\right)\left(5x^{-1} + 3x\right) + 5x^{\frac{2}{3}}\left(-5x^{-2} + 3\right) = \frac{50}{3}x^{-\frac{4}{3}} + 10x^{\frac{2}{3}} - 25x^{-\frac{4}{3}} + 15x^{\frac{2}{3}}$

$= -\frac{25}{3}x^{-\frac{4}{3}} + 25x^{\frac{2}{3}}$

25. Use the Product Rule: $f'(x) = (6\sqrt{x})'(2x^2 - 5) + (6\sqrt{x})(2x^2 - 5)'$

$= \left(6x^{\frac{1}{2}}\right)'\left(2x^2 - 5\right) + (6\sqrt{x})\left(2x^2 - 5\right)' = 3x^{-\frac{1}{2}}\left(2x^2 - 5\right) + 6\sqrt{x}\,(4x)$

$= 6x^{\frac{3}{2}} - 15x^{-\frac{1}{2}} + 24x^{\frac{3}{2}} = 30x^{\frac{3}{2}} - 15x^{-\frac{1}{2}} = \frac{15\sqrt{x}\,(2x^2 - 1)}{x}$

27. Use the Quotient Rule: $g'(x) = \frac{(5x^2 - 11x + 3)(2x)' - (2x)(5x^2 - 11x + 3)'}{(5x^2 - 11x + 3)^2}$

$= \frac{(5x^2 - 11x + 3)(2) - (2x)(10x - 11)}{(5x^2 - 11x + 3)^2} = \frac{(10x^2 - 22x + 6) - (20x^2 - 22x)}{(5x^2 - 11x + 3)^2} = \frac{-10x^2 + 6}{(5x^2 - 11x + 3)^2}$

29. The slope of the tangent line is given by $f'(x)$:

Chapter 2

$$f'(x) = \frac{(x-5)(5x^2+5x)' - (5x^2+5x)(x-5)'}{(x-5)^2} = \frac{(x-5)(10x+5) - (5x^2+5x)(1)}{(x-5)^2}$$

$$= \frac{10x^2 - 45x - 25 - 5x^2 - 5x}{(x-5)^2} = \frac{5x^2 - 50x - 25}{(x-5)^2}$$

In particular, the slope of the tangent line at $x = 1$ is $f'(1) = \frac{5(1)^2 - 50(1) - 25}{(1-5)^2} = -\frac{35}{8}$

31. The slope of the tangent line is given by $f'(x)$:

$$f'(x) = \left(x^{\frac{1}{2}}\right)'(x^2+3) + x^{\frac{1}{2}}(x^2+3)' = \frac{1}{2}x^{-\frac{1}{2}}(x^2+3) + x^{\frac{1}{2}}(2x) = \frac{5}{2}x^{\frac{3}{2}} + \frac{3}{2}x^{-\frac{1}{2}}$$

In particular, the slope of the tangent line at $x = 4$ is

$$f'(4) = \frac{5}{2}(4)^{\frac{3}{2}} + \frac{3}{2}4^{-\frac{1}{2}} = \frac{5}{2}(8) + \frac{3}{2}(\frac{1}{2}) = \frac{83}{4}$$

33. The slope of the tangent line is given by $f'(x)$

$$f'(x) = \frac{(x^2+4x+4)(x^2-5x+1)' - (x^2-5x+1)(x^2+4x+4)'}{(x^2+4x+4)^2}$$

$$= \frac{(x^2+4x+4)(2x-5) - (x^2-5x+1)(2x+4)}{(x^2+4x+4)^2} = \frac{9x^2+6x-24}{(x^2+4x+4)^2} = \frac{3[3x^2+2x-8]}{(x^2+4x+4)^2}$$

$$= \frac{3(3x-4)(x+2)}{[(x+2)(x+2)]^2} = \frac{3(3x-4)}{(x+2)^3}$$

In particular, at $x = 0$ the slope of the tangent line is $f'(0) = \frac{3(-4)}{2^3} = -\frac{3}{2}$

35. The slope of the tangent line is $f'(x) = \frac{5x^2 - 50x - 25}{(x-5)^2}$. (See Problem 29.) In particular, at $x = 0$ $f'(0) = -1$. The point of tangency is at $(0, f(0))$ which is $(0, 0)$. Using the point-slope equation of a line: $y - y_1 = m(x - x_1)$

$$y - 0 = -1(x - 0)$$

$$y = -x \text{ or } x + y = 0$$

37. The slope of the tangent line at $x = 4$ is $f'(4) = \frac{83}{4}$. (See Problem 31.) The point of tangency is $(4, f(4))$ which is $(4, 38)$. Using the point-slope equation of a line:

$$y - y_1 = m(x - x_1)$$

$$y - 38 = \frac{83}{4}(x - 4)$$

$4y - 152 = 83(x - 4)$ Multiply both sides by 4.

$4y - 152 = 83x - 332$

$83x - 4y - 180 = 0$

39. The slope of the tangent line is $f'(x) = \frac{3(3x - 4)}{(x + 2)^3}$. (See Problem 33.) The slope at $x = 1$ is $f'(1) = -\frac{1}{9}$. The point of tangency is $(1, f(1))$ which is $(1, -\frac{1}{3})$. Using the point-slope equation of a line:

$y - y_1 = m(x - x_1)$

$y + \frac{1}{3} = -\frac{1}{9}(x - 1)$

$9y + 3 = -1(x - 1)$ Multiply both sides by 9.

$9y + 3 = -x + 1$

$x + 9y + 2 = 0$

41. The rate of change of D (demand) with respect to x (price) is:

$$D'(x) = \frac{(x^2 + 15x + 25)(100,000)' - (100,000)(x^2 + 15x + 25)'}{(x^2 + 15x + 25)^2} = \frac{0 - 100,000(2x + 15)}{(x^2 + 15x + 25)^2}$$

$$= \frac{-200,000x - 1,500,000}{(x^2 + 15x + 25)^2}$$

43. The rate of change of f (effect) relative to t (time) is:

$$f'(t) = \frac{(.01 + .005t)\, t' - t\,(.01 + .005t)'}{(.01 + .005t)^2} = \frac{(.01 + .005t) - t(.005)}{(.01 + .005t)^2} = \frac{.01}{(.01 + .005t)^2}$$

2.7 THE CHAIN RULE, PAGES 88 - 89

1. $f'(x) = 3(3x + 2)^2 \cdot \frac{d}{dx}(3x + 2) = 3(3x + 2)^2 (3) = 9(3x + 2)^2$

3. $f'(x) = 4(5x - 1)^3 \cdot \frac{d}{dx}(5x - 1) = 4(5x - 1)^3 (5) = 20(5x - 1)^3$

5. $y' = 3(2x^2 + x)^2 \cdot \frac{d}{dx}(2x^2 + x) = 3(2x^2 + x)^2 (4x + 1)$ or $3x^2(2x + 1)^2(4x + 1)$

7. $y' = 2(2x^2 - 3x + 2) \cdot \frac{d}{dx}(2x^2 - 3x + 2) = 2(2x^2 - 3x + 2)(4x - 3)$

9. $g'(x) = 4(x^3 + 5x)^3 \cdot \frac{d}{dx}(x^3 + 5x) = 4(x^3 + 5x)^3 (3x^2 + 5)$ or $4x^3(x^2 + 5)^3(3x^2 + 5)$

11. $m'(x) = -2(2x^2 - 5x)^{-3} \cdot \dfrac{d}{dx}(2x^2 - 5x) = -2(2x^2 - 5x)^{-3}(4x - 5)$

or $-2x^{-3}(2x - 5)^{-3}(4x - 5)$

13. $t'(x) = -1(x^4 + 3x^3)^{-2} \cdot \dfrac{d}{dx}(x^4 + 3x^3) = -1(x^4 + 3x^3)^{-2}(4x^3 + 9x^2)$

$= -1x^{-6}(x + 3)^{-2}x^2(4x + 9) = -x^{-4}(x + 3)^{-2}(4x + 9)$

15. $y' = 15(4x^3 + 3x^2)^2 \cdot (4x^3 + 3x^2)' = 15(4x^3 + 3x^2)^2(12x^2 + 6x)$

$= 15x^4(4x + 3)^2(6x)(2x + 1) = 90x^5(4x + 3)^2(2x + 1)$

17. $y' = \dfrac{1}{4}(x^2 - 3x)^{-\frac{3}{4}} \cdot (x^2 - 3x)' = \dfrac{1}{4}(x^2 - 3x)^{-\frac{3}{4}} \cdot (2x - 3)$ or $\dfrac{2x - 3}{4(x^2 - 3x)^{\frac{3}{4}}}$

19. $y = (x^2 + 16)^{\frac{1}{2}}$

$y' = \dfrac{1}{2}(x^2 + 16)^{-\frac{1}{2}} \cdot (x^2 + 16)' = \dfrac{1}{2}(x^2 + 16)^{-\frac{1}{2}}(2x) = x(x^2 + 16)^{-\frac{1}{2}}$ or $\dfrac{x\sqrt{x^2 + 16}}{x^2 + 16}$

21. $y = 5(x^3 + 8)^{\frac{1}{2}}$

$y' = \dfrac{5}{2}(x^3 + 8)^{-\frac{1}{2}} \cdot (x^3 + 8)' = \dfrac{5}{2}(x^3 + 8)^{-\frac{1}{2}}(3x^2) = \dfrac{15}{2}x^2(x^3 + 8)^{-\frac{1}{2}}$

23. $y = x(3x + 1)^{\frac{1}{2}}$ Using the Product Rule:

$y' = x\left[(3x + 1)^{\frac{1}{2}}\right]' + x'(3x + 1)^{\frac{1}{2}} = x\left[\dfrac{1}{2}(3x + 1)^{-\frac{1}{2}} \cdot \dfrac{d}{dx}(3x + 1)\right] + (1)(3x + 1)^{\frac{1}{2}}$

$= x\left(\dfrac{1}{2}\right)(3x + 1)^{-\frac{1}{2}}(3) + (3x + 1)^{\frac{1}{2}} = \dfrac{3}{2}x(3x + 1)^{-\frac{1}{2}} + (3x + 1)^{\frac{1}{2}}$

$= \dfrac{1}{2}(3x + 1)^{-\frac{1}{2}}[3x + 2(3x + 1)]$ Factor out $\dfrac{1}{2}(3x + 1)^{-\frac{1}{2}}$.

$= \dfrac{1}{2}(3x + 1)^{-\frac{1}{2}}(9x + 2)$

25. $f'(x) = (2x + 1)^2[(3x + 2)^3]' + [(2x + 1)^2]'(3x + 2)^3$

$= (2x + 1)^2[3(3x + 2)^2 \cdot (3x + 2)'] + [2(2x + 1) \cdot (2x + 1)'](3x + 2)^3$

$= (2x + 1)^2[3(3x + 2)^2 \cdot 3] + [2(2x + 1)(2)](3x + 2)^3$

Chapter 2

$\qquad = (2x + 1)(3x + 2)^2 [9(2x + 1) + 4(3x + 2)]$ Factor out $(2x+1)(3x+2)^2$.

$\qquad = (2x + 1)(3x + 2)^2 (30x + 17)$

27. $y' = (5x + 1)^2 [(4x + 3)^{-1}]' + [(5x + 1)^2]' (4x + 3)^{-1}$

$\qquad = (5x + 1)^2 [-1(4x + 3)^{-2} \cdot (4x + 3)'] + [2(5x + 1) \cdot (5x + 1)'](4x + 3)^{-1}$

$\qquad = (5x + 1)^2 [-1(4x + 3)^{-2}(4)] + [2(5x + 1)(5)](4x + 3)^{-1}$

$\qquad = 2(5x + 1)(4x + 3)^{-2} [-2(5x + 1) + 5(4x + 3)]$

$\qquad = 2(5x + 1)(4x + 3)^{-2} (10x + 13)$

29. $g'(x) = \dfrac{(5x + 3)[(2x - 5)^2]' - (2x - 5)^2 [5x + 3]'}{(5x + 3)^2}$

$\qquad = \dfrac{(5x + 3)[2(2x - 5) \cdot (2x - 5)'] - (2x - 5)^2 [5]}{(5x + 3)^2}$

$\qquad = \dfrac{(5x + 3)[2(2x - 5)(2)] - 5(2x - 5)^2}{(5x + 3)^2}$

$\qquad = \dfrac{(2x - 5)[4(5x + 3) - 5(2x - 5)]}{(5x + 3)^2}$ Factor out $2x - 5$.

$\qquad = \dfrac{(2x - 5)(10x + 37)}{(5x + 3)^2}$

31. $t'(x) = \dfrac{(2x - 5)^2 [(x + 5)^4]' - (x + 5)^4 [(2x - 5)^2]'}{(2x - 5)^4}$

$\qquad = \dfrac{(2x - 5)^2 [4(x + 5)^3] - (x + 5)^4 [2(2x - 5)(2)]}{(2x - 5)^4}$

$\qquad = \dfrac{4(2x - 5)(x + 5)^3 [(2x - 5) - (x + 5)]}{(2x - 5)^4}$

$\qquad = \dfrac{4(2x - 5)(x + 5)^3 (x - 10)}{(2x - 5)^4}$

$\qquad = \dfrac{4(x + 5)^3 (x - 10)}{(2x - 5)^3}$

33. $f(x) = 3x(x^2 + 1)^{-\tfrac{1}{2}}$

$\quad f'(x) = (3x)'(x^2 + 1)^{-\tfrac{1}{2}} + 3x[(x^2 + 1)^{-\tfrac{1}{2}}$

Chapter 2

$$= 3(x^2+1)^{-\frac{1}{2}} + 3x[-\frac{1}{2}(x^2+1)^{-\frac{3}{2}} 2x]$$

$$= 3(x^2+1)^{-\frac{1}{2}} - 3x^2(x^2+1)^{-\frac{3}{2}}$$

$$= 3(x^2+1)^{-\frac{3}{2}}[(x^2+1) - x^2] \qquad \text{Factor out } 3(x^2+1)^{-\frac{3}{2}}$$

$$= 3(x^2+1)^{-\frac{3}{2}}$$

35. The rate of change at any t is:

$$N'(t) = 10^9 [(50 - 2t)^3]' = 10^9 [3(50 - 2t)^2 (-2)] = 10^9 (-6)(50 - 2t)^2.$$

At time t = 10: $N'(10) = 10^9 (-6)[(50 - 2(10)]^2 = -5400 \cdot 10^9$. The negative indicates the number is <u>decreasing</u> at the rate of 5,400,000,000,000 bacteria per minute.

37. $\dfrac{f}{g} = fg^{-1}$. Using the Product Rule,

$$(fg^{-1})' = f'(g^{-1}) + f(g^{-1})'$$

$$= f'g^{-1} + f(-g^{-2}g')$$

$$= \frac{f'}{g} - \frac{fg'}{g^2}$$

$$= \frac{gf'}{g^2} - \frac{fg'}{g^2}$$

$$= \frac{gf' - fg'}{g^2}$$

2.8 CHAPTER REVIEW, PAGES 89 - 90

1. 4   **3.** $\lim\limits_{x \to 2} \dfrac{6x+1}{3-x} = 13$ **5.** Look for jumps, holes, or "poles" (vertical asymptotes). There is a hole at x = 0, a jump at x = 3, and a pole at x = 10.

7. Let $f(x) = y = 5 + x - 2x^2$. The average rate of change is $\dfrac{f(3) - f(1)}{3 - 1} = \dfrac{f(3) - f(1)}{3 - 1}$

$= \dfrac{(-10) - (4)}{3 - 1} = -7$

9. The derivative of f(x) is $f'(x) = \lim\limits_{h \to 0} \dfrac{f(x+h) - f(x)}{h}$ provided this limit exists.

11. The derivative of <u>any</u> constant is zero: y' = 0.

Chapter 2

13. $y' = 5(2)x - 2(-1)x^{-2} = 10x + 2x^{-2}$

15. $y' = \dfrac{(4x-5)(15)' - (15)(4x-5)'}{(4x-5)^2} = \dfrac{0 - (15)(4)}{(4x-5)^2} = \dfrac{-60}{(4x-5)^2}$

17. $y' = \dfrac{(1-2x)[(x-5)^3]' - (x-5)^3(1-2x)'}{(1-2x)^2} = \dfrac{(1-2x)[3(x-5)^2] - (x-5)^3(-2)}{(1-2x)^2}$

Factor out $(x-5)^2$: $= \dfrac{(x-5)^2[3(1-2x) + 2(x-5)]}{(1-2x)^2} = \dfrac{(x-5)^2[-4x-7]}{(1-2x)^2}$

Factor out (-1): $= \dfrac{-(x-5)^2(4x+7)}{(1-2x)^2}$

19. $\dfrac{P(25) - P(20)}{25 - 20} = \dfrac{\dfrac{1000 - 25^2}{100 - 25} - \dfrac{1000 - 20^2}{100 - 20}}{25 - 20} = \dfrac{5 - 7.5}{5} = -.5$

The profit is decreasing by $.50 per unit.

2.9 STUDENT'S TEST REVIEW AND ADDITIONAL PRACTICE

OBJECTIVES

The material of this chapter is reviewed in the following list of objectives. After each objective there are some practice questions. Answers to these problems immediately follow. Detailed solutions are given for every third problem. For a sample test select the first question of each set and check your answers. Additional practice is given by the other questions in each set. If you are having trouble with a particular type of problem, or if you want additional practice, look back at the indicated section in the text.

[2.1] Objective 1: *Given a function defined by a graph, find a limit as $x \to a$.*

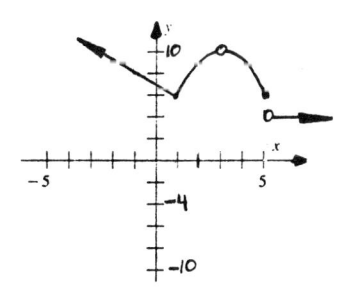

Chapter 2

1. $\lim\limits_{x \to 3} f(x)$ 2. $\lim\limits_{x \to 1} f(x)$ 3. $\lim\limits_{x \to 5} f(x)$ 4. $\lim\limits_{x \to \infty} f(x)$

Objective 2: *Find limits by using a calculator and filling in values on a table.*

5. $\lim\limits_{x \to 3} \dfrac{5x+1}{3-x}$

6. $\lim\limits_{x \to 0} \dfrac{5x+1}{3-x}$

7. $\lim\limits_{|x| \to \infty} \dfrac{5x+1}{3-x}$

8. $\lim\limits_{|x| \to \infty} \dfrac{145 - 2000x + 15x^3}{2x^2 - 3x^3}$

Objective 3: *Evaluate limits.*

9. $\lim\limits_{x \to 1} \dfrac{x^2 + 5x - 6}{x - 1}$

10. $\lim\limits_{x \to 1} \dfrac{x^2 + 7x + 6}{x - 1}$

11. $\lim\limits_{x \to -1} \dfrac{x^2 + 7x + 6}{x - 1}$

12. $\lim\limits_{x \to \infty} \dfrac{3x^2 - 5x + 10}{8x^2 + 2 - x - 5}$

[2.2] Objective 4: *From a graph, find all suspicious points and tell which of those are points of discontinuity.*

13.

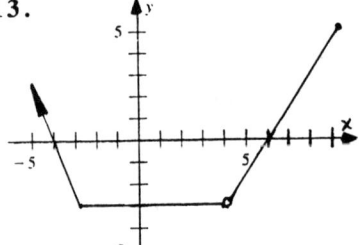

14.

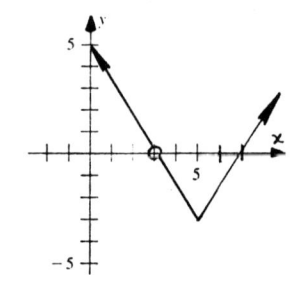

15.

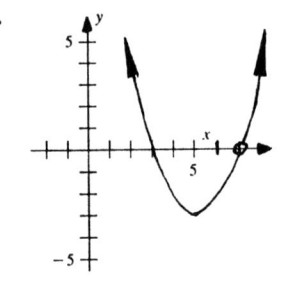

16.

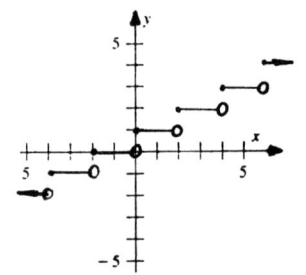

Objective 5: *Decide whether a given applied situation describes a continuous or a discontinuous function.*

17. The number of bacteria in a culture as a function of time.

Chapter 2

18. The distance that a skydiver falls as a function of the time since leaving the aircraft.
19. The odometer reading on an automobile as a function of the distance traveled.
20. The film of a moving picture of a skydiver falling as a function of the time since leaving the aircraft.

Objective 6: Given a function defined over a certain domain, determine whether the function is continuous at all points in the domain. Give the points of discontinuity.

21. $f(x) = \dfrac{x^2 - 15x + 56}{x - 8}$, $-5 \leq x \leq 5$

22. $f(x) = \dfrac{x^2 - 15x + 56}{x - 8}$, $0 \leq x \leq 10$

23. $f(x) = \begin{cases} \dfrac{x^2 - 15x + 56}{x - 8}, & 0 \leq x \leq 10,\ x \neq 8 \\ 4, & x = 8 \end{cases}$

24. $f(x) = \begin{cases} \dfrac{x^2 - 15x + 56}{x - 8}, & 0 \leq x \leq 10,\ x \neq 8 \\ 1, & x = 8 \end{cases}$

[2.3] *Objective 7: Find the average rate of change for a given function on some interval.*

25. $y = 3x^2 + 4x$ for $x = 1$ to $x = 3$
26. $y = 3 + 2x - x^2$ for $x = -1$ to 1
27. $y = \sqrt{3x}$ for $x = 0$ to 6
28. $y = \dfrac{1}{x} - 5$ for $x = 1$ to $x = 4$

Objective 8: Find the instantaneous rate of change for a given function at some point.

29. $y = 3x^2 + 4x$ at $x = 1$
30. $y = 3 + 2x - x^2$ at $x = -1$
31. $y = \sqrt{3x}$ at $x = 0$
32. $y = \dfrac{1}{x} - 5$ at $x = 1$

[2.4] *Objective 9: Know the definition of derivative and also use the definition to find a derivative.*

33. In your own words, state the definition of derivative.
34. $y = 3 - 8x^2$
35. $y - 3\sqrt{x}$
36. $y - \dfrac{1}{(x - 2)}$

[2.5] *Objective 10: Use the power rule, constant rule, constant times a function rule, sum rule, and difference rule to find the derivative of a given function.*

37. $y = x^{14}$
38. $y = x^{-8}$
39. $y = x^{-\frac{7}{9}}$

Chapter 2

40. $y = 2x$ 41. $y = 150$ 42. $y = -\dfrac{23}{25}$

43. $y = 2x^3 - 5x^2 + 12$ 44. $f(x) = 45 - 13x^2 - 5x^3$

[2.6] Objective 11: *Use the product and quotient rules to find derivatives.*

45. $y = (1 - 3x)(2 + 9x)$ 46. $y = 5x^3(x^2 - 3x + 9)$

47. $y = \sqrt{x}\,(x - 1)^{-1}$ 48. $y = 5x\sqrt{9 - x}$

49. $y = \dfrac{1}{(3x^2 + 1)}$ 50. $y = \dfrac{x + 10}{x - 5}$

51. $y = \dfrac{\sqrt{x}}{5x - 3}$ 52. $y = \dfrac{x}{\sqrt{x + 3}}$

[2.7] Objective 12: *Find the derivative of a function using the generalized power rule.*

53. $f(x) = (5x + 9)^4$ 54. $f(x) = (4 - 3x)^8$

55. $y = (5x^2 - 3x)^{\frac{1}{5}}$ 56. $y = \dfrac{(3x + 7)^5}{x^2 - 5}$

ANSWERS TO STUDENT'S TEST REVIEW AND PRACTICE QUESTIONS

1. 10   2. 6   3. As we approach x = 5 from the left, the function approaches 6, but as we approach x = 5 from the right, the function approaches 4; therefore, the limit does not exist.   4. 4

5.
x	2.9	2.99	2.999	3.1	3.01	3.001	3.0001
f(x)	155	1595	15995	-165	-1605	-16005	-160005

Limit does not exist.

6.
x	-.1	-.01	-.001	.1	.01	.001	.0001
f(x)	.1613	.3156	.332	.517	.351	.335	.3335

Limit is .333 ⋯ or $\dfrac{1}{3}$.

7.
x	100	1000	100,000	-100	-1000	-100,000
f(x)	-5.165	-5.02	-5.000	-4.84	-4.98	-4.9998

Limit is -5.

8.
x	100	1000	100,000	-100	-1000	-100,000
f(x)	-4.97	-5.003	-5.000	-4.9	-4,.996	-4.99997

Limit is -5.

Chapter 2

9. $\lim\limits_{x \to 1} \dfrac{x^2 + 5x - 6}{x - 1} = \lim\limits_{x \to 1} \dfrac{(x - 1)(x + 6)}{x - 1} = \lim\limits_{x \to 1} (x + 6) = 7$

10. Limit does not exist.   11. 0

12. $\lim\limits_{x \to \infty} \dfrac{3x^2 - 5x + 10}{8x^2 + 20x - 5} = \lim\limits_{x \to \infty} \dfrac{3 - \dfrac{5}{x} + \dfrac{10}{x^2}}{8 + \dfrac{20}{x} - \dfrac{5}{x^2}} = \dfrac{3 - 0 + 0}{8 + 0 - 0} = \dfrac{3}{8}$

13. Suspicious at $x = -3$ and $x = 4$. Discontinuous at $x = 4$.   14. Suspicious at $x = 3$ and $x = 5$. Discontinuous at $x = 3$.   15. Suspicious at $x = 7$. Discontinuous at $x = 7$.
16. Suspicious at x = even integer. Discontinuous at x = even integer.
17. Discontinuous   18. Continuous   19. Continuous   20. Discontinuous
21. Without the restriction, the function would be undefined at $x = 8$ (division by zero); but since we are limited to x values only between -5 and 5, $f(x)$ is continuous throughout the interval.   22. Discontinuous at $x = 8$ 23. Discontinuous at $x = 8$

24. First, $\lim\limits_{x \to 8} f(x) = \lim\limits_{x \to 8} \dfrac{x^2 - 15x + 56}{x - 8} = \lim\limits_{x \to 8} \dfrac{(x - 7)(x - 8)}{x - 8} = \lim\limits_{x \to 8} (x - 7) = 1$. The only suspicious point is at $x = 8$, but $\lim\limits_{x \to 8} f(x) = 1 = f(8)$, so $f(x)$ is continuous throughout $0 \leq x \leq 10$.

25. 16   26. 2   27. Let $y = f(x) = \sqrt{3x}$. The average rate of change is:

$\dfrac{f(6) - f(0)}{6} = \dfrac{\sqrt{18} - \sqrt{0}}{6} = \dfrac{\sqrt{2}}{2}$ 28. $-\dfrac{1}{4}$ 29. 10

30. The instantaneous rate of change at any x is:

$\lim\limits_{h \to 0} \dfrac{[3 + 2(x + h) - (x + h)^2] - [3 + 2x - x^2]}{h}$

$= \lim\limits_{h \to 0} \dfrac{[3 + 2x + 2h - x^2 - 2xh - h^2] - 3 - 2x + x^2}{h} = \lim\limits_{h \to 0} \dfrac{2h - 2xh - h^2}{h}$

$= \lim\limits_{h \to 0} (2 - 2x - h) = 2 - 2x$. At $x = -1$, we get $2 - 2(-1) = 4$.

31. 0   32. -1

33. For a function $f(x)$, the derivative is $f'(x) = \lim\limits_{h \to 0} \dfrac{f(x + h) - f(x)}{h}$ provided this limit exists.

34. $\lim\limits_{h \to 0} \dfrac{[3 - 8(x + h)^2] - [3 - 8x^2]}{h} = -16x$ 35. $\lim\limits_{h \to 0} \dfrac{3\sqrt{x + h} - 3\sqrt{x}}{h} = \dfrac{3\sqrt{x}}{2x}$

Chapter 2   58

36. $\lim\limits_{h \to 0} \dfrac{\left[\dfrac{1}{(x+h)-2}\right]-\left[\dfrac{1}{x-2}\right]}{h} = \lim\limits_{h \to 0} \dfrac{\dfrac{x-2}{(x-2)(x+h-2)} - \dfrac{x+h-2}{(x-2)(x+h-2)}}{h}$

$= \lim\limits_{h \to 0} \dfrac{-h}{(x-2)(x+h-2)} \cdot \dfrac{1}{h} = \lim\limits_{h \to 0} \left[\dfrac{-1}{(x-2)(x+h-2)}\right] = \dfrac{-1}{(x-2)^2}$

37. $14x^{13}$ 38. $-8x^{-9}$ 39. $-\dfrac{7}{9}x^{-\frac{16}{9}}$ 40. 2   41. 0

42. The derivative of <u>any</u> constant is 0.   43. $6x^2 - 10x$ 44. $-26x - 15x^2$

45. $y' = (1 - 3x)(2 + 9x)' + (1 - 3x)'(2 + 9x)$ 46. $y' = 25x^4 - 60x^3 + 135x^2$
$= (1 - 3x)(9) + (-3)(2 + 9x)$
$= 3 - 54x$

47. $\dfrac{\sqrt{x} - x\sqrt{x}}{2x(x-1)^2}$

48. $y' = 5x(\sqrt{9-x})' + (5x)'\sqrt{9-x} = 5x\left[(9-x)^{\frac{1}{2}}\right]' + (5x)'\sqrt{9-x}$

$= 5x\left[\dfrac{1}{2}(9-x)^{-\frac{1}{2}}(-1)\right] + 5\sqrt{9-x} = \dfrac{5}{2}(9-x)^{-\frac{1}{2}}[-x + 2(9-x)] = \dfrac{5}{2}(9-x)^{-\frac{1}{2}}(18 - 3x)$

$= -\dfrac{15}{2}(9-x)^{-\frac{1}{2}}(x - 6)$

49. $\dfrac{-6x}{(3x^2+1)^2}$ 50. $\dfrac{-15}{(x-5)^2}$

51. $y' = \dfrac{(5x-3)(\sqrt{x})' - (\sqrt{x})(5x-3)'}{(5x-3)^2} = \dfrac{(5x-3)\left(\dfrac{1}{2\sqrt{x}}\right) - \sqrt{x}(5)}{(5x-3)^2}$

$= \dfrac{(5x-3)\left(\dfrac{\sqrt{x}}{2x}\right) - 5\sqrt{x}\left(\dfrac{2x}{2x}\right)}{(5x-3)^2} = \dfrac{\dfrac{\sqrt{x}}{2x}[(5x-3) - 10x]}{(5x-3)^2} = \dfrac{\sqrt{x}(-5x-3)}{2x(5x-3)^2}$

52. $\dfrac{\sqrt{x+3}(x+6)}{2(x+3)^2}$ 53. $20(5x+9)^3$

54. $f'(x) = 8(4-3x)^7(4-3x)' = 8(4-3x)^7(-3) = -24(4-3x)^7$

Chapter 2   59

55. $\dfrac{1}{5}(5x^2 - 3x)^{-\frac{4}{5}}(10x - 3) = \dfrac{(10x - 3)\sqrt[5]{5x^2 - 3x}}{5(5x^2 - 3x)}$ 56. $\dfrac{(3x + 7)^4(9x^2 - 14x - 75)}{(x^2 - 5)^2}$

2.10 MODELING APPLICATION 2: SAMPLE ESSAY

Instantaneous Acceleration; A case study of the Mazda 626

The performance of the 1982 Mazda 626 Sport Coup is shown in the following advertisement.*

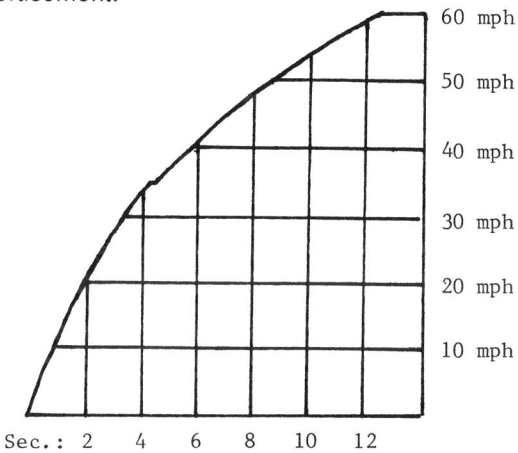

Figure 1 Acceleration for the Mazda 626 Sport Coup

Using this advertisement as a basis for discussion, we can consider the following concepts:
1. Average rate of travel
2. Distance traveled
3. Velocity (instantaneous rate of travel)
4. Acceleration**

As we saw in Section 2.1, the distance formula (d = rt) can be used only when the rate is a constant. However, if we want to approximate the distance for a small period of time, say t = 8 to t = 10, we can do so by using the advertisement shown in Figure 1. At time t = 8, the rate is about 48 mph and at t = 10 it is about 54 mph. The <u>average rate</u> for this interval can be found:

$$\dfrac{48 + 52}{2} = 50 \text{ mph}$$

---

*From <u>Time,</u> January 4, 1982.
**This extended application is adapted from Peter A. Lindstrom, "A Beginning Calculus Project," UMAP Journal, Vol. 5, No. 3, pp. 271–276

Chapter 2

Thus, for the time interval under consideration,
$$d = rt$$
$$= 50(\frac{1}{1800}) \qquad r = 50 \text{ and } t = 2 \text{ sec}$$
$$= \frac{1}{36} \qquad\qquad = 2(\frac{1}{3600}) \text{ hours}$$
$$\qquad\qquad\qquad = \frac{1}{1800}$$

If we convert $\frac{1}{36}$ mile to feet, we find d ≈ 147 feet. (1 mile is 5280 feet.)

Take another look at Figure 1. Notice that the distance can also be found by finding the area of a rectangle (which we have shaded in Figure 2).

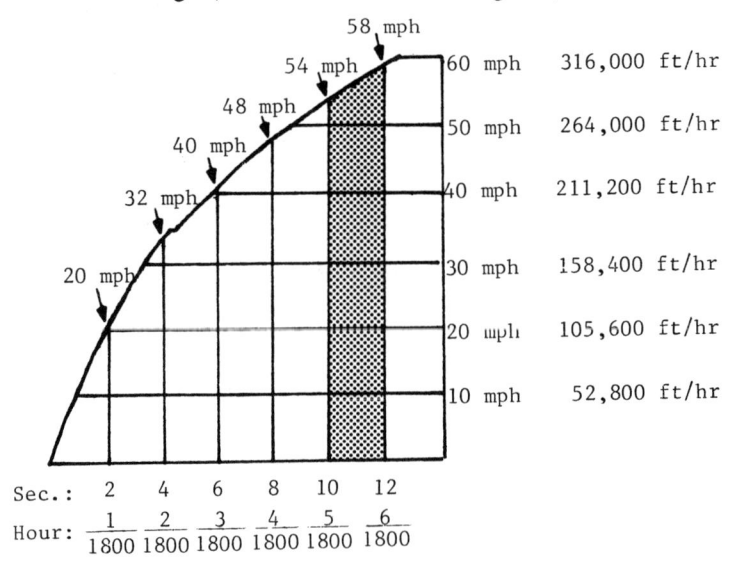

```
                                58 mph
                        54 mph          60 mph    316,000 ft/hr
                 48 mph                  50 mph    264,000 ft/hr
           40 mph                        40 mph    211,200 ft/hr
      32 mph                             30 mph    158,400 ft/hr
  20 mph                                 20 mph    105,600 ft/hr
                                         10 mph     52,800 ft/hr

Sec.:  2   4   6   8   10  12
Hour:  1   2   3   4   5   6
      1800 1800 1800 1800 1800 1800
```

Figure 2 Finding the distance traveled using areas

The area (A) of the shaded rectangle is:
$$A = bh$$
$$= (\frac{1}{1800})(50) \qquad \text{Notice that } b = 2 \text{ sec or } \frac{1}{1800} \text{ hr and } h = 50$$
$$= \frac{1}{36}$$

If you wish to work in feet instead of miles, use the scale shown in color in Figure 2:
$$A = bh$$
$$= (\frac{1}{1800})(264{,}000)$$

≈ 147

This area can also be approximated by using a trapezoid. Recall the formula for the area of a trapezoid is $A = \frac{1}{2}b(h_1 + h_2)$. Using this formula instead,

$$A = \frac{1}{2}(\frac{1}{1800})(253{,}440 + 274{,}560)$$ Note: $h_1 = 48$ mph ≈ 253,440 ft/hr
$$\phantom{A = \frac{1}{2}(\frac{1}{1800})(253{,}440 + 274{,}560)\ \text{Note:}\ } h_2 ≈ 52$$ mph ≈ 274,560 ft/hr
≈ 146.6

We can see, therefore, that the instantaneous rate, or velocity at a time t_1 is

$$\lim_{h \to 0} \frac{d(t_1 + h) - d(t_1)}{(t_1 + h) - t_1} = \lim_{h \to 0} \frac{d(t_1 + h) - d(t_1)}{h}$$

From the definition of derivative, we see that since the velocity function (v) is the rate of change of the distance function with respect to time, that

$v(t) = d'(t)$.

Next, consider the average acceleration. The acceleration is defined at the rate at which the velocity is changing with respect to time. To find the average acceleration, you can use

$$\text{average acceleration} = \frac{\text{change in velocity}}{\text{change in time}}$$

For example, assume, from Figure 2, $v(8) ≈ 48$ mph (this is 253,448 ft/hr ≈ 4224 ft/sec) and $v(10) ≈ 4752$ ft/sec, then

$$\text{average acceleration} = \frac{(4752 - 4224)\ \text{ft/sec}}{(10 - 8)\ \text{sec}}$$

$$= 264\ \text{ft/sec}^2 ≈ 3\ \text{mph}$$

How does this average acceleration compare to the slope of the line segment joining two points on the graph of $v = v(t)$?

To approximate the instantaneous acceleration, we will first rewrite the average acceleration using functional notation.

$$\text{average acceleration} = \frac{v(t_2) - v(t_1)}{t_2 - t_1}$$

If we let h be the length of time from t_1 to t_2, then $t_2 - t_1 = h$ and $t_2 = t_1 + h$ so

$$\text{average acceleration} = \frac{v(t_1 + h) - v(t_1)}{h}$$

Then, the instantaneous acceleration at $t = t_1$ is

Chapter 2

$$\lim_{h \to 0} \frac{v(t_1 + h) - v(t_1)}{h}$$

For $0 \le t \le 12$ let the function $a = a(t)$ be the instantaneous acceleration for the Mazda at time t, we see from the definition of the derivative that $a(t) = v'(t)$. That is, the acceleration function is the rate of change of the velocity function with respect to time.

CHAPTER 3
ADDITIONAL DERIVATIVE TOPICS

3.1 IMPLICIT DIFFERENTIATION, PAGE 96

1. $y' + (5x^2)' + 12' = 0'$
$y' + 10x + 0 = 0$
$y' = -10x$

3. $(xy)' = 5'$
$xy' + x'y = 5'$
$xy' + (1)y = 0$
$xy' = -y$
$y' = \dfrac{-y}{x}$

5. $(x^2)' - (3xy)' = 50'$
$2x - 3(xy' + x'y) = 0$
$2x - 3(xy' + (1)y) = 0$
$2x - 3xy' - 3y = 0$
$2x - 3y = 3xy'$
$\dfrac{2x - 3y}{3x} = y'$

7. $(x^2) + (y^2)' = 4'$
$2x + 2yy' = 0$
$2yy' = -2x$
$y' = -\dfrac{2x}{2y} = -\dfrac{x}{y}$

9. $(2x^4)' - (5y^3)' = 3'$
$8x^3 - (15y^2 y') = 0$
$8x^3 = 15y^2 y'$
$\dfrac{8x^3}{15y^2} = y'$

11. $(x^2)' - (xy)' + (y^2)' = 1'$
$2x - (xy' + x'y) + 2yy' = 0$
$2x - xy' - y + 2yy' = 0$
$2yy' - xy' = y - 2x$
$y'(2y - x) = y - 2x$
$y' = \dfrac{y - 2x}{2y - x}$
which is also $\dfrac{2x - y}{x - 2y}$

13.
$(3x^2 y^3)' - (3xy^2)' + (5xy)' = 2'$
$3[x^2(y^3)' + (x^2)'y^3] - 3[x(y^2)' + x'y^2] + 5(xy' + x'y) = 2'$
$3[x^2(3y^2 y') + 2xy^3] - 3[x(2yy') + y^2] + 5(xy' + y) = 0$
$9x^2 y^2 y' + 6xy^3 - 6xyy' - 3y^2 + 5xy' + 5y = 0$
$9x^2 y^2 y' - 6xyy' + 5xy' = -6xy^3 + 3y^2 - 5y$
$y'(9x^2 y^2 - 6xy + 5x) = -6xy^3 + 3y^2 - 5y$
$y' = \dfrac{-6xy^3 + 3y^2 - 5y}{9x^2 y^2 - 6xy + 5x}$

15. $(x^3)' + (2x^2)' + (xy)' - 4' = 0'$
$3x^2 + 4x + (xy' + x'y) - 0 = 0$
$3x^2 + 4x + xy' + y = 0$
$xy' = -3x^2 - 4x - y$
$y' = \dfrac{-3x^2 - 4x - y}{x}$

Chapter 3   64

17. First, the <u>slope</u> of the tangent line is given by y':

$(x^2)' + (y^2)' - 4' = 0'$

$2x + 2yy' - 0 = 0$

$y' = -\dfrac{2x}{2y} = -\dfrac{x}{y}.$

At (0, 2) $y' = -\dfrac{0}{2} = 0$. The slope is 0 and the line passes through (0, 2):

$y - 2 = 0(x - 0)$

$y - 2 = 0$

19. First, the <u>slope</u> of the tangent line is given by y':

$(2x^4)' - (5y^3)' - 7' = 0'$

$8x^3 - 15y^2 y' - 0 = 0$

$8x^3 = 15y^2 y'$

$\dfrac{8x^3}{15y^2} = y'.$

At (-1, -1) $y' = -\dfrac{8}{15}$

The slope is 0 and the line passes through (-1, -1): $y + 1 = -\dfrac{8}{15}(x + 1)$

Multiply by 15: $15y + 15 = -8(x + 1)$

$15y + 15 = -8x - 8$

$8x + 15y + 23 = 0$

21. First, the <u>slope</u> of the tangent line is given by y':

$(x^2)' + (xy)' + (y^2)' - 7' = 0'$

$2x + (xy' + x'y) + 2yy' - 0 = 0$

$2x + xy' + y + 2yy' = 0$

$xy' + 2yy' = -2x - y$

$y'(x + 2y) = -2x - y$

$y' = \dfrac{-2x - y}{x + 2y}$

At (1, -3) $y' = \dfrac{1}{-5}$

The slope is $-\dfrac{1}{5}$ and the line passes through (1, -3): $y + 3 = -\dfrac{1}{5}(x - 1)$

$5y + 15 = -(x - 1)$

$x + 5y + 14 = 0$

23. First, the <u>slope</u> of the tangent line is given by y':

$((x - 2)^2)' + ((y - 1)^2)' = 9'$

Chapter 3

65

$$2(x - 2) + 2(y - 1) \cdot (y - 1)' = 9'$$
$$2x - 4 + 2(y - 1) \cdot y' = 0$$
$$2(y - 1) \cdot y' = 4 - 2x$$
$$y' = \frac{4 - 2x}{2(y - 1)} = \frac{2 - x}{y - 1}.$$

At x = 2 the slope is 0. To find y when x = 2, substitute into the equation of the curve:

$$(2 - 2)^2 + (y - 1)^2 = 9$$
$$(y - 1)^2 = 9$$
$$y - 1 = \pm 3$$
$$y = 4 \text{ or } -2$$

Since the slope is 0, the line passing through (2,4) has equation: y - 4 = 0. The line passing through (2,-2) has equation y + 2 = 0.

25. The <u>slope</u> of the tangent line is given by y':

$$(x^2)' - (2xy)' + (y^2)' = 0'$$
$$2x - 2(xy' + x'y) + 2yy' = 0$$
$$x - (xy' + y) + yy' = 0 \qquad \text{Divide by 2}$$
$$x - xy' - y + yy' = 0$$
$$yy' - xy' = y - x$$
$$y'(y - x) = y - x$$
$$y' = \frac{y - x}{y - x} = 1. \text{ At } x = -1$$

At x = -1 (or anywhere else) the slope is 1. To find y when x = 1, substitute into the equation of the curve:

$$(-1)^2 - 2(-1)y + y^2 = 0$$
$$1 + 2y + y^2 = 0$$
$$(y + 1)^2 = 0$$
$$y = -1.$$

The line with slope 1, passing through (-1, -1) has the equation y + 1 = 1(x + 1)

x - y = 0

27. The slope of the tangent line is given by y':

$$\left[\frac{(x + 1)^2}{25}\right]' + \left[\frac{(y - 1)^2}{4}\right]' = 1'$$

Chapter 3

$$\frac{1}{25}[2(x+1)] + \frac{1}{4}[2(y-1) \cdot y'] = 0 \qquad \text{Multiply by 100}$$

$$8(x+1) + 50(y-1)y' = 0$$

$$50(y-1)y' = -8(x+1)$$

$$y' = \frac{-8(x+1)}{50(y-1)} = \frac{-4(x+1)}{25(y-1)}.$$

At x = -1 the slope is 0.
To find y when x = -1, substitute into the equation of the curve:

$$\frac{(-1+1)^2}{25} + \frac{(y-1)^2}{4} = 1$$

$$0 + \frac{(y-1)^2}{4} = 1$$

$$(y-1)^2 = 4$$

$$y - 1 = \pm 2$$

$$y = 3 \text{ or } -1.$$

The line with slope 0, passing through (-1, 3) is y - 3 = 0. The line with slope 0, passing through (-1,-1) is y + 1 = 0.

29. $x' = (p^2)' - (5p)' + 500'$

$$1 = 2pp' - 5p' + 0$$

$$1 = p'(2p - 5)$$

$$\frac{1}{2p-5} = p'$$

3.2 DIFFERENTIALS, PAGES 100 - 101

1. $\frac{dy}{dx} = 15x^2$, so $dy = 15x^2 \, dx$ 3. $\frac{dy}{dx} = -10x^{-2}$, so $dy = -10x^{-2} \, dx$

5. $\frac{dy}{dx} = 300x^2 - 50$, so $dy = (300x^2 - 50)dx$

7. $y = 3(x-1)^{\frac{1}{2}}$; $\frac{dy}{dx} = \frac{3}{2}(x-1)^{-\frac{1}{2}}$, so $dy = \frac{3}{2}(x-1)^{-\frac{1}{2}} dx$

9. $\frac{dy}{dx} = 5$, so $dy = 5dx$

11. $\frac{dy}{dx} = (5x-3)^2(x^2-3)' + [(5x-3)^2]'(x^2-3) = (5x-3)^2(2x) + 10(5x-3)(x^2-3)$
$$= (5x-3)[(5x-3)(2x) + 10(x^2-3)]$$
$$= (5x-3)(20x^2 - 6x - 30)$$
$$= 2(5x-3)(10x^2 - 3x - 15)$$

So $dy = 2(5x-3)(10x^2 - 3x - 15)dx$

Chapter 3

13. $\dfrac{dy}{dx} = \dfrac{(x-2)(3x+1)' - (3x+1)(x-2)'}{(x-2)^2} = \dfrac{(x-2)(3) - (3x+1)}{(x-2)^2} = \dfrac{-7}{(x-2)^2}$.

So $dy = -7(x-2)^{-2}$

15. $y = x^2 - x + 3$; $\dfrac{dy}{dx} = 2x - 1$, so $dy = (2x-1)dx$

17. First expand: $y = \left(5 - \dfrac{1}{x^2}\right)\left(1 + \dfrac{1}{x}\right) = 5 + 5x^{-1} - x^{-2} - x^{-3}$.

$\dfrac{dy}{dx} = -5x^{-2} + 2x^{-3} + 3x^{-4}$, so $dy = (-5x^{-2} + 2x^{-3} + 3x^{-4})dx$

19. $\dfrac{dy}{dx} = \dfrac{(x^2 + 3x - 2)(x^2 - 5x + 1)' - (x^2 - 5x + 1)(x^2 + 3x - 2)'}{(x^2 + 3x - 2)^2}$

$= \dfrac{(x^2 + 3x - 2)(2x - 5) - (x^2 - 5x + 1)(2x + 3)}{(x^2 + 3x - 2)^2}$

$= \dfrac{8x^2 - 6x + 7}{(x^2 + 3x - 2)^2}$. So $dy = \dfrac{8x^2 - 6x + 7}{(x^2 + 3x - 2)^2} dx$

21. $dy = f'(x)dx = (2x - 2)dx$. Substituting $x = 10$ and $dx = \Delta x = .1$,

$dy = (2(10) - 2)(.1) = 1.8$.

$\Delta y = f(x + \Delta x) - f(x)$

$= (x + \Delta x)^2 - 2(x + \Delta x) + 5 - (x^2 - 2x + 5)$

$= 2x \Delta x + \Delta x^2 - 2 \Delta x$. Substituting $x = 10$, $\Delta x = .1$,

$\Delta y = 2(10)(.1) + (.1)^2 - 2(.1) = 1.81$

23. $dy = f'(x)dx = \dfrac{5}{2}(5x)^{-\frac{1}{2}} dx$. Substituting $x = 5$ and $dx = \Delta x = .15$,

$dy = \dfrac{5}{2}(5(5))^{-\frac{1}{2}}(.15) = \dfrac{5}{2}(\dfrac{1}{5})(.15) = .075$

$\Delta y = f(x + \Delta x) - f(x) = \sqrt{5x(x + \Delta x)} - \sqrt{5x}$. Substituting $x = 5$ and $\Delta x = .15$,

$\Delta y = \sqrt{5(5.15)} - \sqrt{5(5)} \approx .0744457825$

25. $dy = \dfrac{(5x+2)(2x-3)' - (2x-3)(5x+2)'}{(5x+2)^2} dx = \dfrac{(5x+2)(2) - (2x-3)(5)}{(5x+2)^2} dx = \dfrac{19}{(5x+2)^2} dx$

Chapter 3

For $x = 100$, $dx = \Delta x = 3$: $\Delta y \approx dy = \dfrac{19}{[5(100) + 2]^2}(3) = \dfrac{57}{502^2} \approx .0002261869$

27. $\dfrac{dy}{dx} = 20 \left(3 - \dfrac{1}{x^2}\right)' = 20 \left(\dfrac{2}{x^3}\right) = \dfrac{40}{x^3}$. So $dy = \dfrac{40}{x^3} dx$.

At $x = 10$, $dx = \Delta x = .02$: $\Delta y \approx dy = \dfrac{40}{10^3}(.02) = .0008$

29. $dy = 450 \left[\dfrac{\sqrt{x - 50}\,(x + 300)' - (x + 300)\sqrt{x - 50}\,'}{\sqrt{x - 50}^2} \right] dx$

$= 450 \left[\dfrac{\sqrt{(x - 50)}\,(1) - (x + 300)\left[\dfrac{1}{2}(x - 50)^{-\frac{1}{2}}\right]}{x - 50} \right] dx$

$= 450 \left[\dfrac{\sqrt{x - 50} - \dfrac{x + 300}{2\sqrt{x - 50}}}{x - 50} \right] \cdot \dfrac{2\sqrt{x - 50}}{2\sqrt{x - 50}} dx$

$= 450 \left[\dfrac{2(x - 50) - (x + 300)}{2(x - 50)^{\frac{3}{2}}} \right] dx$

At $x = 1000$, $dx = \Delta x = 10$: $\Delta y \approx dy = 450 \left[\dfrac{2(950) - (1300)}{2(950)^{\frac{3}{2}}} \right](10) = 46.105036$

31. First, $c = 100$ and $\Delta c = 110 - 100 = 10$. $\dfrac{ds}{dc} = 500 - 2c$, so $ds = (500 - 2c)\,dc$.

$\Delta s \approx ds = (500 - 2c)\,dc = (500 - 2(100))(10) = 3000$.

So the sales will increase by about 3000 units.

33. First, $x = 3$ and $\Delta x = 3.5 - 3 = .5$.

$\dfrac{dA}{dx} = \dfrac{(100 + x^2)(3x)' - (3x)(100 + x^2)'}{(100 + x^2)^2} = \dfrac{(100 + x^2)(3) - (3x)(2x)}{(100 + x^2)^2} = \dfrac{300 - 3x^2}{(100 + x^2)^2}$

So $\Delta A \approx dA = \dfrac{300 - 3x^2}{(100 + x^2)^2} dx = \dfrac{300 - 3(3)^2}{(100 + 3^2)^2}(.5) = .011489$

The alcohol will increase by about .01 percent.

Chapter 3 69

35. The area of a circle is given by $A = \pi r^2$. $\dfrac{dA}{dr} = 2\pi r$, so $dA = 2\pi r\, dr$.

At $r = 2$ and $\Delta r = 2.1 - 2 = .1$, $\Delta A \approx dA = 2\pi (2)(.1) = .4\pi \approx 1.256637$ square miles.

37. $dR = 60 dx$. For $x = 100$ and $\Delta x = 110 - 100 = 10$: $dR = 60(10) = 600$, so the change in revenue is $\Delta R \approx dR = \$600$.

$P = R(x) - C(x) = 60x - (400 + 30x) = 30x - 400$

$dP = (30)dx$, so the change in profit is $\Delta P \approx dP = 30 dx = 30(10) = \$300$.

39. $\dfrac{dN}{dx} = \dfrac{1}{5}x - \dfrac{3}{50{,}000}x^2$; $dN = \left(\dfrac{1}{5}x - \dfrac{3}{50{,}000}x^2\right)dx$. The change in N for $x = 1000$

and $x = 1100 - 1000 = 100$ is:

$\Delta N \approx dN = \left(\dfrac{1}{5}(1000) - \dfrac{3}{50{,}000}(1000)^2\right)(100) = 14{,}000$ votes.

3.3 BUSINESS MODELS USING DIFFERENTIATION, PAGE 106

1. Use the ordered pairs (p,n): (15, 20,000) and (25, 5,000). The line passing through these has slope $\dfrac{20{,}000 - 5{,}000}{15 - 25} = -1{,}500$. Using the point-slope form of a line:

$n - 5{,}000 = -1{,}500\,(p - 25)$

$n - 5{,}000 = -1{,}500p + 37{,}500$

$1{,}500p + n - 42{,}500 = 0$

3. See Problem 1. Isolate n in $1{,}500p + n - 42{,}500 = 0$, $n = 42{,}500 - 1{,}500p$. Substitute for n: $C(n) = 15{,}000 + 8.5n$

$C(p) = 15{,}000 + 8.5\,(42{,}500 - 1{,}500p)$

$= 15{,}000 + 361{,}250 - 12{,}750p$

$= 376{,}250 - 12{,}750p$

5. $R(p) = 29{,}750p - 1{,}050p^2$, so $R'(p) = 29{,}750 - 2{,}100p$. This gives the rate of change of revenue per unit change in price.

7. $P(p) = 1{,}050p^2 + 42{,}500p - 376{,}250$, so $P'(p) = -2{,}100p + 42{,}500$. This gives the rate of change of profit per unit change in price.

9. $\overline{C}(x) = \dfrac{50{,}000 + 50x}{x} = 50{,}000x^{-1} + 50$ so $\overline{C}\,'(x) = -50{,}000x^{-2}$

Chapter 3

11. $R(x) = 140x - .028x^2$, so $R'(x) = 140 - 0.056x$

13. See Problem 11. $R'(1,000) = 140 - .056(1,000) = 84$; at a production level of 1,000 items, the revenue is increasing at a rate of $84 per item. $R'(2,500) = 0$; at a production level of 2,500 items the revenue does not change per unit change in the number of items.

15. $P(x) = (140x - .028x^2) - (50,000 + 50x) = 90x - .028x^2 - 50,000$
$P'(x) = 90 - .056x$, so $P'(1,000) = 34$ which means that at a production level of 1,000 items, the profit is increasing at $34 per item. Similarly, $P'(2,500) = -50$ which means the profit is decreasing at $50 per item.

17. The graph of $C(x) = 50,000 + 50x$ is a straight line passing through $(0, C(0)) = (0, 50,000)$ and $(1,000, C(1,000)) = (1,000, 100,000)$. The graph of $R(x) = 140x - .028x^2$ is a parabola, some of the ordered pairs are $x = 0$, $R(0) = 0$

$$x = 1,000, R(1,000) = 140(1,000) - .028(1,000)^2$$
$$= 112,000$$

$$x = 2,000, R(2,000) = 140(2,000) - .028(2,000)^2$$
$$= 168,000$$

You should calculate several other ordered pairs. The graphs are below:

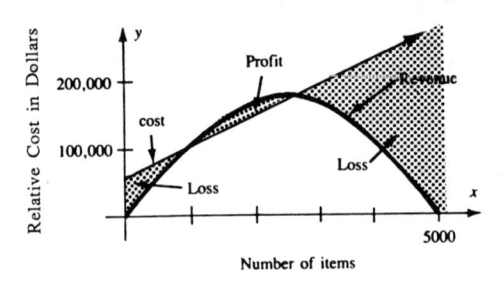

19. $R'(x) = (5x))' - (.001x^2)' = 5 - .002x$

3.4 RELATED RATES, PAGES 111 - 112

1. Differentiate both sides with respect to t: $x^2 + y^2 = 25$, $2x\dfrac{dx}{dt} + 2y\dfrac{dy}{dt} = 0$

Divide both sides by 2, then substitute $\dfrac{dx}{dt} = 4$ and $x = 3$:

$x\dfrac{dx}{dt} + y\dfrac{dy}{dt} = 0$, $3(4) + y\dfrac{dy}{dt} = 0$ or $\dfrac{dy}{dt} = \dfrac{-12}{y}$.

We need to know y when $x = 3$: $x^2 + y^2 = 25$, $3^2 + y^2 = 25$, $y^2 = 16$, $y = 4$

Chapter 3   71

So, $\dfrac{dy}{dt} = \dfrac{-12}{y} = \dfrac{-12}{4} = -3$

3. Differentiate both sides with respect to t: $5x^2 - y = 100$, $10x\dfrac{dx}{dt} - \dfrac{dy}{dt} = 0$,

   Isolate $\dfrac{dt}{dt}$: $\dfrac{dy}{dt} = 10x\dfrac{dx}{dt}$

   Substitute $\dfrac{dx}{dt} = 10$ and $x = 10$: $\dfrac{dy}{dt} = 10(10)(10) = 1000$

5. Differentiate both sides with respect to t: $\dfrac{dy}{dt} = x^{-\tfrac{1}{2}}\dfrac{dx}{dt}$

   Isolate $\dfrac{dx}{dt}$: $x^{\tfrac{1}{2}}\dfrac{dy}{dt} = \dfrac{dx}{dt}$

   Substitute $\dfrac{dy}{dt} = 5$ and $x = 9$: $9^{\tfrac{1}{2}}(5) = \dfrac{dx}{dt}$, $15 = \dfrac{dx}{dt}$

7. Differentiate both sides with respect to t: $xy = 10$

   (Remember to use the Product Rule) $x\dfrac{dy}{dt} + y\dfrac{dx}{dt} = 0$

   Substitute $\dfrac{dx}{dt} = -2$ and $x = 5$: $(5)\dfrac{dy}{dt} + y(-2) = 0$

   Isolate $\dfrac{dy}{dt}$: $\dfrac{dy}{dt} = \dfrac{2y}{5}$

   We need to know y (when $x = 5$), $xy = 10$, $(5)y = 10$, $y = 2$

   So $\dfrac{dy}{dt} = \dfrac{2y}{5} = \dfrac{2(2)}{5} = \dfrac{4}{5}$

9. Differentiate both sides with respect to t (remember to use the Product Rule on the xy term):

   $x^2 + xy - y^2 = 10$, $2x\dfrac{dx}{dt} + x\dfrac{dy}{dt} + y\dfrac{dx}{dt} - 2y\dfrac{dy}{dt} = 0$

   To isolate $\dfrac{dx}{dt}$, put all terms with $\dfrac{dx}{dt}$ on one side of the equation and all other terms on the other side: $2x\dfrac{dx}{dt} + y\dfrac{dx}{dt} = 2y\dfrac{dy}{dt} - x\dfrac{dy}{dt}$

   Now factor out $\dfrac{dx}{dt}$: $\dfrac{dx}{dt}(2x + y) = 2y\dfrac{dy}{dt} - x\dfrac{dy}{dt}$

   Divide by $(2x + y)$: $\dfrac{dx}{dt} = \dfrac{2y\dfrac{dy}{dt} - x\dfrac{dy}{dt}}{2x + y}$

   To finish we will need y (when $x = 4$). So, using the original equation:

Chapter 3

$x^2 + xy - y^2 = 10$, $4^2 + 4y - y^2 = 10$, $0 = y^2 - 4y - 6$

Quadratic Formula: $y = \dfrac{4 \pm \sqrt{4^2 - 4(1)(-6)}}{2} = \dfrac{4 \pm 2\sqrt{10}}{2} = 2 \pm \sqrt{10} = 2 + \sqrt{10}$

since y is positive. So $\dfrac{dx}{dt} = \dfrac{2(2+\sqrt{10})(5) - (4)(5)}{2(4) + (2+\sqrt{10})} = \dfrac{10[(2+\sqrt{10}) - 2]}{10 + \sqrt{10}} = \dfrac{10\sqrt{10}}{10 + \sqrt{10}}$

So $\dfrac{dx}{dt} = \dfrac{10\sqrt{10}}{10 + \sqrt{10}} \cdot \dfrac{10 - \sqrt{10}}{10 - \sqrt{10}} = \dfrac{100\sqrt{10} - 100}{100 - 10} = \dfrac{10(\sqrt{10} - 1)}{9}$

11. Differentiate with respect to t: $\dfrac{dp}{dt} = 125\dfrac{dx}{dt} - \dfrac{1}{100} x \dfrac{dx}{dt}$. We are given $\dfrac{dx}{dt} = -1$

    production is decreasing 1 unit per week) and that x = 200:

    $\dfrac{dp}{dt} = 125(-1) - \dfrac{1}{100}(200)(-1) = -123$.

    So profit is decreasing at a rate of $123 per week.

13. Let x represent the distance the product has moved along the conveyor belt. By the Pythagorean Theorem, $d^2 = 8^2 + x^2$, $d^2 = 64 + x^2$

    Differentiate both sides with respect to time (t): $2d\dfrac{dd}{dt} = 2x\dfrac{dx}{dt}$

    when x = 9, $d^2 = 8^2 + 9^2$, so $d = \sqrt{145}$. Also, $\dfrac{dx}{dt} = 3$ (ft/sec): $2d\dfrac{dd}{dt} = 2x\dfrac{dx}{dt}$,

    $2(\sqrt{145})\dfrac{dd}{dt} = 2(9)(3)$. Solving for $\dfrac{dd}{dt}$: $\dfrac{dd}{dt} = \dfrac{2(9)(3)}{2\sqrt{145}} = \dfrac{27}{\sqrt{145}} \approx 2.24$.

    So the distance is changing at about 2.24 feet per second.

15. If the other details remain the same, we only need to change $\dfrac{dv}{dt} = 500$:

    $\dfrac{dv}{dt} = 400\pi\dfrac{dh}{dt}$, $500 = 400\pi\dfrac{dh}{dt}$, $\dfrac{500}{400\pi} = \dfrac{dh}{dt}$, $\dfrac{dh}{dt} = \dfrac{5}{4\pi} \approx .398$ (feet per minute).

17. Start with $A = \pi r^2$ and differentiate both sides with respect to t: $\dfrac{dA}{dt} = 2\pi r \dfrac{dr}{dt}$.

    The radius is changing at 1 foot per second, so $\dfrac{dr}{dt} = 1$: $\dfrac{dA}{dt} = 2\pi(10)(1) = 20\pi$.

    So the area is growing at a rate of about 62.8 square feet per second.

19. Differentiate both sides with respect to t (time): $d = \dfrac{-5x}{x + 1000} + 70$

Chapter 3

(Quotient Rule): $\dfrac{dd}{dt} = \dfrac{(x+1000)(-5\frac{dx}{dt}) - (-5x)(\frac{dx}{dt})}{(x+1000)^2} + 0$

$= \dfrac{5\frac{dx}{dt}(-(x+1000)+x)}{(x+1000)^2} = \dfrac{(-5000)\frac{dx}{dt}}{(x+1000)^2}$

We are given that $x = 2000$ and $\dfrac{dx}{dt} = -200$: $\dfrac{dd}{dt} = \dfrac{-5000(-200)}{(2000+1000)^2} = \dfrac{1,000,000}{9,000,000} \approx .11$

So the price is increasing at $.11 per day.

21. Differentiate both sides of $v = (\frac{4}{3})\pi r^3$ with respect to t (time): $\dfrac{dv}{dt} = 4\pi r^2 \dfrac{dr}{dt}$

We are given that $\dfrac{dr}{dt} = .3$ and $r = 4$: $\dfrac{dv}{dt} = 4\pi(4)^2(.3) \approx 60.32$ (cm^3 per minute).

23. First draw a picture:

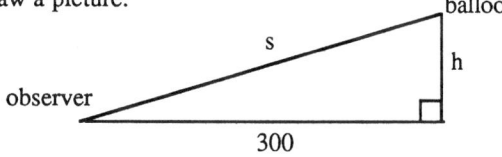

[Diagram: right triangle with observer at left, balloon at upper right, hypotenuse labeled s, vertical side labeled h, horizontal base labeled 300]

By the Pythagorean Theorem, $s^2 = 300^2 + h^2$, $s^2 = 90,000 + h^2$

Differentiating with respect to t (time): $2s\dfrac{ds}{dt} = 2h\dfrac{dh}{dt}$

Isolate $\dfrac{ds}{dt}$: $\dfrac{ds}{dt} = \dfrac{2h\frac{dh}{dt}}{2s} = \dfrac{h\frac{dh}{dt}}{s}$

We know $\dfrac{dh}{dt} = 10$ (feet per second) and that $h = 400$. We need to know s:

$s^2 = 300^2 + h^2 = 300^2 + 400^2$, $s^2 = 250,000$, $s = 500$.

So $\dfrac{ds}{dt} = \dfrac{h(\frac{dh}{dt})}{s} = \dfrac{(400)(10)}{500} = 8$ (feet per second)

3.5 CHAPTER REVIEW, PAGES 112 - 113

1. $y' - (10x^2)' + (6x)' - 15' = 0'$, $\dfrac{dy}{dx} - 20x + 6 = 0$, $\dfrac{dy}{dx} = 20x - 6$

3. $(x^4)' + (5xy)' - (3x^2 y)' + (9xy^2) - 155' = 0'$

$4x^3 + 5(x\dfrac{dy}{dx} + y) - 3(x^2\dfrac{dy}{dx} + y(2x)) + 9(x(2y\dfrac{dy}{dx}) + y^2) = 0$

Chapter 3 74

$$4x^3 + 5x\frac{dy}{dx} + 5y - 3x^2\frac{dy}{dx} - 6xy + 18xy\frac{dy}{dx} + 9y^2 = 0$$

$$5x\frac{dy}{dx} - 3x^2\frac{dy}{dx} + 18xy\frac{dy}{dx} = -4x^3 - 5y + 6xy - 9y^2$$

$$\frac{dy}{dx}(5x - 3x^2 + 18xy) = -4x^3 - 5y + 6xy - 9y^2$$

$$\frac{dy}{dx} = \frac{-4x^3 - 5y + 6xy - 9y^2}{5x - 3x^2 + 18xy} = \frac{4x^3 - 6xy + 5y + 9y^2}{3x^2 - 18xy - 5x}$$

5. $9x^2 + 16y^2 = 145$, $18x\frac{dx}{dt} + 32y\frac{dy}{dt} = 0$, $32y\frac{dy}{dt} = -18x\frac{dx}{dt}$, $\frac{dy}{dt} = \frac{-18x\frac{dx}{dt}}{32y} = \frac{-9x\frac{dx}{dt}}{16y}$

To find y, use x = 3 in $9x^2 + 16y^2 = 145$, $9(3)^2 + 16y^2 = 145$, $16y^2 = 64$,

$y = \pm 2$, so $\frac{dy}{dt} = \frac{-9x\frac{dx}{dt}}{16y} = \frac{-9(3)(4)}{16(\pm 2)} = \pm\frac{27}{8}$

7. $y = 25x^2$, $\frac{dy}{dt} = 50x\frac{dx}{dt}$. $(-4) = 50(3)\frac{dx}{dt}$, $\frac{dx}{dt} = \frac{-4}{150} = \frac{-2}{75}$

9. Let x be the number of $50 price raises

Rent = 500 + 50x

Number = 100 - 2x

R(x) = Revenue = Rent · Number

= (500 + 50x)(100 - 2x)

= 50,000 + 4,000 - 100x²

R'(x) = 4,000 - 200x

The marginal revenue is 4,000 - 200x

3.6 STUDENT'S TEST REVIEW AND ADDITIONAL PRACTICE
OBJECTIVES

The material of this chapter is reviewed in the following list of objectives. After each objective there are some practice questions. Answers to these problems immediately follow. Detailed solutions are given for every third problem. For a sample test select the first question of each set and check your answers. Additional practice is given by the other questions in each set. If you are having trouble with a particular type of problem, or if you want additional practice, look back at the indicated section in the text.

[3.1] Objective 1: *Find the derivative of y with respect to x implicitly.*

Chapter 3   75

1. $x^5 + 2y^2 + y + 10 = 0$ 2. $3x^8y^4 = 1$

3. $(x - 3)^2 - (y + 1)^2 = 1$ 4. $x^3 - 3y^2 + 40 = 0$

Objective 2: *Find the slope of a tangent line at a given point.*

5. $f(x) = \dfrac{5x + 3}{x - 2}$ at $x = 1$ 6. $f(x) = 5x - 6x^3$ at $x = 0$

7. $f(x) = 5x^2(6x - 5)^3$ at $x = 1$ 8. $x^3 - 3y^2 + 40 = 0$ at $(2, 4)$

Objective 3: *Find the standard form equation of a line tangent to a given curve at a given point.*

9. $f(x) = \dfrac{5x + 3}{x - 2}$ at $x = 1$ 10. $f(x) = 5x - 6x^3$ at $x = 0$

11. $f(x) = 5x^2(6x - 5)^3$ at $x = 1$ 12. $x^3 - 3y^2 + 40 = 0$ at $(2, 4)$

[3.2] Objective 4: *Find dy for the given functions.*

13. $y = 6\sqrt{3x^2 + 5}$ 14. $y = \dfrac{2x + 1}{x - 3}$

15. $y = \dfrac{2x^2 + x - 1}{3x^2 + x - 4}$ 16. $y = (1 - \dfrac{1}{x^2})(2 - \dfrac{1}{x})$

Objective 5: *Find dy and Δy for the indicated values.*
17. $y = f(x) = 5x^2$, $x = 10, \Delta x = .1$
18. $y = f(x) = 2x^2 + 5x - 3$, $x = 5, \Delta x = .1$
19. $y = f(x) = \sqrt{6x}$, $x = 6, \Delta x = .2$

20. $y = f(x) = \dfrac{x + 1}{x - 5}$, $x = 1, \Delta x = .01$

Objective 6: *Estimate Δy by using dy.*

21. $y = 5(10 - \dfrac{1}{x^3})$, $x = 10, \Delta x = .03$

22. $y = \dfrac{1 - 2x}{\sqrt{x}}$, $x = 50, \Delta x = .1$

23. $y = \dfrac{x^2 - 1}{x + 2}$, $x = 20, \Delta x = 1$

Chapter 3

24. $y = \dfrac{50(10 - 8x)}{\sqrt{x^2 + 10}}$, $x = 40$, $\Delta x = 2$

[3.3] Objective 7: *Solve applied problems based on the preceding objectives. For specific examples of the types of applications look at the list of applications in this chapter on page 92.*

25. <u>Profit Function</u> Suppose that the profit, P, for a manufacturer is a function of the number of units produced and behaves according to the model
$$P(x) = \dfrac{200 - x^3}{10 - x^2}$$
If the current level of production is 10 units, what is the per unit increase in profit if production is increased from 10 to 15 units?

26. <u>Marginal Profit</u> Find the marginal profit for the function given in Problem 25.

27. <u>Cost Function</u> Suppose that the cost, C, in dollars for producing x items is given by the formula
$$C(x) = 20x^3(5x - 100)^2$$
Find average rate of change of cost as x increases from 50 to 75 units.

28. <u>Marginal Cost</u> Find the marginal cost for the function given in Problem 27.

29. <u>Rate of Change</u> An object moving in a straight line travels d centimeters in t minutes according to the formula $d(t) = .005t^3 + 20t$. What is the object's velocity?

[3.4] Objective 8: *Find the indicated rate, given the other information. Assume that both x and y are positive.*

30. Find $\dfrac{dy}{dt}$ where $4x^2 + 9y^2 = 36$ and $\dfrac{dx}{dt} = 3$ when $x = 2$.

31. Find $\dfrac{dx}{dt}$ where $y = 4x^2$ and $\dfrac{dy}{dt} = -5$ when $x = 2$.

32. Find $\dfrac{dx}{dt}$ where $xy = 4$ and $\dfrac{dy}{dt} = -4$ when $x = 1$.

33. Find $\dfrac{dy}{dt}$ where $x^2 + 2xy + y^2 = 20$ and $\dfrac{dx}{dt} = 3$ when $x = 4$.

ANSWERS TO STUDENT'S TEST REVIEW AND PRACTICE QUESTIONS

1. $5x^4 + 4yy' + y' = 0 \Rightarrow y' = \dfrac{-5x^4}{4y + 1}$

2. $24x^7y^4 + 12x^8y^3y' = 0 \Rightarrow y' = -\dfrac{2y}{x}$

Chapter 3

3. $[(x-3)^2]' - [(y+1)^2]' = 1' \Rightarrow 2(x-3) - 2(y+1)y' = 0 \Rightarrow y' = \dfrac{2(x-3)}{2(y+1)} = \dfrac{x-3}{y+1}$

4. $3x^2 - 6yy' = 0 \Rightarrow y' = \dfrac{x^2}{2y}$ **5.** $f'(1) = -13$ **6.** $f'(x) = 5 - 18x^2;\ f'(0) = 5$

7. $f'(1) = 100$ **8.** $\dfrac{1}{2}$

9. First, find the derivative to calculate the slope:

$$f'(x) = \dfrac{(x-2)(5x+3)' - (5x+3)(x-2)'}{(x-2)^2} = \dfrac{(x-2)5 - (5x+3)(1)}{(x-2)^2} = \dfrac{-13}{(x-2)^2}$$

$f'(1) = -13$. The point of tangency if $(1, f(1)) = (1, -8)$. Using $y - y_1 = m(x - x_1)$ we have $y + 8 = -13(x - 1)$. In standard form this is $13x + y - 5 = 0$.

10. $5x - y = 0$ **11.** $100x - y - 95 = 0$

12. The slope of the tangent is $\dfrac{1}{2}$. (See Problem 8.) Using $y - y_1 = m(x - x_1)$ we have

$y - 4 = \dfrac{1}{2}(x - 2)$. In standard form this is $x - 2y + 6 = 0$.

13. $dy = \dfrac{18x}{\sqrt{3x^2 + 5}}\,dx = \dfrac{18x\sqrt{3x^2+5}}{3x^2+5}\,dx$ **14.** $dy = \dfrac{-7}{(x-3)^2}\,dx$

15. $\dfrac{dy}{dx} = \dfrac{(3x^2 + x - 4)(2x^2 + x - 1)' - (2x^2 + x - 1)(3x^2 + x - 4)'}{(3x^2 + x - 4)^2}$

$= \dfrac{(3x^2 + x - 4)(4x + 1) - (2x^2 + x - 1)(6x + 1)}{(3x^2 + x - 4)^2}$

$= \dfrac{-x^2 - 10x - 3}{(3x^2 + x - 4)^2}$. So $dy = \dfrac{-x^2 - 10x - 3}{(3x^2 + x - 4)^2}\,dx$

16. $dy = (\dfrac{1}{x^2} + \dfrac{4}{x^3} - \dfrac{3}{x^4})dx = \dfrac{x^2 + 4x - 3}{x^4}\,dx$ **17.** $dy = 10;\ \Delta y = 10.05$

18. $dy = f'(x)dx = (4x + 5)dx = [4(5) + 5](.1) = 2.5$
$\Delta y = f(x + \Delta x) - f(x) = [2(5.1)^2 + 5(5.1) - 3] - [2(5)^2 + 5(5) - 3] = 2.52$

19. $dy = .1;\ \Delta y \approx .09918$ **20.** $dy = -.00375;\ \Delta y \approx -.0037594$

21. $\Delta y \approx dy = 15x^{-4}dx = 15(10)^{-4}(.03) = .000045$

22. $\Delta y \approx dy = (-\dfrac{1}{2}(50)^{-\tfrac{3}{2}} + 50^{-\tfrac{1}{2}})(.1) \approx .0140007$ **23.** $\Delta y \approx dy \approx .9938$

24. $\Delta y \approx dy = 50 \dfrac{\sqrt{x^2+10}\,(10-8x)' - (10-8x)\sqrt{x^2+10}\,'}{\sqrt{x^2+10}\,^2}\,dx$

$= 50 \dfrac{\sqrt{x^2+10}\,(-8) - (10-8x)\left[x(x^2+10)^{-\frac{1}{2}}\right]}{x^2+10}\,dx$

$= 50 \dfrac{\sqrt{40^2+10}\,(-8) - \left[10-8(40)\right]\left[40(40^2+10)^{-\frac{1}{2}}\right]}{40^2+10}(2) \approx -.743023$

25. $\dfrac{P(15)-P(10)}{5} \approx 1.17571$, about $1.18. **26.** $P'(x) = \dfrac{x^4 - 30x^2 + 400x}{(10-x^2)^2}$

27. $\dfrac{C(75)-C(50)}{75-50} = \dfrac{[20(75)^3(5(75)-100)^2] - [20(50)^3(5(50)-100)^2]}{25} = 23{,}273{,}433{,}500$

28. $C'(x) = 2{,}500x^2(x-20)(x-4)$
29. $d'(5) = .015t^2 + 20$ cm per minute

30. $\dfrac{d}{dt}(4x^2) + \dfrac{d}{dt}(9y^2) = \dfrac{d}{dt}(36)\ \Rightarrow\ 8x\dfrac{dx}{dt} + 18y\dfrac{dy}{dt} = 0$

So $\dfrac{dy}{dt} = \dfrac{-4x}{9y}\cdot\dfrac{dx}{dt}$. To find y when $x = 2$ use $4x^2 + 9y^2 = 36\ \Rightarrow\ 4(2)^2 + 9y^2 = 36$

$\Rightarrow y = \dfrac{\sqrt{20}}{3}$. Therefore, $\dfrac{dy}{dt} = \dfrac{-4(2)}{9\left(\frac{\sqrt{20}}{3}\right)}\cdot(3) = \dfrac{-4}{\sqrt{5}} = \dfrac{-4\sqrt{5}}{5}$.

31. $-\dfrac{5}{16}$ **32.** 1

33. To find $\dfrac{dy}{dt}$: $\dfrac{dx^2}{dt} + \dfrac{d}{dt}(2xy) + \dfrac{d}{dt}y^2 = \dfrac{d}{dt}(20)$

$2x\dfrac{dx}{dt} + \left(2x\dfrac{dy}{dt} + 2y\dfrac{dx}{dt}\right) + 2y\dfrac{dy}{dt} = 0$

Divide by 2: $x\dfrac{dx}{dt} + x\dfrac{dy}{dt} + y\dfrac{dx}{dt} + y\dfrac{dy}{dt} = 0$

To find y, substitute $x = 4$ into the original equation:

$x^2 + 2xy + y^2 = 20\ \Rightarrow\ (x+y)^2 = 20\ \Rightarrow\ (4+y)^2 = 20\ \Rightarrow\ y = \sqrt{20} - 4$

Now, substitute known values into the last equation:

$$x\frac{dx}{dt} + x\frac{dy}{dt} + y\frac{dx}{dt} + y\frac{dy}{dt} = 0$$

$$4(3) + 4\frac{dy}{dt} + (\sqrt{20} - 4)(3) + (\sqrt{20} - 4)\frac{dy}{dt} = 0$$

$$\sqrt{20}\frac{dy}{dt} = -3\sqrt{20}$$

$$\frac{dy}{dt} = -3$$

3.7 MODELING APPLICATION 3: SAMPLE ESSAY
Publishing; An Economic Model

Karlin Press sells its World Dictionary at a list price of $20 and it presently sells 5000 copies per year. Suppose you are being considered for an editorial position and are asked to do an analysis of the price and sales of this book in order to assess your competency for the position. What questions would you ask, and what conclusions might be reached? Among other things, this report should reach a conclusion about pricing the book for maximum revenue and/or maximum profit. Some consideration should also be given to inventory vs reprint schedule.

The annual cost associated with this book are sumarized in Table 1 and the book has a net price of 80% the list price.

Table 1 Costs Associated with Publishing World Dictionary (Karlin Press)

Advertising	$1750.00
Author's Royalty	0.00
Binding	1.85 per book
Composition	3600.00
Computer Services	750.00
Investment Return	3000.00
Operating Overhead	9000.00*
Printing	4.90 per book
Set-up Charges	400.00
Storage	.50 per book
Taxes	1.20 per book

*Salaries and office expenses prorated for this title.

The first question might be to assess the market demand, which is assumed to be linear. Market research shows that at a list price of $15 the company will sell 7500 copies, but at $25 sales would drop to 2500. This means that the number of copies sold is dependent on the price. Let

p = list price, and
n = number of copies sold.

Chapter 3

We have the following information.

p price	n number sold
$20	5000
$15	7500
$25	2500

Since this relationship is linear,

$$m = \frac{7500 - 5000}{15 - 20} = -500$$

For the demand equation, use the form $y - y_1 = m(x - x_1)$ where the independent variable (x) for this problem is p and the dependent variable (y) is n. Therefore, we know m = -500 and a point (20, 5000). Note any of the three known points could be used. The demand equation is

$$n - 5000 = -500(p - 20)$$
$$n = -500p + 15{,}000$$

There are four items for analysis: revenue, cost, profit, and reprint schedule.

REVENUE

The revenue function, R(p) is found by

REVENUE = (NET PRICE)(NUMBER SOLD)

$$R(p) = (.8p)(n)$$
$$= .8p(-500p + 15{,}000) \quad \text{Substitute } n = -500p + 15{,}000$$
$$= -400p^2 + 12{,}000$$

The maximum revenue is the value of R for a price p for which $R'(p) = 0$ and $R''(p) < 0$.

$$R'(p) = -800p + 12{,}000$$
$$R''(p) = -800 \quad \text{Negative value, so all critical values are relative minima.}$$

Critical values: $R'(p) = 0$ if $-800p + 12{,}000 = 0$
$$12{,}000 = 800p$$
$$15 = p$$

This means that revenue will be maximized if you reduce the list price $5.

The derivative of the revenue is the **marginal revenue** and Table 2.

Table 2 Marginal Revenue

Price	Production Level	Marginal Revenue
10	10,000	4,000
12	9,000	2,400
14	8,000	800
16	7,000	-800
18	6,000	-2,400
20	5,000	-4,000
22	4,000	-5,600
24	3,000	-7,200

From Table 2 you can see that at a production level of 10,000 ($10 list price), a $1 increase in price would increase revenue $4,000, but at a production level of 8,000 books ($14 list) a $1 increase in price would increase revenue only $800. Notice that at a production level of 5,000 books ($20 list) a $1 increase in price would actually decrease revenue by $4,000.

COST

The **cost function** C is found by

COST = (FIXED COSTS) + (VARIABLE COSTS)(NUMBER OF ITEMS)

The fixed costs are those in Table 1 <u>not</u> indicated by a "per book" charge; these amounts total $18,500. The variable costs are those listed in Table 1 as "per book" charges. These total $8,45 per book. Thus,

$C(n) = 18,500 + 8.45n$

The **marginal cost** is

$C'(n) = 8.45$

which means that it costs an additional $8.45 to product one more book at all production levels.

PROFIT

The **profit function** P is found by

PROFIT = REVENUE - COST

We need to consider the profit as a function of the number of items produced, but production is determined by the price. So in this example, the profit P is ultimately a function of price p. Thus,

$P(p) = R(p) - C(p)$

where $R(p) = -400p^2 + 12,000p$ and $C(n) = 18,500 + 8.45n$. But $n = -500p + 15,000$ so that

$C(p) = 18,500 + 8.45(-500p + 15,000)$
$= 142,250 - 422p$

Thus,

$P(p) = (-400p^2 + 12,000p) - (142,250 - 4225p)$
$= -400p^2 + 16,225p - 142,250$

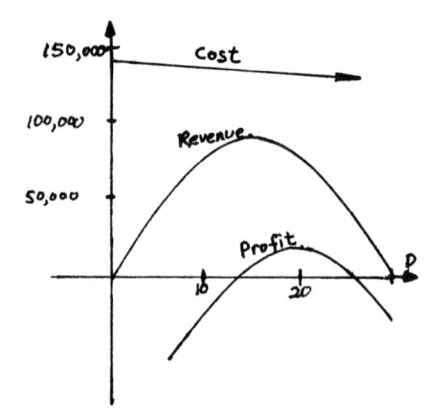

Figure 1 Cost, Revenue, and Profit Functions

The **marginal profit** is

P'(p) = -800p + 16,225

Table 3 gives some values of this function.

Table 3 MARGINAL PROFIT

Price	Production Level	Marginal Profit
10	10,000	8,225
12	9,000	6,625
14	8,000	5,025
16	7,000	3,425
18	6,000	1,825
20	5,000	225
22	4,000	-1,375
24	3,000	-2,975

From Table 3, you can see that at a production level of 10,000 ($10 list) a $1 increase in price would increase profit $8,225, at a production level of 5000 copies ($20 list) a $1 increase in price would increase profit $225, but at a production level of 4,000 ($22 list), a unit increase in price would <u>decrease</u> profit by $1,375.

In Section 3.3 we showed that maximum profit would occur when the marginal revenue and marginal cost are equal. That is,

R'(p) = C'(p)

-800p + 12,000 = -4,225

-800p = -16,225

p = 20.28125

This means that for a maximum profit the list price should be raised to $20.28. At this price, the production level should be

n = -500p + 15,000

= -5,000(20.28) + 15,000

= 4,860

Chapter 3

REPRINT SCHEDULE AND INVENTORY

The annual inventory (for maximum profit) should be 4,860. Allowing for some office and sample copies, suppose we assume an inventory of 7,000 copies. How many of the same sized printings of this book should the publisher make during the year? The cost of storing the books is $.50 per book, but the number of books stored is not the same as the number of books produced. Suppose that the total storage costs are

$.50 (AVERAGE NUMBER OF BOOKS STORED)

where we define the average number of books stored as the total number printed divided by 2. This concept is illustrated by Figure 2.

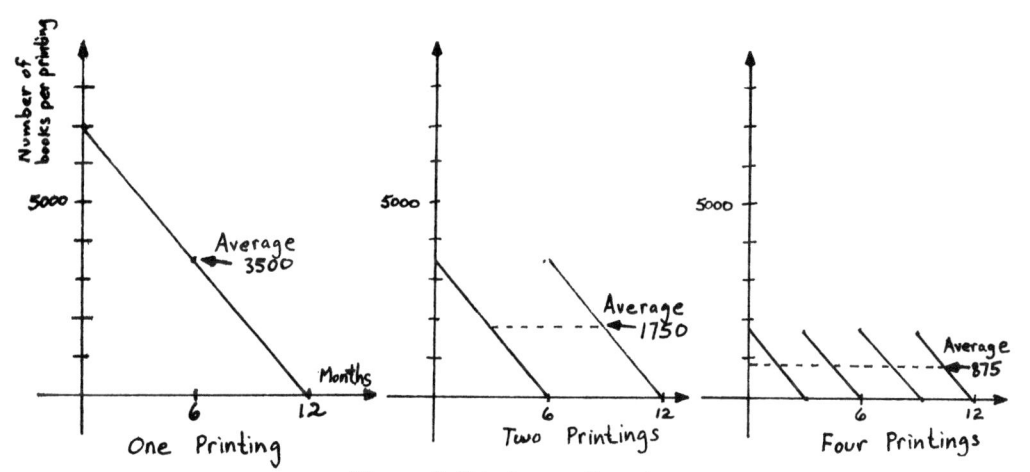

Figure 2 Printings vs Inventory; Average Number of Books Stored

Suppose you decide to do x printings per year. The known factors are:

s = storage cost per book; s = $.50 in this example
c = set-up charges; c = $400 in this example from Table 1
v = variable costs per book; v = 8.45 from Table 1
N = number of books produced per year; N = 7,000

There are two costs associated with production: (1) manufacturing costs and (2) storage costs. If D is the function representing these production costs then

D(x) = MANUFACTURING COSTS + STORAGE COSTS

1. Manufacturing Costs

$$\begin{pmatrix} \text{COST PER} \\ \text{PRINTING} \end{pmatrix} = \begin{pmatrix} \text{SET UP} \\ \text{CHARGE} \end{pmatrix} + \begin{pmatrix} \text{COST PER} \\ \text{BOOK} \end{pmatrix} \begin{pmatrix} \text{NUMBER OF BOOKS} \\ \text{IN EACH PRINTING} \end{pmatrix}$$

Chapter 3

$$= 400 + 8.45 \left(\frac{700}{x}\right)$$

$$= 400 + \frac{59{,}150}{x}$$

MANUFACTURING COSTS = (COST PER PRINTING)(NUMBER OF PRINTINGS)

$$= \left(400 + \frac{59{,}150}{x}\right)(x)$$

$$= 400x + 59{,}150$$

2. Storage Costs

$$\text{AVERAGE INVENTORY} = \binom{\text{NUMBER OF BOOKS}}{\text{IN EACH PRINTING}} \left(\frac{1}{2}\right)$$

$$= \frac{7{,}000}{x} \cdot \frac{1}{2}$$

$$= \frac{3{,}500}{x}$$

STORAGE COSTS = (STORAGE COST PER UNIT)(AVERAGE INVENTORY)

$$= .50\left(\frac{3{,}500}{x}\right)$$

$$= \frac{1{,}750}{x}$$

Therefore,

$$D(x) = 400x + 59{,}150 + \frac{1{,}750}{x}$$

$$D'(x) = 400 - 1750x^{-2}$$

Critical values for D:

$$D'(x) = \text{if } 400 - 1750x^{-2} = 0$$

$$\frac{1{,}750}{x^2} = 400$$

$$x^2 = \frac{1{,}750}{400}$$

$$x = \sqrt{\frac{1{,}750}{400}} \quad \text{x is positive, so reject the negative value}$$

$$\approx 2.09$$

$D''(x) = 3500x^{-3} > 0$ so $x = 2.09$ is a minimum. The cost will be minimized when $x = 2$. This means that if you were the editor, you should order two printings of 3500 copies each.

SUMMARY

Let n be the number of units sold in some time interval.

REVENUE = (NET PRICE)(NUMBER SOLD)

    Maximum revenue, $R'(n) = 0$ and $R''(n) < 0$

    Marginal revenue, $R'(n)$

COST = FIXED COSTS + (VARIABLE COSTS)(NUMBER OF ITEMS)

    Minimum cost, $C'(n) = 0$ and $C''(n) > 0$

    Marginal cost, $C'(n)$

PROFIT = REVENUE - COST

    Marginal profit, $P'(n)$

    Maximum profit, $R'(n) = C'(n)$

INVENTORY PROBLEM

    Constants

        s = storage cost per unit

        c = fixed set-up cost to manufacture the product

        v = variable costs to manufacture a single unit

        N = number of units produced annually

    Variable

        x^* = number of printings to produce the annual number N

    Formula

        $x^* = \sqrt{\dfrac{sN}{2c}}$ You might try to derive this formula. The steps are identical to those we showed in this section.

CHAPTER 4
APPLICATIONS AND DIFFERENTIATION

4.1 FIRST DERIVATIVES AND GRAPHS, PAGES 124 - 125

1. $f'(x) = 16x$; $16x = 0$ when $x = 0$. The vertex is at $(0, f(0)) = (0, 0)$. A few points are:
$x = \pm 1, f(\pm 1) = 8(1)^2 = 8$
$x = \pm 1.5, f(\pm 1.5) = 8(\pm 1.5)^2 = 18$
You should calculate a few more points.
The graph is shown at right:

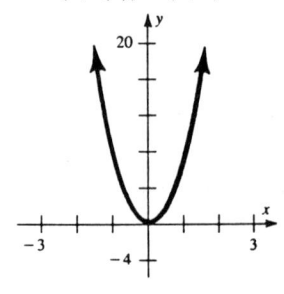

3. $f'(x) = -40x$; $-40x = 0$ when $x = 0$. The vertex is at $(0, f(0)) = (0, 0)$. A few points are:
$x = \pm 1, f(\pm 1) = -20(\pm 1)^2 = -20$
$x = \pm 1.5, f(\pm 1.5) = -20(\pm 1.5)^2 = -45$
You should calculate a few more points.
The graph is shown at right:

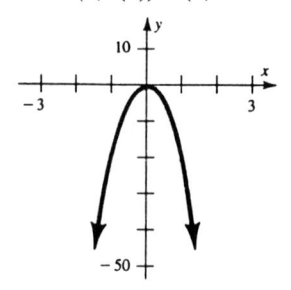

5. $(5y)' + (15x^2)' = 0'$, $5y' + 30x = 0$, $5y' = -30x$, $y' = -6x$, $-6x = 0$ when $x = 0$
The vertex and a few other points are:
$5y + 15x^2 = 0$, $y = -3x^2$
$x = 0, y = -3(0)^2 = 0$
$x = \pm 1, y = -3(\pm 1)^2 = -3$
$x = \pm 2, y = -3(\pm 2)^2 = -12$

The graph is shown at right:

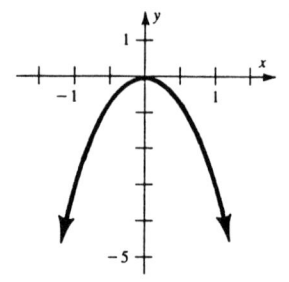

7. $f'(x) = 10x - 20$; $10x - 20 = 0$ when $x = 2$. The vertex is at $(2, f(2)) = (2, -18)$. A few points are:
$x = 1, f(1) = 5(1)^2 - 20(1) + 2 = -13$
$x = 0, f(0) = 5(0)^2 - 20(0) + 2 = 2$
$x = 3, f(3) = 5(0)^2 - 20(3) + 2 = -13$
You should calculate a few more points.
The graph is shown at right:

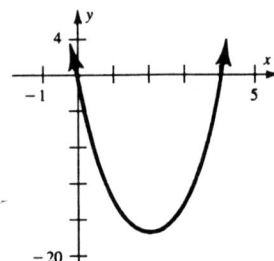

Chapter 4

9. $(x^2)' + (4y)' - (3x)' + 1' = 0'$

$2x + 4y' - 3 = 0$

$4y' = 3 - 2x$

$y' = \dfrac{3 - 2x}{4}; \dfrac{3 - 2x}{4} = 0$ when $x = \dfrac{3}{2}$ The vertex and a few other points are:

$x^2 + 4y - 3x + 1 = 0$

$4y = 3x - 1 - x^2$

$x = 1, y = \dfrac{3(1) - 1 - (1)^2}{4} = \dfrac{1}{2}$

$y = \dfrac{3x - 1 - x^2}{4}$

$x = 0, y = \dfrac{3(0) - 1 - (0)^2}{4} = -\dfrac{1}{4}$

$x = \dfrac{3}{2}, y = \dfrac{3(\tfrac{3}{2}) - 1 - (\tfrac{3}{2})^2}{4} = \dfrac{5}{16}$

You should calculate a few more points.

The graph is shown at right:

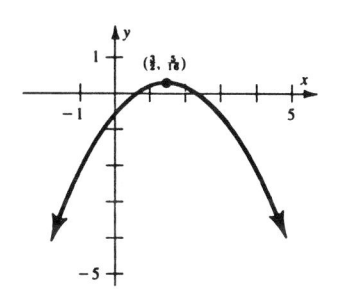

11. $(9x^2)' + (6x)' + (18y)' - 23' = 0'$

$18x + 6 + 18y' = 0$

$y' = \dfrac{-18x - 6}{18} = \dfrac{-3x - 1}{3}; \dfrac{-3x - 1}{3} = 0$ when $x = -\dfrac{1}{3}$. The vertex and a few other points are:

$9x^2 + 6x + 18y - 23 = 0$

$18y = -9x^2 - 6x + 23$

$x = 0, y = \dfrac{-9(0)^2 - 6(0) + 23}{18} = \dfrac{23}{18}$

$y = \dfrac{-9x^2 - 6x + 23}{18}$

$x = -\dfrac{1}{3}, y = \dfrac{-9(-\tfrac{1}{3})^2 - 6(-\tfrac{1}{3}) + 23}{18} = \dfrac{4}{3}$

$x = -1, y = \dfrac{-9(-1)^2 - 6(-1) + 23}{18} = \dfrac{10}{9}$

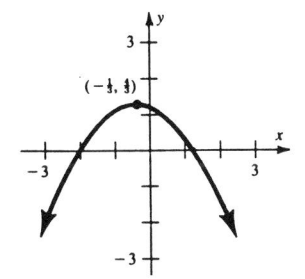

13. $y' = 10x - 3$; $10x - 3 = 0$ when $x = \dfrac{3}{10}$. The vertex is:

$x = \dfrac{3}{10}, y = 5(\dfrac{3}{10})^2 - 3(\dfrac{3}{10}) + 1 = \dfrac{11}{20}$

You should calculate a few more points.

The graph is shown at right:

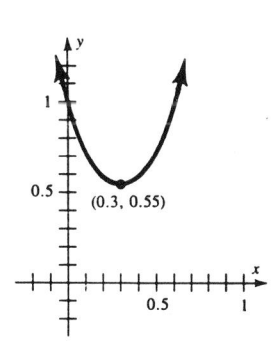

Chapter 4

15. $y' = 14x + 2$; $14x + 2 = 0$ when $x = -\frac{1}{7}$. The vertex is:

$x = -\frac{1}{7}$, $y = 7(-\frac{1}{7})^2 + 2(-\frac{1}{7}) - 3 = -\frac{22}{7}$

You should calculate a few more points.
The graph is shown at right:

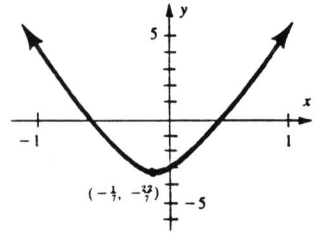

17. Increasing on $(-\infty, 2)$ and decreasing on $(2, \infty)$. Horizontal tangent at $(2, 4)$.

19. Decreasing on $(-\infty, 0)$ and on $(5, \infty)$; increasing on $(0, 3)$; constant on $(3, 5)$.
Horizontal tangent at $(0, 3)$ and for all $3 < x < 5$.

21. Increasing on $(-\infty, 2)$ and $(2, \infty)$; no horizontal tangents.

23. **a.** $f'(x) = 10 - 2x$ which is always defined and $f'(x) = 0$ when $x = 5$, so the only critical value is at $x = 5$. **b.** The function is increasing when $f'(x) > 0$: $10 - 2x > 0$, $10 > 2x$, $5 > x$. So $f(x)$ is increasing on $(-\infty, 5)$. The function is decreasing when $f'(x) < 0$: $10 - 2x < 0$, $10 < 2x$, $5 < x$. So $f(x)$ is decreasing on $(5, \infty)$.

25. **a.** $f'(x) = 2x - 12$ is always defined and $f'(x) = 0$ when $x = 6$, which is the only critical value. **b.** The function is increasing when $f'(x) > 0$: $2x - 12 > 0$, $2x > 12$, $x > 6$. So $f(x)$ is increasing on $(6, \infty)$. The function is decreasing when $f'(x) < 0$: $2x - 12 < 0$, $x < 6$. So the function is decreasing on $(-\infty, 6)$.

27. **a.** $g'(x) = 6x^2 - 6x - 36$ which is always defined. Setting $g'(x) = 0$:
$6x^2 - 6x - 36 = 0$
$x^2 - x - 6 = 0$
$(x - 3)(x + 2) = 0$
$x = 3$ or $x = -2$

So the only critical values are at $x = 3$ and $x = -2$. **b.** It might help to draw a number line with the critical points $x = 3$ and $x = -2$ on it. There are three areas to check:
1. If $x < -2$ then both $(x - 3)$ and $(x + 2)$ are negative, so
$g'(x) = (x - 3)(x + 2) = $ (negative)(negative) > 0.
$g'(x) > 0$

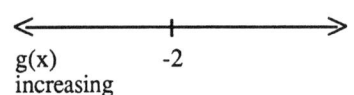

←――――――+――――――→
g(x)   -2
increasing

2. If $-2 < x < 3$ then $(x - 3)$ is negative and $(x + 2)$ is positive, so
$g'(x) = (x - 3)(x + 2) = $ (negative)(positive) < 0.

Chapter 4

$$g'(x) < 0$$
⟵——+————————+——⟶
    -2   g(x)   3
            decreasing

3. If x > 3 then both (x - 3) and (x + 2) are positive, so g'(x) = (x - 3)(x + 2) > 0.

$$g'(x) > 0$$
⟵——————+—————————⟶
           3   g(x)
               increasing

Summarizing, g(x) is increasing on $(-\infty, -2)$ and $(3, \infty)$; decreasing on $(-2, 3)$.

29. a. $y' = 3x^2 + 10x + 8 = (3x + 4)(x + 2)$, which is always defined and equals zero when $x = -2$ or $-\frac{4}{3}$, the only critical values. **b.** It might help to draw a number line with the critical values -2 and $-\frac{4}{3}$ on it. There are three areas to check:

1. If $x < -2$ then both $(3x + 4)$ and $(x + 2)$ are negative, so $y' = (3x + 4)(x + 2)$ = (negative)(negative) > 0.

$$y' > 0$$
⟵—————————+————⟶
y   -2
increasing

2. If $-2 < x < -\frac{4}{3}$ then $(3x + 4)$ is negative and $(x + 2)$ is positive, so $y' = (3x + 4)(x + 2)$ = (negative)(positive) < 0.

$$y' < 0$$
⟵——+———————+——⟶
    -2   y $-\frac{4}{3}$
       decreasing

3. If $x > -\frac{4}{3}$ both $(3x + 4)$ and $(x + 2)$ are positive so $y' = (3x + 4)(x + 2) > 0$.

$$y' > 0$$
⟵——+————————————⟶
  $-\frac{4}{3}$ y
      increasing

Summarizing, y is increasing on $(-\infty, -2)$ and $(-\frac{4}{3}, \infty)$; decreasing on $(-2, -\frac{4}{3})$.

31. Average cost $= \frac{c(x)}{x} = .25x + \frac{12,100}{x}$. To minimize this, take the derivative and set equal to zero:

$.25 - 12,100x^{-2} = 0$

$.25x^2 - 12,100 = 0$

$x^2 = \frac{12,100}{.25} = 48,400$

$x = \sqrt{48,400} = 220$

Average Cost $= \frac{c(x)}{x} = .25x + \frac{12,100}{x}$

Marginal Cost $= c'(x) = .50x$

Chapter 4

These are equal when: $.25x + \dfrac{12{,}100}{x} = .50x$

$\dfrac{12{,}100}{x} = .25x$

$12{,}100 = .25x^2$

$x = 220$

33. $S(x)$ is increasing when $S'(x) > 0$: $S'(x) = 5000 - 50x - 3x^2 > 0$
Multiply by -1: $3x^2 + 50x - 5000 < 0$

$(3x - 100)(x + 50) < 0$

which has the solution $-50 < x < \dfrac{100}{3}$. Since x must be positive, when $x < \dfrac{100}{3} = 33.\overline{33}$ (thousands of dollars). So $S(x)$ is increasing when the amount spent on advertising is under \$33,333.

4.2 SECOND DERIVATIVES AND GRAPHS, PAGE 132

1. $f'(x) = 10x^4 - 12x^3 + 3x^2 - 10x + 19$; $f''(x) = 40x^3 - 36x^2 + 6x - 10$;
$f'''(x) = 120x^2 - 72x + 6$

3. $y = \sqrt{5}x^{\frac{1}{2}}$; $y' = \dfrac{\sqrt{5}}{2}x^{-\frac{1}{2}}$; $y'' = -\dfrac{\sqrt{5}}{4}x^{-\frac{3}{2}}$; $y''' = \dfrac{3\sqrt{5}}{8}x^{-\frac{5}{2}}$; $\dfrac{d^4y}{dx^4} = y'''' = -\dfrac{15\sqrt{5}}{16}x^{-\frac{7}{2}}$

$= \dfrac{-15\sqrt{5}}{16\sqrt{x^7}} \cdot \dfrac{\sqrt{x}}{\sqrt{x}} = \dfrac{-15\sqrt{5x}}{16x^4}$

5. $g' = 12x^2 + 2x^{-2}$; $g'' = 24x - 4x^{-3}$; $g''' = 24 + 12x^{-4}$; $g^{(4)}(x) = -48x^{-5}$

7. $y' = 5x^{\frac{2}{3}}$; $y'' = \dfrac{10}{3}x^{-\frac{1}{3}}$; $\dfrac{d^3y}{dx^3} = y''' = \dfrac{-10}{9}x^{-\frac{4}{3}} = \dfrac{-10}{9} \cdot \dfrac{\sqrt[3]{x^2}}{\sqrt[3]{x^4}\sqrt[3]{x^2}} = \dfrac{-10\sqrt[3]{x^2}}{9x^2}$

9. $y = 3(x - 1)^{-1}$; $y' = -3(x - 1)^{-2}$; $y'' = 6(x - 1)^{-3}$; $\dfrac{d^3y}{dx^3} = y''' = -18(x - 1)^{-4}$

11. $y' = \dfrac{(x + 4)(x^2 - 1)' - (x^2 - 1)(x + 4)'}{(x + 4)^2} = \dfrac{(x + 4)(2x) - (x^2 - 1)(1)}{(x + 4)^2}$

$= \dfrac{2x^2 + 8x - x^2 + 1}{(x + 4)^2} = \dfrac{x^2 + 8x + 1}{(x + 4)^2}$

$\dfrac{d^2y}{dx^2} = y'' = \dfrac{(x + 4)^2(x^2 + 8x + 1)' - (x^2 + 8x + 1)[(x + 4)^2]'}{(x + 4)^4}$

Chapter 4

$$= \frac{(x+4)^2(2x+8) - (x^2+8x+1)[2(x+4)]}{(x+4)^4} = \frac{(x+4)^2 2(x+4) - (x^2+8x+1)2(x+4)}{(x+4)^4}$$

Factor $2(x+4)$

$$= \frac{2(x+4)[(x+4)^2 - (x^2+8x+1)]}{(x+4)^4}$$

Reduce and simplify

$$= \frac{2[x^2 + 8x + 16 - x^2 - 8x - 1]}{(x+4)^3}$$

$$= \frac{2[15]}{(x+4)^3} = \frac{30}{(x+4)^3}$$

13. $f'(x) = 4(2x+1)^3(2) = 8(2x+1)^3$, which is defined for all x, so our candidates will result from setting $f'(x) = 0$: $8(2x+1)^3 = 0$, $2x + 1 = 0$, $x = -\frac{1}{2}$. Trying the second derivative test, $f''(x) = 24(2x+1)^2(2) = 48(2x+1)^2$; $f''(-\frac{1}{2}) = 0$, so the test fails (i.e. no information).

Using the first derivative test we will try $c_- = -1$ and $c_+ = 0$: $f'(-1) = 8(2(-1)+1)^3 = -8 < 0$, so curve is decreasing at $x = -1$. $f'(0) = 8(2(0)+1)^3 = 8 > 0$, so curve is increasing at $x = 0$. Therefore, a relative minimum at $(-\frac{1}{2}, 0)$.

15. $f(x) = \frac{x^2+9}{x} = \frac{x^2}{x} + \frac{9}{x} = x + 9x^{-1}$

$f'(x) = 1 - 9x^{-2}$, which is defined for all $x \neq 0$. Since $f(0)$ is also undefined, the only critical values result from solving $f'(x) = 0$:

$$1 - 9x^{-2} = 0, \ 1 = 9x^{-2}, \ x^2 = 9, \ x = \pm 3$$

Using the second derivative test, $f''(x) = 18x^{-3}$:

$f''(3) = \frac{18}{3^3} = \frac{2}{3} > 0$ Curve is concave up at $x = 3$; meaning a relative minimum at $(3, 6)$.

$f''(-3) = \frac{18}{(-3)^3} = -\frac{2}{3} < 0$ Curve is concave down at $x = -3$, meaning a relative maximum at $(-3, -6)$.

17. $f'(x) = \frac{1}{2}x^{-\frac{1}{2}} = \frac{1}{2\sqrt{x}}$ which is undefined at $x = 0$. Since $f(0)$ is defined, $x = 0$ is a critical value. To find any others, set $f'(x) = 0$. But since $\frac{1}{2\sqrt{x}} \neq 0$, the only critical value is (the original) $x = 0$. But $f'(x)$ is not real for any negative x, so there are no relative maxima or

# Chapter 4	92

minima at $x = \dfrac{1}{2\sqrt{x}}$. Checking the undefined nature $x = 0$, we see that there is a relative minimum at $(0, 0)$.

19. $f'(x) = 16x - 4x^3$ which is defined for all x, so the only critical values result from solving
$f'(x) = 0$: $16x - 4x^3 = 0$ (Do not divide both sides by x)
$\qquad\qquad\quad 4x - x^3 = 0$
$\qquad\qquad\quad x(4 - x^2) = 0$
$\qquad\qquad\quad x(2 - x)(2 + x) = 0$; so $x = 0$, $x = 2$, or $x = -2$ are critical values.
Using the second derivative test, $f''(x) = 16 - 12x^2$:
$f''(0) = 16 > 0$ Curve is concave up, relative minimum at $x = 0$; this is point $(0, 0)$.
$f''(-2) = -32 < 0$ Curve is concave down, relative maximum at $x = -2$; point $(-2, 16)$.
$f''(2) = -32 < 0$ Curve is concave down, relative maximum at $x = 2$; point $(2, 16)$.

21. $f'(x) = 3x^2 + 10x - 8$ which is defined for all x so the only critical values result from solving $f'(x) = 0$: $3x^2 + 10x - 8 = 0$, $(3x - 2)(x + 4) = 0$; so $x = \dfrac{2}{3}$ or $x = -4$.
Using the second derivative test, $f''(x) = 6x + 10$:
$f''(\dfrac{2}{3}) = 14 > 0$ Curve is concave up, relative minimum at $x = \dfrac{2}{3}$; this is point $(\dfrac{2}{3}, \dfrac{194}{27})$.
$f''(-4) = -14 < 0$ Curve is concave down, relative maximum at $x = -4$; point $(-4, 58)$.

23. $y' = 12x^2 - 54x - 30$. The only critical values result from solving $y' = 0$:
$12x^2 - 54x - 30 = 0$
$2x^2 - 9x - 5 = 0$
$(2x + 1)(x - 5) = 0$, so $x = -\dfrac{1}{2}$ or 5.
Using the second derivative test, $y'' = 24x - 54$:
y'' at $x = -\dfrac{1}{2}$: $24(-\dfrac{1}{2}) - 54 = -66 < 0$, relative maximum at $(-\dfrac{1}{2}, \dfrac{7}{4})$.
y'' at $x = 5$: $24(5) - 54 = 66 < 0$, relative minimum at $(5, -331)$.

25. $g(x) = x + x^{-1}$; $g'(x) = 1 - x^{-2}$. The only critical value results from solving $g'(x) = 0$:
$1 - x^{-2} = 0$
$x^2 - 1 = 0$
$x = \pm 1$.
Using the second derivative test, $g''(x) = 2x^{-3}$:
$g''(1) = 2 > 0$ Curve concave up, relative minimum at $x = 1$; this is point $(1, 2)$.
$g''(-1) = -2 < 0$ Curve concave down, relative maximum at $x = -1$; point $(-1, -2)$.

27. $g'(x) = \dfrac{(x-1)(x^2)' - x^2(x-1)'}{(x-1)^2} = \dfrac{(x-1)(2x) - x^2}{(x-1)^2} = \dfrac{x^2 - 2x}{(x-1)^2}$

Chapter 4

The only critical values will result from setting g'(x) = 0:

$$\frac{x^2 - 2x}{(x-1)^2} = 0$$

$x^2 - 2x = 0$

$x(x - 2) = 0$; so $x = 0$ or 2.

Using the first derivative test, g'(-1) = (-1)(-1 - 2) = 3 > 0 Curve is increasing left of x = 0.

g'(1) = 1(1 - 2) = -1 < 0 Curve is decreasing for 0 < x < 2.

g'(3) = 3(3 - 2) = 3 > 0 Curve is increasing right of x = 2.

Therefore, a relative maximum at (0, 0) and a relative minimum at (2, 4).

29. $f'(x) = \frac{2}{3}(x + 1)^{-\frac{1}{3}} = \frac{2}{3\sqrt[3]{x+1}}$ which is never zero but undefined at x = -1 and since f(-1) is defined, x = -1 is a critical value. Using the first derivative test:

$f'(-2) = -\frac{2}{3} < 0$ Curve is decreasing left of x = -1.

$f'(0) = \frac{2}{3} > 0$ Curve is increasing right of x = -1.

Therefore, a relative minimum occurs at (-1, 0).

31. **a.** f'(x) = 12 - 2x = 0 when x = 6. **b.** f'(x) > 0: 12 - 2x > 0 when x < 6; increasing for x < 6. f'(x) < 0: 12 - 2x < 0 when x > 6; decreasing for x > 6. **c.** f"(x) = -2 which is always negative; curve is concave down for all x.

33. **a.** $g'(x) = 3x^2 - 14x - 5 = 0$, $(3x + 1)(x - 5) = 0$, $x = -\frac{1}{3}$ or 5

b. Pick values left of $-\frac{1}{3}$, between $-\frac{1}{3}$ and 5, and right of 5:

g'(-1) = 12 > 0 Curve is increasing left of $x = -\frac{1}{3}$.

g'(0) = -5 < 0 Curve is decreasing $-\frac{1}{3} < x < 5$.

g'(6) = 19 > 0 Curve is increasing right of x = 5.

c. g"(x) = 6x - 14 > 0 when $x > \frac{7}{3}$ Curve is concave up for $x > \frac{7}{3}$.

g"(x) = 6x - 14 < 0 when $x < \frac{7}{3}$ Curve is concave down for $x < \frac{7}{3}$.

35. **a.** $y' = 3x^2 + 22x - 45 = 0$, $(3x - 5)(x + 9) = 0$, $x = \frac{5}{3}$ or -9

b. Pick values left of -9, between -9 and $\frac{5}{3}$, and right of $\frac{5}{3}$:

y' at x = -10: $3(-10)^2 + 22(-10) - 45 = 35 > 0$ Curve is increasing left of -9.

y' at x = 0: $3(0)^2 + 22(0) - 45 = -45 < 0$ Curve is decreasing for $-9 < x < \frac{5}{3}$.

y' at x = 2: $3(2)^2 + 22(2) - 45 = 11 > 0$ Curve is increasing right of $\frac{5}{3}$.

Chapter 4

c. $y'' = 6x + 22 > 0$ when $x > -\frac{11}{3}$ Curve is concave up for $x > -\frac{11}{3}$.

$y'' = 6x + 22 < 0$ when $x < -\frac{11}{3}$ Curve is concave down for $x < -\frac{11}{3}$.

37. a. Differentiate implicitly (with respect to x): $36x^2 - 10x - 4 - 2y' = 0$

$$-2y' = -36x^2 + 10x + 4$$
$$y' = 18x^2 - 5x - 2$$

Set $y' = 0$: $18x^2 - 5x - 2 = 0$

$$(9x + 2)(2x - 1) = 0$$
$$x = -\frac{2}{9} \text{ or } \frac{1}{2}$$

b. Pick values of x left of $-\frac{2}{9}$, between $-\frac{2}{9}$ and $\frac{1}{2}$, and right of $\frac{1}{2}$:

y' at $x = -1$: $18(-1)^2 - 5(-1) - 2 = 21 > 0$ Curve is increasing left of $-\frac{2}{9}$.

y' at $x = 0$: $18(0)^2 - 5(0) - 2 = -2 < 0$ Curve is decreasing $-\frac{2}{9} < x < \frac{1}{2}$.

y' at $x = 1$: $18(1)^2 - 5(1) - 2 = 11 > 0$ Curve is increasing right of $\frac{1}{2}$.

c. $y'' = 36x - 5 > 0$ when $x > \frac{5}{36}$ Curve is concave up for $x > \frac{5}{36}$.

$y'' = 36x - 5 < 0$ when $x < \frac{5}{36}$ Curve is concave down for $x < \frac{5}{36}$.

39. Revenue = (price)(number of items)

a. $R(x) = xP(x) = \left(5 - \left(\frac{x}{100}\right)^2\right)(x) = 5x - \frac{1}{10,000}x^3$

b. $R'(x) = 5 - \frac{3}{10,000}x^2$

c. Look at derivative of marginal revenue:

$R''(x) = -\frac{3}{5,000}x$ at $x = 100$: $R''(100) = -\frac{3}{50} < 0$ so R'(x), the marginal revenue, is decreasing.

4.3 CURVE SKETCHING--RELATIVE MAXIMA AND MINIMA, PAGES 135 - 136

1. $f'(x) > 0$ means the curve must be increasing, and $f''(x) > 0$ means the curve must be concave up. One possibility is shown at right.

Chapter 4

3. f '(x) < 0 means the curve must be decreasing, and f "(x) > 0 means the curve must be concave up. One possibility is shown at right.

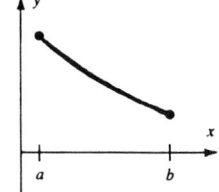

5. f '(-2) = 0 and f '(2) = 0 means the curve must have a horizontal tangent at x = -2 and 2 (curve is not increasing or decreasing). f "(x) > 0 if x < 0 means for negative values of x the curve is concave up; f "(x) < 0 if x > 0 means for positive values of x the curve is concave down. One possibility is shown at right.

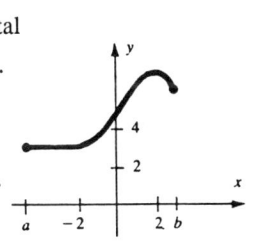

7. y' = 6 - 2x which is positive for x < 3 and negative for x > 3; hence the curve is increasing for x < 3 and decreasing for x > 3. The only critical point is x = 3. y" = -2 < 0 for all x; hence the curve is concave down for all x. Some points are:
(vertex) x = 3, y = 6(3) - (3)² = 9
x = 5, y = 6(5) - 5² = 5
x = 1, y = 6(1) - 1² = 5
You should calculate a few more points. The curve, a parabola, is shown at right.

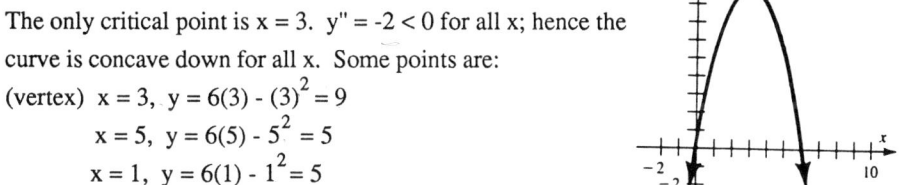

9. y' = 4x - 3 which is positive for x > $\frac{3}{4}$; hence the curve is decreasing for x < $\frac{3}{4}$ and increasing for x > $\frac{3}{4}$. The only critical point is x = $\frac{3}{4}$. y" = 4 > 0 for all x; hence the curve is concave up for all x. Some points are:
(vertex) x = $\frac{3}{4}$, y = 2($\frac{3}{4}$)² - 3($\frac{3}{4}$) + 5 = 3$\frac{7}{8}$
x = 1, y = 2(1)² - 3(1) + 5 = 4
x = 0, y = 5
You should calculate several other ordered pairs. The curve, a parabola, is shown at right.

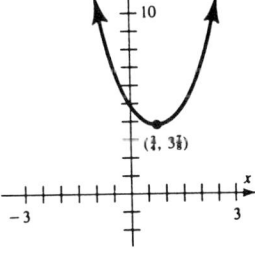

Chapter 4

11. $y' = 3x^2 + 6x - 9 = 3(x + 3)(x - 1) = 0$ when $x = -3$ or $x = 1$. It might help to draw a number line (the x-axis). Pick a number to the left of -3, like -4. y' at $x = -4$:
$3(-4 + 3)(-4 - 1) = 15 > 0$, so the curve is increasing for $x < -3$:

```
        y' < 0
  <------+------>
      y  -3
  increasing
```

Now pick a number between -3 and 1, like 0. y' at $x = 0$: $3(0 + 3)(0 - 1) = -9 < 0$, so the curve is decreasing for $-3 < x < 1$:

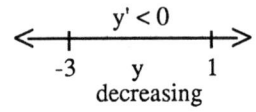

```
          y' < 0
  <---+----------+--->
     -3    y     1
        decreasing
```

Finally, pick a number greater than 1, like 2. y' at $x = 2$: $3(2 + 3)(2 - 1) = 15 > 0$, so the curve is increasing for $x > 1$:

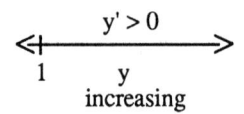

```
              y' > 0
    <-+-------------->
      1       y
           increasing
```

$y'' = 6x + 6$ which is positive for $x > -1$ and negative for $x < -1$; hence the curve is concave up for all $x > -1$ and concave down for $x < -1$. A few points are:

$x = -4$, $y = -4^3 + 3(-4)^2 - 9(-4) + 5 = 25$
$x = -3$, $y = (-3)^3 + 3(-3)^2 - 9(-3) + 5 = 32$
$x = -2$, $y = (-2)^3 + 3(-2)^2 - 9(-2) + 5 = 27$

You should calculate several others. The graph is shown at right.

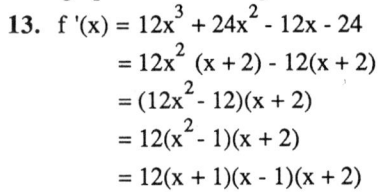

13. $f'(x) = 12x^3 + 24x^2 - 12x - 24$
$= 12x^2(x + 2) - 12(x + 2)$
$= (12x^2 - 12)(x + 2)$
$= 12(x^2 - 1)(x + 2)$
$= 12(x + 1)(x - 1)(x + 2)$

$f'(x) = 0$ when $x = -1, 1,$ or -2. If we pick a value for x left of -2 like -3:
$f'(-3) = 12(-3 + 1)(-3 - 1)(-3 + 2) = -96 < 0$, the curve is decreasing for $x < -2$. If we pick a value for x between -2 and -1, like -1.5:
$f'(-1.5) = 12(-1.5 + 1)(-1.5 - 1)(-1.5 + 2) = 7.5 > 0$, curve is increasing for $-2 < x < -1$.
If we pick a value for x between -1 and 1, like 0: $f'(0) = -24 < 0$, curve is decreasing for

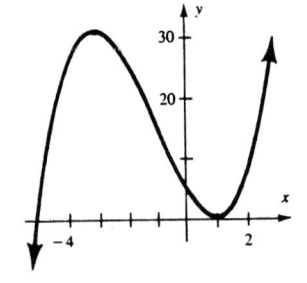

Chapter 4   97

-1 < x < 1. If we pick a value for x to the right of 1, like 2: f '(2) = 144 > 0, curve is
increasing for x > 1. A few points are
(-3, f(-3)) = (-3, 40); (-2, f(-2)) = (-2, 3);
(-1, f(-1)) = (-1, 8). The graph is shown at right.

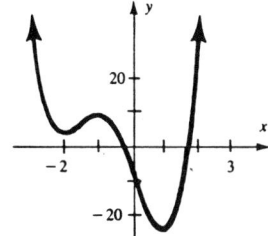

15. f '(x) = 3x² which is never negative, and f '(x) = 0 only when
x = 0, so the curve is increasing for all x ≠ 0 with a horizontal
tangent line at x = 0. f "(x) = 6x which is negative when
x < 0 and f "(x) is positive when x > 0, so the curve is
concave down for x < 0, concave up for x > 0. A few points
are: (-2, f(-2)) = (-2, -7); (-1, f(-1)) = (-1, 0); (1, f(1))
= (1, 2); (2, f(2)) = (2, 9). The graph is shown at right.

17. g'(x) = 4x² which is negative when x < 0 and positive when
x > 0; so the curve is decreasing for x < 0 and increasing for
x > 0. g'(0) = 0, so there is a horizontal tangent at x = 0.
g"(x) = 12x² which is never negative, so for all x ≠ 0 the
curve is concave up. A few points are (0, g(0)) = (0, -1);
(± 1, g(± 1)) = (± 1, 0); (± 2, g(± 2)) = (± 2, 15). The graph
is shown at right.

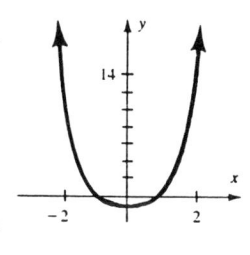

19. $y' = \frac{(x-1)(x^2)' - (x^2)(x-1)'}{(x-1)^2} = \frac{(x-1)(2x) - x^2}{(x-1)^2} = \frac{x^2 - 2x}{(x-1)^2}$, which is zero if and only if
$x^2 - 2x = 0$; which occurs when x = 0 or x = 2. Since the denominator is always
positive, y' is positive when $x^2 - 2x = x(x - 2)$ is positive: x(x - 2) > 0 when x > 2 and
when x < 0. Hence the curve is increasing when x > 2 and when x < 0 and decreasing
when 0 < x < 2.

$$y'' = \frac{(x-1)^2(x^2-2x)' - (x^2-2x)[(x-1)^2]'}{(x-1)^4}$$

$$= \frac{(x-1)^2(2x-2) - (x^2-2x)[2(x-1)]}{(x-1)^4}$$

$$= \frac{(x-1)^2 2(x-1) - 2(x^2-2x)(x-1)}{(x-1)^4}$$

$$= \frac{(x-1)[2(x-1)^2 - 2(x^2-2x)]}{(x-1)^4}$$

$$= \frac{[2(x^2-2x+1) - 2x^2 + 4x]}{(x-1)^3}$$

$$= \frac{2}{(x-1)^3}$$

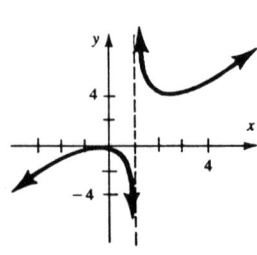

which is negative if $x < 1$ and positive when $x > 1$, so the curve is concave down for $x < 1$, concave up for $x > 1$. The function is undefined at $x = 1$, and the graph is shown at right.

21. It might be helpful to divide: $y = \frac{x+1}{x-1}$:

$$\begin{array}{r} 1 \\ x-1\overline{\smash{)}x+1} \\ \underline{x-1} \\ 2 \end{array}, \text{ so } y = \frac{x+1}{x-1} = 1 + \frac{2}{x-1} = 1 + 2(x-1)^{-1}$$

$y' = -2(x-1)^{-2} = \frac{-2}{(x-1)^2}$ which is negative for all $x \neq 1$, hence the curve is always decreasing. $y'' = 4(x-1)^{-3}$ which is negative if $x < 1$, positive when $x > 1$; hence the curve is concave down for $x < 1$, concave up for $x > 1$. Also, the function is undefined at $x = 1$ and $\lim_{|x| \to \infty} \frac{x+1}{x-1} = 1$ which give the horizontal and vertical asymptotes. The graph is shown at right.

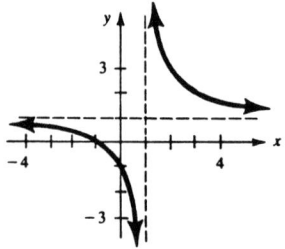

23. $y' = 2x^{-\frac{1}{3}} = \frac{2}{\sqrt[3]{x}}$ which is undefined at $x = 0$. Since y is defined at $x = 0$ a vertical tangent line occurs at $x = 0$. Also, if $x < 0$ y' is negative, and if $x > 0$ y' is positive; hence the curve is decreasing for $x < 0$ and increasing for $x > 0$.

Chapter 4

$y'' = -\frac{2}{3}x^{-\frac{4}{3}} = \frac{-2}{3\sqrt[3]{x^4}}$ which is negative for all

$x \neq 0$; hence the curve is concave down for all $x \neq 0$.
The graph is shown at right.

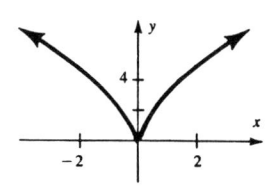

25. $f'(x) = 3x^{-\frac{1}{2}} - 6x^{\frac{1}{2}}$, set equal to zero:

$3x^{-\frac{1}{2}} - 6x^{\frac{1}{2}} = 0$

$3 - 6x = 0$ (multiply by $x^{\frac{1}{2}}$)

$\frac{1}{2} = x$

If $x < \frac{1}{2}$ (but $x > 0$) $f'(x)$ is positive. If $x > \frac{1}{2}$ $f'(x)$ is negative; hence the curve is increasing for $0 < x < \frac{1}{2}$ and decreasing for $x > \frac{1}{2}$.

$f''(x) = -\frac{3}{2}x^{-\frac{3}{2}} - 3x^{-\frac{1}{2}}$

$= \frac{-3}{2\sqrt{x^3}} - \frac{3}{\sqrt{x}}$

$= \frac{-3}{2x\sqrt{x}} - \frac{6x}{2x\sqrt{x}} = \frac{-3 - 6x}{2x\sqrt{x}}$, which is always

negative (since $x > 0$). Hence, the curve is concave down. The graph is shown at right.

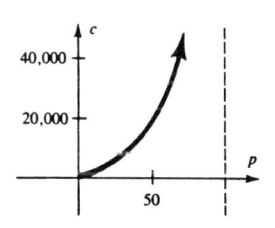

27. $C'(p) = \dfrac{(100 - p)(20,000\,p)' - (20,000\,p)(100 - p)'}{(100 - p)^2}$

$= \dfrac{(100 - p)(20,000) - (20,000\,p)(-1)}{(100 - p)^2}$

$= \dfrac{20,000[(100 - p) + p]}{(100 - p)^2}$

$= \dfrac{2,000,000}{(100 - p)^2} = 2,000,000\,(100 - p)^{-2}$

which is always positive, so the curve is increasing.
$C''(p) = 4,000,000\,(100 - p)^{-3}$ which is positive for
$0 < p < 100$; hence the curve is concave up. The graph is shown at right.

4.4 ABSOLUTE MAXIMUM AND MINIMUM, PAGES 142 - 143

1. $f'(x) = 15x^4 - 150x^2 + 135 = 15[x^4 - 10x^2 + 9] = 15(x^2 - 9)(x^2 - 1)$
   $= 15(x + 3)(x - 3)(x + 1)(x - 1)$
   $f'(x) = 0$ when $x = -3, 3, -1,$ or 1. Compute $f(x)$ for all critical values on $[-5, 0]$, and also at the endpoints: $f(-5) = -3050$; $f(-3) = 966$; $f(-1) = 662$; $f(0) = 750$. Maximum = 966 occurs at $x = -3$; minimum = -3050 occurs at $x = -5$.

3. $g'(x) = 3x^2 - 4x - 15 = (3x + 5)(x - 3)$. $g'(x) = 0$ when $x = -\frac{5}{3}$ or 3. Compute $g'(x)$ for all critical values on $[0, 5]$, and also the endpoints: $g(0) = 42$, $g(3) = 6$, $g(5) = 42$. Maximum = 42 occurs at $x = 0$ and $x = 5$; minimum = 6 occurs at $x = 3$.

5. $h'(x) = 15x^4 - 75x^2 + 60 = 15[x^4 - 5x^2 + 4] = 15(x^2 - 1)(x^2 - 4)$
   $= 15(x - 1)(x + 1)(x - 2)(x + 2)$. $h'(x) = 0$ when $x = 1, -1, 2,$ or -2. Compute $h(x)$ for all critical values on $[0, 3]$, and also the endpoints: $h(0) = -200$, $h(1) = -162$, $h(2) = -184$, $h(3) = 34$. Maximum = 34 occurs at $x = 3$; minimum = -200 occurs at $x = 0$.

7. $f'(x) = \frac{1}{3}x^{-\frac{2}{3}} = \frac{1}{3\sqrt[3]{x^2}}$ which is undefined at $x = 0$. Since $f'(x) \neq 0$, $x = 0$ is the only critical value. Calculate $f(x)$ at $x = 0$ and the endpoints: $f(-1) = 0$; $f(0) = 1$; $f(8) = 3$. Maximum = 3 occurs at $x = 8$; minimum = 0 occurs at $x = -1$.

9. $f'(x) = \frac{1}{2}x^{-\frac{1}{2}} + 1$ which is undefined at $x = 0$. Set $f'(x) = 0$ to find other critical values:
   $$\frac{1}{2}x^{-\frac{1}{2}} + 1 = 0$$
   (Multiply both sides by $2x^{\frac{1}{2}}$) $1 + 2\sqrt{x} = 0$ which has no solution since $\sqrt{x}$ is non-negative. Calculate $f(x)$ at $x = 0$ and the endpoints: $f(0) = 0$; $f(4) = 6$. Maximum = 6 occurs at $x = 4$; minimum = 0 occurs at $x = 0$.

11. $f'(x) = (x + 1)^2(x - 3)' + (x - 3)[(x + 1)^2]' = (x + 1)^2 + (x - 3)2(x + 1)$
    $= (x + 1)[(x + 1) + 2(x - 3)]$
    $= (x + 1)[3x - 5]$
    $f'(x) = 0$ when $x = -1$ or $\frac{5}{3}$. Since $x = -1$ is not in the interval, calculate $f(x)$ at $\frac{5}{3}$ and the endpoints: $f(0) = -3$; $f(\frac{5}{3}) = -\frac{256}{27}$; $f(5) = 72$. Maximum = 72 occurs at $x = 5$; minimum = $-\frac{256}{27}$ occurs at $x = \frac{5}{3}$.

Chapter 4   101

13. $f'(x) = -2(x-3)^{-2} = \dfrac{-2}{(x-3)^2}$ (which is undefined at $x = 3$, but $f(x)$ is also undefined at $x = 3$, so it is not a critical value). A fraction is zero if and only if the numerator is zero, hence $f'(x) \neq 0$. Since we have no critical values, calculate $f(x)$ at the endpoints: $f(0) = -\dfrac{2}{3}$; $f(5) = 1$. These are <u>not</u> maximums or minimums, because the graph of $f(x)$ has a vertical asymptote at $x = 3$, and $\lim_{x \to 3^+} f(x) = \infty$; $\lim_{x \to 3^-} f(x) = -\infty$.

15. $f'(x) = \begin{cases} 1 \text{ if } x > 0 \\ -1 \text{ if } x < 0 \end{cases}$ and $f'(x)$ is undefined at $x = 0$ Calculate $f(x)$ at the only critical value, $x = 0$, and the endpoints: $f(-3) = 3$; $f(0) = 0$; $f(3) = 3$. Maximum = 3 occurs at $x = 3$ or $x = -3$; minimum = 0 occurs at $x = 0$.

17. $f'(x) = \dfrac{(x+1)^2(x-1)' - (x-1)[(x+1)^2]'}{(x+1)^4} = \dfrac{(x+1)^2 - (x-1)2(x+1)}{(x+1)^4} = \dfrac{x+1-2(x-1)}{(x+1)^3}$

$= \dfrac{3-x}{(x+1)^3}$

The critical values are (setting $f'(x) = 0$) at $x = 3$ and $x = -1$. Calculate $f(x)$ at $x = 0$ and 3: $f(0) = -1$; $f(3) = \dfrac{1}{8}$. Maximum = $\dfrac{1}{8}$ occurs at $x = 3$; minimum = -1 occurs at $x = 0$.

19. Revenue = (price)(number of items) = $(p)(x) = (1000 - 10x)(x) = 1000x - x^2$.

$P(x)$ = Profit = Revenue - Cost = $(1000x - 10x^2) - (30{,}000 - 500x)$

$= -10x^2 + 1500x - 30{,}000$.

$P'(x) = -20x + 1500 = 0$ when $x = 75$. By the first derivative test, this is a maximum because $P'(x) > 0$ if $x < 75$ and $P'(x) < 0$, if $x > 75$. So the maximum profit is

$P(75) = \$26{,}250$, at a price of $p = 1000 - 10x$

$= 1000 - 10(75)$

$= 250$

21. Set $R'(x) = C'(x)$: $35 - .06x = 5$, solving gives $x = 500$.

23. Set $R'(x) = C'(x)$: $5 - \dfrac{64}{(x-64)^2} = 1$

$-\dfrac{64}{(x-64)^2} = -4$

$-64 = -4(x-64)^2$

$$16 = (x - 64)^2$$
$$\pm 4 = x - 64 \text{ so } x = 60 \text{ or } 68. \text{ But since } x < 62, x = 60.$$

25. Let x be (as before) the distance from Plant P_1. The domain will be [1, 19] and
$$P(x) = \frac{60}{x} + \frac{240}{20 - x} = 60x^{-1} + 240(20 - x)^{-1}$$

$P'(x) = -60x^{-2} + 240(20 - x)^{-2}$; set equal to zero and solve:

$$\frac{-60}{x^2} + \frac{240}{(20 - x)^2} = 0$$

$$\frac{1}{x^2} - \frac{4}{(20 - x)^2} = 0 \quad \text{Divide by -60}$$

$$(20 - x)^2 - 4x^2 = 0 \quad \text{Multiply by } x^2(20 - x)^2$$

$$400 - 40x - 3x^2 = 0 \quad \text{Expand and simplify}$$

$$3x^2 + 40x - 400 = 0$$

$$(3x - 20)(x + 20) = 0$$

Since x can not be negative, the only critical value is $x = \frac{20}{3}$. $P(1) = 72.6$; $P(\frac{20}{3}) = 27$; $P(19) = 243.2$. So the minimum is 27 which occurs when $x = \frac{20}{3}$ or $6\frac{2}{3}$ miles from plant P_1.

27. Let x represent the number of new people signed up. The price of each ticket will be
2000 if x = 100
1995 if x = 101
1990 if x = 102
.
.
.

In general, the price will always be 2000 - 5x (if x new people are signed up)
Revenue = R(x) = (price)(number)
$$= (2000 - 5x)(100 + x)$$
$$= 200,000 + 150x - 5x^2$$
Cost = C(x) = fixed costs + variable costs
$$= 125,000 + 500(100 + x)$$
$$= 175,000 + 500x$$
Profit = P(x) = R(x) - C(x) = $(200,000 + 1500x - 5x^2) - (175,000 + 500x)$
$$= -5x^2 + 1000x + 25,000$$
$P'(x) = -10x + 1000$ so $P'(x) = 0$ when x = 100.
The interval for x is [0, 50], so we evaluate P only at the endpoints: P(0) = 25,000 and

P(5) = 62,500. Maximum is $62,500 which occurs when x = 50. (That is, there are 150 people on the flight.) In this case the price is 2000 - 5(50) = 1750.

29. Let x represent the number of grapevines per acre. The number of pounds will be

x = 50, 150 pounds per grapevine

x = 51, 148 pounds per grapevine

x = 52, 146 pounds per grapevine, etc.

In general, if x vines are planted per acre, each vine will produce 250 - 2x pounds. The yield per acre will be $Y(x) = (250 - 2x)(x) = 250x - 2x^2$. $Y'(x) = 250 - 4x = 0$ when x = 62.5, but since there must be a whole number of vines, plant 62 per acre.

31. Volume, V, = (length)(width)(height) = $(10 - 2x)(5 - 2x)x = 4x^3 - 30x^2 + 50x$. Taking the derivative and setting equal to zero: $12x^2 - 60x + 50 = 0$

$$6x^2 - 30x + 25 = 0$$

$$x = \frac{30 \pm \sqrt{30^2 - 4(6)(25)}}{2(6)} = \frac{30 \pm \sqrt{300}}{12} \approx 3.94 \text{ or } 1.057$$

The interval is [0, 2.5] so we calculate the volume at the endpoints and at x = 1.06:

V(0) = 0, V(1.06) = 24.06, V(2.5) = 0. So the maximum volume is about 24.06 cubic inches which occurs when the height (x) is 1.06, width (5 - 2x) is 2.89, and the length (10 - 2x) is 7.89. For all practical purposes, the box has dimensions 1" × 3" × 8" with a volume of 24 cubic inches.

4.5 CHAPTER REVIEW, PAGE 144

1. **a.** $y' = -9x^2 + 5x^4$; $y'' = -18x + 20x^3$; $y''' = -18 + 60x^2$; $y^{(4)} = 120x$; $y^{(5)} = 120$; $y^{(6)} = y^{(7)} = \cdots = 0$.

   b. $g' = 15x^4 + 3x^{-2}$; $g'' = 60x^3 - 6x^{-3}$; $g''' = 180x^2 + 18x^{-4}$; $g^{(4)} = 360x - 72x^{-5}$

3. $g'(x) = 3x^2 + 6x - 9 = 3(x - 1)(x + 3)$. $g'(x) = 0$ when x = 1 or x = -3. $g'(x) > 0$ when x < -3 or x > 1; $g'(x) < 0$ when -3 < x < 1. A few points are: (-4, g(-4)) = (-4, 25); (-3, g(-3)) = (-3, 32); (-2, g(-2)) = (-2, 27). You should calculate more points; the graph is shown at right.

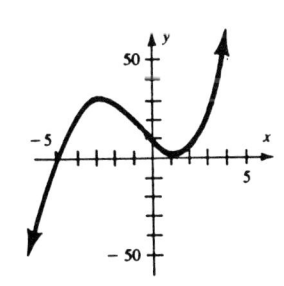

Chapter 4 104

5. $y' = \dfrac{(x+1)(x-2)' - (x-2)(x+1)'}{(x+1)^2} = \dfrac{3}{(x+1)^2}$ which is always positive, so the curve is always increasing. $y'' = \dfrac{-6}{(x+1)^3}$ which is negative if $x > -1$ and positive if $x < -1$, so the curve is concave down to the right of $x = -1$ and concave up to the left of $x = -1$. Also, there is a vertical asymptote at $x = -1$, and since $\lim_{|x| \to \infty} \dfrac{x-2}{x+1} = 1$, $y = 1$ is a horizontal asymptote. The graph is shown at right.

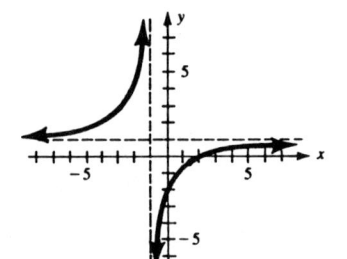

7. $g'(x) = 1 - 8x^{-2}$. Set g' equal to 0: $1 - 8x^{-2} = 0$

$$x^2 - 8 = 0$$
$$x = \pm\sqrt{8} = \pm 2\sqrt{2}.$$

$g''(x) = 16x^{-3}$; $g''(2\sqrt{2}) > 0$ and $g''(-2\sqrt{2}) < 0$. By the second derivative test, there is a relative maximum at $x = -2\sqrt{2}$ and a relative minimum at $x = 2\sqrt{2}$.

9. Revenue = (price)(number of items) = $\left(50 - \dfrac{x}{1000}\right)(x)$. $R(x) = 50x - \dfrac{x^3}{1000^2}$.

Marginal revenue = $R'(x) = 50 - \dfrac{3}{1000^2}x^2$. Take the derivative of marginal revenue to see if increasing or decreasing: $R''(x) = -\dfrac{6}{1000^2}x$; $R''(100) = -\dfrac{6}{1000^2}(100) < 0$

So $R'(x)$, the marginal revenue, is decreasing when $x = 100$.

4.6 CUMULATIVE REVIEW CHAPTERS 1-4, PAGES 144 - 145

1. As x approaches (but never equals) 3, the graph gets closer and closer to a y-value of 5. The limit is 5.

3. $\lim\limits_{x \to 5} \dfrac{2x^2 - 7x - 15}{x - 5} = \lim\limits_{x \to 5} \dfrac{(2x+3)(x-5)}{x-5} = \lim\limits_{x \to 5} (2x+3) = 13.$

5. $\lim\limits_{x \to \infty} \dfrac{(3x+1)(5x-2)}{x^2} = \lim\limits_{x \to \infty} \dfrac{15x^2 - x - 2}{x^2} \cdot \dfrac{\frac{1}{x^2}}{\frac{1}{x^2}} = \lim\limits_{x \to \infty} \dfrac{15 - \frac{1}{x} - \frac{2}{x^2}}{1} = \dfrac{15 - 0 - 0}{1} = 15.$

7. We should be suspicious of $x = 8$, so check to see if $\lim\limits_{x \to 8} f(x) = f(8)$:

Chapter 4

$$\lim_{x \to 8} f(x) = \lim_{x \to 8} \frac{x^2 - 15x + 56}{x - 8} = \lim_{x \to 8} \frac{(x - 8)(x - 7)}{x - 8} = \lim_{x \to 8} (x - 7) = 1.$$

But $f(8) = 7 \neq 1 = \lim_{x \to 8} f(x)$, so $f(x)$ is discontinuous at $x = 8$.

9. $y = 4 - x^{-1}; \dfrac{dy}{dx} = x^{-2} = \dfrac{1}{x^2}$. At $x = 1$, $\dfrac{dy}{dx} = \dfrac{1}{1^2} = 1$.

11. $y = -24x - 15x^2$

13. $y' = 5(2 - 5x)^4 (2 - 5x)' = 5(2 - 5x)^4 (-5) = -25(2 - 5x)^4$

15. Differentiate implicitly with respect to x (remember to use the Product Rule):

$$3x^5 y^4 = 10$$

$$x^5 y^4 = \frac{10}{3}$$

$$(x^5 y^4)' = \left(\frac{10}{3}\right)'$$

$$x^5 (4y^3 y') + (5x^4) y^4 = 0$$

$$4x^5 y^3 y' = -5x^4 y^4$$

$$y' = \frac{-5x^4 y^4}{4x^5 y^3} = \frac{-5y}{4x}$$

17. The slope is given by $f'(x) = 2 - 18x^2$. At $x = 1$, $f'(1) = 2 - 18(1)^2 = -16$. Now use the point-slope form of the equation for a line: $y - y_1 = m(x - x_1)$

$$y - (-4) = -16(x - 1)$$
$$y + 4 = -16(x - 1).$$
$$y + 4 = -16x + 16$$
$$16x + y - 12 = 0$$

19. $y' = 8x^3 - 3x^2 + 6x$; $y'' = 24x^2 - 6x + 6$; $y''' = 48x - 6$; $y^{(4)} = 48$; $y^{(5)} = y^{(6)} = \cdots = 0$.

21. First find the critical values by setting $y' = 0$:

$$y = x^{\frac{1}{2}} - x; \quad y' = \frac{1}{2} x^{-\frac{1}{2}} - 1; \quad \frac{1}{2} x^{-\frac{1}{2}} - 1 = 0$$

Multiply by $2x^{\frac{1}{2}}$: $1 - 2x^{\frac{1}{2}} = 0$

$$1 - 2\sqrt{x} = 0$$

Chapter 4

$$\frac{1}{2} = \sqrt{x}$$

$$\frac{1}{4} = x.$$

Now calculate f(x) here and at the endpoints: $f(0) = 0$, $f(9) = -6$; $f(\frac{1}{4}) = \frac{1}{4}$

The maximum is $\frac{1}{4}$, which occurs at $x = \frac{1}{4}$; the minimum is -6, which occurs at x = 9.

23. Revenue = (price)(number of items)

$$R(x) = (25 - \frac{x}{100})(x)$$

$$= 25x - \frac{1}{100}x^2$$

Marginal revenue = $R'(x) = 25 - \frac{1}{50}x$. To see if marginal revenue, R'(x), is increasing or decreasing we examine the derivative of marginal revenue, R''(x): $R''(x) = -\frac{1}{50}$ which is always negative; hence R'(x) is always decreasing. Marginal revenue is decreasing at x = 100.

25. $C'(x) = \dfrac{(100 - p)(50,000) - 50,000p(-1)}{(100 - p)^2} = \dfrac{5,000,000}{(100 - p)^2}$ which is positive for all values of p; hence C(x) is increasing. The minimum value for C(x) will be when x = 0 (the leftmost x in the domain. The minimum cost is C(0) = 0.

4.7 STUDENT'S TEST REVIEW AND ADDITIONAL PRACTICE
OBJECTIVES

The material of this chapter is reviewed in the following list of objectives. After each objective there are some practice questions. Answers to these problems immediately follow. Detailed solutions are given for every third problem. For a sample test select the first question of each set and check your answers. Additional practice is given by the other questions in each set. If you are having trouble with a particular type of problem, or if you want additional practice, look back at the indicated section in the text.

[4.1] Objective 1: *Graph a parabola by using calculus to find its vertex and how it opens.*
  1. $y = 2x^2 - 8x + 5$
  2. $x^2 - 6x + 3y - 4 = 0$
  3. $x^2 + 4x + 2y + 3 = 0$
  4. $2x^2 - 16x - 3y + 23 = 0$

Objective 2: *Graph a curve by finding the critical values, finding when it is increasing or decreasing, and by deciding upon the concavity.*

5. $f(x) = x^3 - 27x$ 6. $f(x) = 2x^3 - 3x^2 - 36x$
7. $g(x) = x^3 + 3x^2 - 9x + 5$ 8. $g(x) = 4x^3 + 5x^2 + 2x + 3$

[4.2] Objective 3: *Find successive derivatives of a given function.*

9. Find all derivatives of $y = 3x^4 - x^3 + 5x^2 + 79$.

10. Find all derivatives of $y = 1 - 3x^3 + x^5$.

11. Find the first four derivatives of $f(x) = \dfrac{1}{\sqrt{x}}$.

12. Find the first four derivatives of $g(x) = 3x^5 - 3x^{-1}$.

[4.3] Objective 4: *Find all relative maxima and minima for a given function.*

13. $f(x) = x^3 - x^2 - 5x$ 14. $g(x) = x - \dfrac{2}{x}$

15. $t(x) = 4x^2 - x^4$ 16. $s(x) = x^3 - 2x^2 - 4x - 8$

Objective 5: *Graph a given function.*

17. $f(x) = 4x^3 - 30x^2 + 48x$ 18. $g(x) = 3x^5 - 50x^3 + 135x + 12$

19. $y = x^3 - 1$ 20. $y = \dfrac{x-2}{x+1}$

[4.4] Objective 6: *Find the absolute maximum and minimum for a function defined on a closed interval.*

21. $y = 3x^5 - 85x^3 + 240x$ on $[-2, 5]$

22. $y = \sqrt{x} - x$ on $[0, 9]$

23. $y = (x-1)(x+3)^3$ on $[-5, 5]$

24. $y = 1 - x^{\frac{3}{2}}$ on $[0, 4]$

Objective 7: *If you are given revenue and cost functions, find the number of items to produce a maximum profit.*

25. $R(x) = 50x - .05x^2$; $C(x) = 5000 + 2x$

26. $R(x) = 10x - 3x^2 + .01x^3$; $C(x) = 15{,}000 + x$; $x \geq 100$

27. $R(x) = 4x - .01x^2 - .005x^3$; $C(x) = 4000 + x$

28. $R(x) = 4x + \dfrac{x}{(x-49)}$; $C(x) - 3000 + x$

Objective 8: *Solve applied problems based on the preceeding objectives. For specific examples of the types of applications look at the list of applications in this chapter on page 115.*

29. <u>Marginal Revenue</u> A manufacturer has determined that the price of selling x items is given by the formula

$$p(x) = 25 - \left(\dfrac{x}{500}\right)^2$$

Chapter 4

Is the marginal revenue increasing or decreasing when x = 100 items?

30. <u>Cost Benefit Model</u> The cost benefit model relating the cost, C, of removing p% of the pollutants from the atmosphere is
$$C(p) = \frac{50,000p}{110 - p}$$
where the domain of p is [0, 100]. Find the minimum cost.

31. <u>Property Management</u> A property management company manages 100 apartments renting for $500 with all the apartments rented. For each $50 per month increase in rent there will be two vacancies with no possibility of filling them. What rent per apartment will maximize the monthly revenue?

32. A rectangular cardboard poster is to have 208 sq. in. for printed matter. It is to have a 3-inch margin at the top and a 2-inch margin at the sides and bottom.

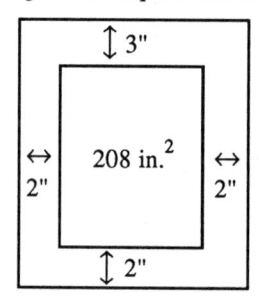

Find the length and width (to the nearest inch) of the poster so that the amount of cardboard used is minimized.

ANSWERS TO STUDENT'S TEST REVIEW AND PRACTICE QUESTIONS

1. $y' = 0$ when $x = 2$; $y'' = 4 > 0$

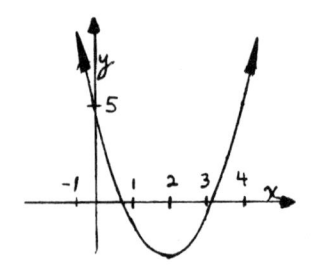

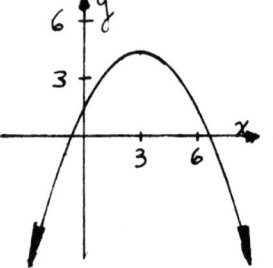

2. $y' = 0$ when $x = 3$; $y'' = \frac{2}{3} > 0$

3. Differentiate implicitly: $2x + 4 + 2y' = 0$
   Set $y' = 0$: $\quad 2x + 4 = 0$
   $\qquad\qquad\qquad x = -2$ which is where the vertex occurs.
   Differentiate again implicitly: $(2x)' + 4' + (2y')' = 0'$

Chapter 4 109

$$2 + 0 + 2y'' = 0$$
$$y'' = -1 < 0,$$

so the graph is concave down. A few points are:

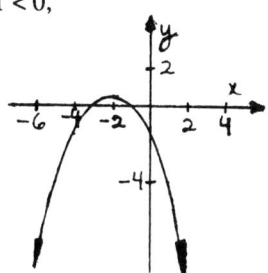

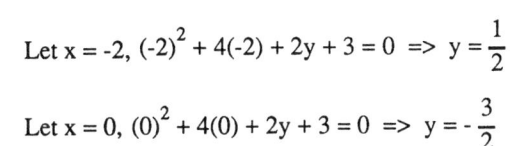

Let $x = -2$, $(-2)^2 + 4(-2) + 2y + 3 = 0 \Rightarrow y = \dfrac{1}{2}$

Let $x = 0$, $(0)^2 + 4(0) + 2y + 3 = 0 \Rightarrow y = -\dfrac{3}{2}$

Let $x = 1$, $(1)^2 + 4(1) + 2y + 3 = 0 \Rightarrow y = -4$

4.

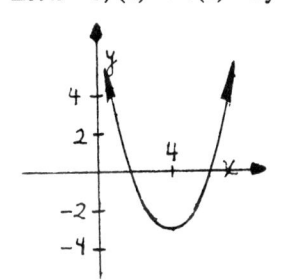

5.

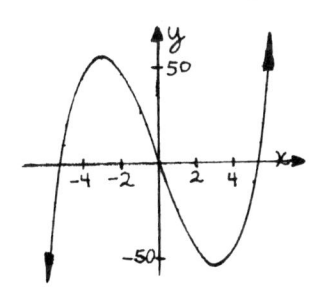

6. Set $f'(x)$ equal to zero to find critical points:
$$f'(x) = 6x^2 - 6x - 36 = 0$$
$$x^2 - x - 6 = 0$$
$$(x - 3)(x + 2) = 0$$
$$x = -2 \text{ or } 3$$

$f''(x) = 12x - 6$; $f''(-2) = -30$, $f''(3) = 30$, so we have a relative maximum at $x = -2$, and a relative minimum at $x = 3$. Also, since $f''(x) < 0$ for $x < \dfrac{1}{2}$ and $f''(x) > 0$ for $x > \dfrac{1}{2}$, concavity changes at $x = \dfrac{1}{2}$ (i.e., point of inflection at $x = \dfrac{1}{2}$). Some points are:

Let $x = -3$, $y = 2(-3)^3 - 3(-3)^2 - 36(-3) = 27$
Let $x = -2$, $y = 2(-2)^3 - 3(-2)^2 - 36(-2) = 44$
Let $x = -1$, $y = 2(-1)^3 - 3(-1)^2 - 36(-1) = 31$
Let $x = 3$, $y = 2(3)^3 - 3(3)^2 - 36(3) = -81$

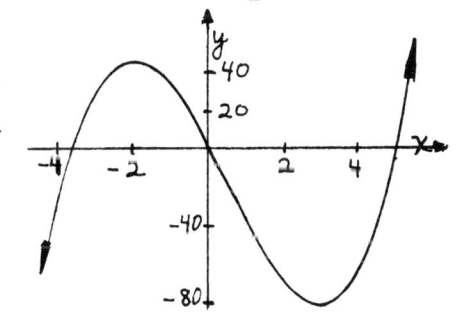

Chapter 4

7.

8.

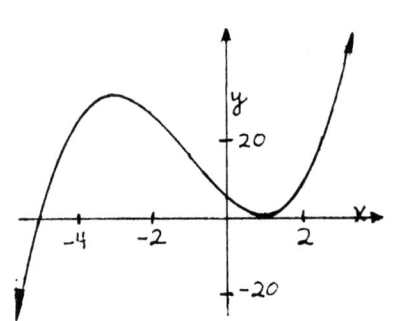

9. $y' = 12x^3 - 3x^2 + 10x$
    $y'' = 36x^2 - 6x + 10$
    $y''' = 72x - 6$
    $y^{(4)} = 72$
    $y^{(N)} = 0$ for $N > 4$

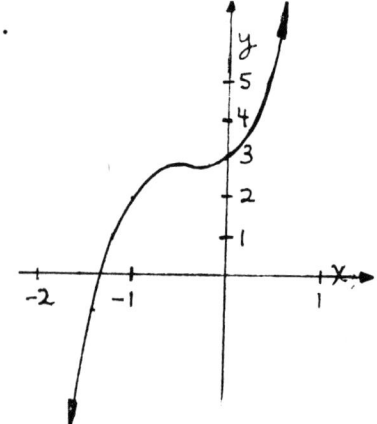

10. $y' = -9x^2 + 5x^4$
    $y'' = -18x + 20x^3$
    $y''' = -18 + 60x^2$
    $y^{(4)} = 120x$
    $y^{(5)} = 120$
    $y^{(N)} = 0$ for $N > 5$

11. $f'(x) = -\dfrac{1}{2} x^{-\frac{3}{2}}$

    $f''(x) = \dfrac{3}{4} x^{-\frac{5}{2}}$

    $f'''(x) = -\dfrac{15}{8} x^{-\frac{7}{2}}$

    $f^{(4)}(x) = \dfrac{105}{6} x^{-\frac{9}{2}}$

12. $g'(x) = 15x^4 + 3x^{-2}$

    $g''(x) = 60x^3 - 6x^{-3}$

    $g'''(x) = 180x^2 + 18x^{-4}$

    $g^{(4)}(x) = 360x - 72x^{-5}$

13. Relative maximum = 3, occurs at $x = -1$; relative minimum = $-\dfrac{175}{27}$, occurs at $x = \dfrac{5}{3}$.

14. No relative extrema.

15. Set the first derivative equal to zero to find critical points
    $t'(x) = 8x - 4x^3 = 0 \Rightarrow 2x - x^3 = 0 \Rightarrow x(x - x^2) = 0 \Rightarrow x = 0, \sqrt{2}, -\sqrt{2}$
    $t''(x) = 8 - 12x^2$
    $t''(0) = 8 - 12(0)^2 = 8 > 0$
    $t''(\pm\sqrt{2})^2 = -16 < 0$

Chapter 4

So we have a relative maximum = 4, which occurs at $x = \pm \sqrt{2}$
and a relative minimum = 0, which occurs at $x = 0$.

16. Relative maximum $= -\dfrac{176}{27}$, occurs at $x = -\dfrac{2}{3}$; relative minimum $= -16$, occurs at $x = 2$.

17.

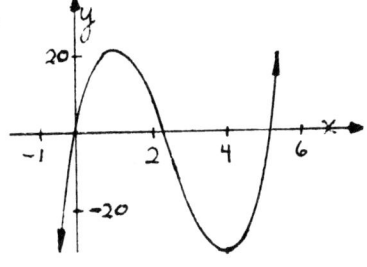

18. Find the critical points:

$g'(x) = 15x^4 - 150x^2 + 135 = 0 \implies x^4 - 10x^2 + 9 = 0 \implies (x^2 - 1)(x^2 - 9) = 0$
$\implies (x + 1)(x - 1)(x + 3)(x - 3) = 0 \implies x = -3, -1, 1, 3$

$g''(x) = 60x^3 - 300x$

$g''(-3) = 60(-3)^3 - 300(-3) = -720 < 0$ (relative maximum)

$g''(-1) = 60(-1)^3 - 300(-1) = 240 > 0$ (relative minimum)

$g''(1) = 60(1)^3 - 300(1) = -240 < 0$ (relative maximum)

$g''(3) = 60(3)^3 - 300(3) = 720 > 0$ (relative minimum)

You should calculate $g(x)$ for $x = -4, -3, -2, \ldots, 2, 3, 4$.

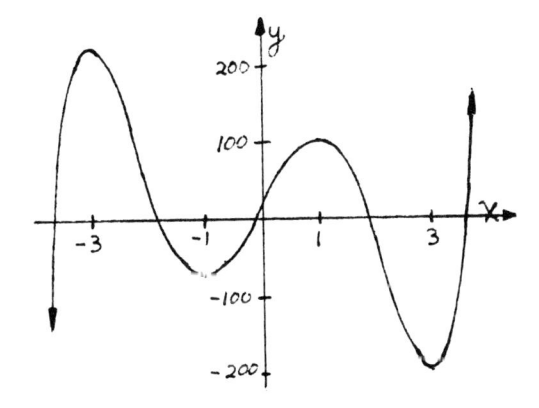

19.

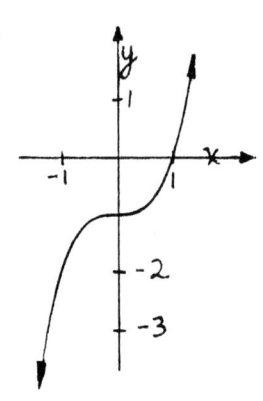

20.

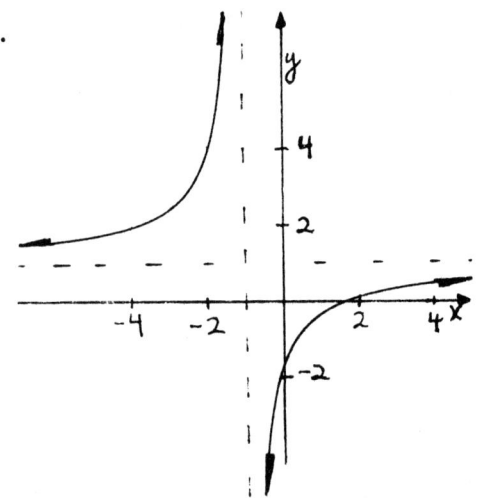

21. Find critical values:

$y' = 15x^4 - 255x^2 + 240 = 0 \Rightarrow x^4 - 17x^2 + 16 = 0 \Rightarrow (x^2 - 16)(x^2 - 1) = 0 \Rightarrow x = 4, -4, 1, -1$

Evaluate the function at the endpoints and the critical points within [-2, 5]:

At $x = -2$, $y = 3(-2)^5 - 85(-2)^3 + 240(-2) = 104$

At $x = 5$, $y = 3(5)^5 - 85(5)^3 + 240(5) = 0$

At $x = 4$, $y = (4)^4 - 85(4)^3 + 240(4) = -1408$ minimum

At $x = 1$, $y = (1)^4 - 85(1)^3 + 240(1) = 156$

At $x = -1$, $y = (-1)^4 - 85(-1)^3 + 240(-1) = -154$

The maximum = 104, which occurs at $x = -2$; the minimum = -1408, which occurs at $x = 4$.

22. The maximum = $\frac{1}{4}$, which occurs at $x = \frac{1}{4}$; the minimum = -6, which occurs at $x = 9$.

23. The maximum = 2048, which occurs at $x = 5$; the minimum = -27, which occurs at $x = 0$.

24. Find the critical points: $y' = -\frac{3}{2}x^{-\frac{1}{2}} = = \frac{3}{2\sqrt{x}}$. $y' \neq 0$, but at $x = 0$, y' is undefined and y is defined. Evaluate y at the critical point $x = 0$ and the endpoints:

At $x = 0$, $y = 1 - 0^{\frac{3}{2}} = 1$

At $x = 4$, $y = 1 - 4^{\frac{3}{2}} = -7$

The maximum = 1, which occurs at $x = 0$; the minimum = -7, which occurs at $x = 4$.

25. 480 items   **26.** No maximum profit

Chapter 4

$P(x) = -.005x^3 - .01x^2 + 3x - 4000$

$P'(x) = -.-15x^2 - .02x + 3$. Set this equal to zero and solve:

$-.015x^2 - .02 + 3 = 0 \Rightarrow 15x^2 + 2\text{-}x - 3000 = 0 \Rightarrow 3x^2 + 4x - 600 = 0$

Using the Quadratic Formula: $x \approx 13.49$ or -14.82. Rejecting the negative value gives us $x \approx 13.49$, or we should produce 13 items.

28. 50 items

29. $R(x) = xp(x) = 25x - \dfrac{x^3}{(500)^2} = 25x - (.000004)x^3$

$R'(x) = 25 - .000012x^2$

$R'(100) = 25 - .000012(100)^2 = 24.88 > 0$, thus marginal revenue is increasing at $x = 100$.

30. Minimum cost = $0   **31.** $1,500 (60 will stay rented)

32. Let x represent the height of the printed matter. Since the area of printed matter is 208, the width of the printed matter is $\dfrac{208}{x}$. The height of the poster is $(x + 5)$, and the width of the poster is $(\dfrac{208}{x} + 4)$. The amount of cardboard used is:

$$A(x) = (x + 5)(\dfrac{208}{x} + 4)$$

$$= 4x + 1040x^{-1} + 228$$

The critical points:

$A' = 4 - 1040x^{-2} = 0 \Rightarrow 4 = \dfrac{1040}{x^2} \Rightarrow x^2 = \dfrac{1040}{4} = 260 \Rightarrow x \approx 16$

Since $A'' = \dfrac{2080}{x^3} > 0$ for $x > 0$, this is a minimum. The height (length) should be 21 inches and the width should be 17 inches.

CHAPTER 5
EXPONENTIAL FUNCTIONS, LOGARITHMIC FUNCTIONS

5.1 EXPONENTIAL FUNCTIONS, PAGE 153

1. $25^{\frac{1}{2}} = \sqrt{25} = 5$ 3. $(-25)^{\frac{1}{2}} = \sqrt{-25}$ which is not a real number   5. $-27^{\frac{1}{3}} = -\sqrt[3]{27} = -3$

7. $7^{\frac{1}{3}} \cdot 7^{\frac{2}{3}} = 7^{\frac{1}{3}+\frac{2}{3}} = 7^1 = 7$ 9. $1000^{-\frac{1}{3}} = \dfrac{1}{1000^{\frac{1}{3}}} = \dfrac{1}{\sqrt[3]{1000}} = \dfrac{1}{10}$

11. $100^{-\frac{3}{2}} = \dfrac{1}{100^{\frac{3}{2}}} = \dfrac{1}{(\sqrt{100})^3} = \dfrac{1}{10^3} = \dfrac{1}{1000}$ 13. $x^{\frac{1}{2}}\left(x^{\frac{1}{2}} + x^{\frac{1}{2}}\right) = x^1 + x^1 = 2x$

15. $x^{\frac{2}{3}}\left(x^{-\frac{2}{3}} + x^{\frac{1}{3}}\right) = x^0 + x^1 = 1 + x$ 17. $\left(x^{\frac{1}{2}} + y^{\frac{1}{2}}\right)^2 = \left(x^{\frac{1}{2}} + y^{\frac{1}{2}}\right)\left(x^{\frac{1}{2}} + y^{\frac{1}{2}}\right) = x + 2x^{\frac{1}{2}}y^{\frac{1}{2}} + y$

19. Some points are:

x	3	2	1	0	-1	-2
y	$3^3 = 27$	$3^2 = 9$	3	$3^0 = 1$	$3^{-1} = \dfrac{1}{3}$	$3^{-2} = \dfrac{1}{9}$

The graph is shown at right:

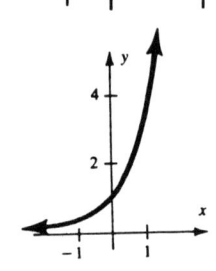

21. Some points are:

x	3	2	1	0	-1	-2
y	$\left(\frac{1}{3}\right)^3=\frac{1}{27}$	$\left(\frac{1}{3}\right)^2=\frac{1}{9}$	$\frac{1}{3}$	$\left(\frac{1}{3}\right)^0=1$	$\left(\frac{1}{3}\right)^{-1}=3$	$\left(\frac{1}{3}\right)^{-2}=9$

The graph is shown at right.

23. Some points are:

x	2	1	0	-1	-2
y	$e^{-2} \approx .14$	$e^{-1} \approx .37$	$e^0 = 1$	$e^1 \approx 2.7$	$e^2 \approx 7.4$

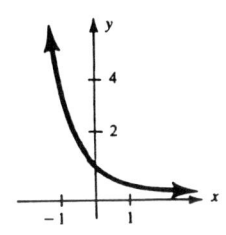

The graph is shown at right.

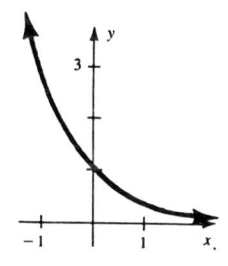

25. 20.085537   **27.** 1.0512711   **29.** 1.046028

31. $A = P(1+i)^N = 1000(1+.07)^{25} = \$5,427.43$

33. $A = Pe^{rt} = 1000\, e^{(.16)(25)} = 1000\, e^4$
$= 1000\,(54.59815)$
$= \$54,598.15$

35. $A = P(1+i)^N = 8500\left(1+\frac{.18}{12}\right)^{(12)(4)} = 8500\,(1.015)^{48}$
$= 8500\,(2.043478)$
$= \$17,369.57$

37. $A = P(1+i)^N = 9400\left(1+\frac{.14}{360}\right)^{(\frac{1}{2})(360)} = 9400\,(1.000388889)^{180}$

Chapter 5

$$= 9400 \, (1.0724936)$$
$$= \$10,081.44$$

5.2 LOGARITHMIC FUNCTIONS, PAGE 158

1. Remember a logarithm is an <u>exponent</u>; since the exponent = 6, logarithm = 6
   base = 2
   $\log_2 64 = 6$

3. The exponent = -2, logarithm = -2
   base = $\frac{1}{3}$
   $\log_{\frac{1}{3}} 9 = -2$

5. The exponent = c, logarithm = c
   base = b
   $\log_b a = c$

7. logarithm = $\frac{1}{2}$, exponent = $\frac{1}{2}$
   base = 4
   $4^{\frac{1}{2}} = 2$

9. logarithm = -2, exponent = -2
   base = 10
   $10^{-2} = .01$ (Log with no base is understood to be base 10.)

11. logarithm = 2, exponent = 2
    base = e
    $e^2 = e^2$ (ln x = $\log_e x$)

13. Let $x = \log_b b^2$. In exponential form this is $b^x = b^2$, hence $\log_b b^2 = x = 2$.

15. Let $x = \log_\pi \sqrt{\pi}$. In exponential form this is $\pi^x = \sqrt{\pi} = \pi^{\frac{1}{2}}$, hence $x = \log_\pi \sqrt{\pi} = \frac{1}{2}$.

17. Let $x = \log 1000$. In exponential form this is $10^x = 1000$, hence $x = \log 1000 = 3$.

19. Let $x = \log_{16} 1$. In exponential form this is $16^x = 1 = 16^0$, hence $x = \log_{16} 1 = 0$.

21. Using a calculator enter 1.08 on display, then press the [log] key, which should give .033423755.

23. Using a calculator enter 9760 on display, then press the [log] key, which should give 3.989449818.

25. Enter .321 on display, then press the [log] key, which should give -.493494967.

27. Enter 2.27 on display, then press the [ln] key, giving .81977983.

29. Enter 2 on display, then press the [ln] key, giving .69314718.

31. A few points are:

x	x ≤ 0	$\frac{1}{3}$	1	$\sqrt{3}$	3	9
y	undefined	-1	0	$\frac{1}{2}$	1	2

Chapter 5

The graph is at right.

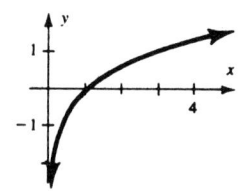

33. Using a calculator, a few points are:

x	x ≤ 0	.5	1	1.5	2	2.5	3
y	undefined	-.69	0	.4	.69	.92	1.1

The graph is at right.

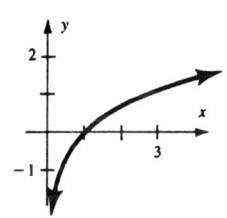

35. Using a calculator, a few points are:

x	x ≤ 0	.5	1	1.5	2	2.5	3
y	undefined	-.35	0	.2	.35	.46	.55

Notice the y values are half those of Problem 33.
The graph is at right.

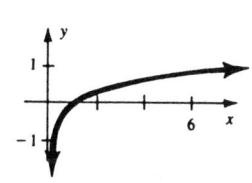

37. a. $N = 1500 + 300 \ln(1000) \approx 1500 + 300(6.9078) \approx 3{,}572$
   b. $N = 1500 + 300 \ln(50{,}000) \approx 1500 + 300(10.8198) \approx 4{,}746$

39. a. $M = \dfrac{\log 15^{15} - 11.8}{1.5} \approx \dfrac{17.64137 - 11.8}{1.5} \approx 3.9$

   b. $M = \dfrac{\log 10^{25} - 11.8}{1.5} = \dfrac{25 - 11.8}{1.5} = 8.8$

41. Let $x = \log_b 1$. This is equivalent to $b^x = 1$, which implies $x = 0$. Thus, $\log_b 1 = x = 0$.

43. Let $y = \ln e^x$. This is equivalent to $e^y = e^x$, which implies $x = y = \ln e^x$.

Chapter 5

5.3 LOGARITHMIC AND EXPONENTIAL EQUATIONS, PAGES 163 - 164

1. The equation is equivalent to $5^x = 25$, so $x = 2$. Another way would be to use the change of base formula (with a calculator): $x = \log_5 25 = \dfrac{\log 25}{\log 5} = \dfrac{1.39794}{.69897} = 2$.

3. The equation is equivalent to $10^x = \dfrac{1}{10} = 10^{-1}$, so $x = -1$. Another way would be to use a calculator: $x = \log(0.1) = -1$.

5. The equation is equivalent to $x^2 = 28$, so $x = \sqrt{28} = 2\sqrt{7}$.

7. The equation is equivalent to $e^3 = x$. Using a calculator press $\boxed{3}$ $\boxed{e^x}$ giving a display of 20.08553692.

9. Since $\ln 9.3 = \ln x$, $9.3 = x$. If you want to "do" something to both sides to see this, raise e to the power $\ln 9.3$ and $\ln x$: $e^{\ln 9.3} = e^{\ln x}$, giving $9.3 = x$

11. Since $\ln x^2 = \ln 12$; $x^2 = 12$, so $x = \pm\sqrt{12} = \pm 2\sqrt{3}$.

13. The equation is equivalent to $3^x = 27\sqrt{3} = 3^3 \, 3^{\frac{1}{2}} = 3^{3.5}$, so $x = 3.5$. Another way would be to use the change of base formula (with a calculator):

    $x = \log_3 27\sqrt{3} = \dfrac{\log 27\sqrt{3}}{\log 3} \approx \dfrac{1.6699244}{.47712125} = 3.5$.

15. The equation is equivalent to $x^0 = 10$ which implies $1 = 10$. Since this is a contradiction, there is no solution.

17. If you know $2^7 = 128$, the answer is $x = 7$. Another way is to use the definition of log:

    $x = \log_2 128 = \dfrac{\log 128}{\log 2} \approx \dfrac{2.10721}{.30103} = 7$

19. Since $125 = 5^3$ and $25 = 5^2$, the given equation is equivalent to $(5^3)^x = 5^2$, so $5^{3x} = 5^2$ which leads to $3x = 2$; hence $x = \dfrac{2}{3}$.

21. Since $\dfrac{1}{9} = 3^{-2}$, the given equation is equivalent to $3^{4x-3} = 3^{-2}$, so $4x - 3 = -2$. Solving for x gives $x = \dfrac{1}{4}$.

23. $\log_8 5 + \dfrac{1}{2}\log_8 9 = \log_8 x$ Given

    $\log_8 5 + \log_8 9^{\frac{1}{2}} = \log_8 x$ Power rule for logs

Chapter 5

$$\log_8 5 + \log_8 3 = \log_8 x$$
$$\log_8 15 = \log_8 x \qquad \text{Product rule for logs}$$
$$15 = x$$

25. $\ln 10 - \dfrac{1}{2}\ln 25 = \ln x \qquad \text{Given}$

$\ln 10 - \ln 25^{\frac{1}{2}} = \ln x \qquad \text{Power rule for logs}$

$\ln 10 - \ln 5 = \ln x$

$\ln \left(\dfrac{10}{5}\right) = \ln x \qquad \text{Quotient rule for logs}$

$\ln 2 = \ln x$

$2 = x$

27. The given equation in logarithmic form is:

$\log 515 = 2x - 1$

$2.711807229 \approx 2x - 1 \qquad \text{Calculator}$

$3.711807229 \approx 2x$

$1.855903615 \approx x$

29. The given equation in logarithmic form is:

$\log_4 .82 = x$

$\dfrac{\log .82}{\log 4} = x \qquad \text{Change of base formula}$

$-\dfrac{.0806186}{.60206} \approx x \qquad \text{Calculator}$

$-.143152 \approx x$

31. Using $A = P(1 + i)^N$, $2000 = 1000\left(1 + \dfrac{.12}{2}\right)^{2t}$ where t is the number of years. Divide both sides by 1000:

$2 = (1.06)^{2t}$

$\log_{1.06} 2 = 2t \qquad \text{Write as a logarithm}$

$\dfrac{\log 2}{\log 1.06} = 2t \qquad \text{Change of base formula}$

$\dfrac{1}{2} \cdot \dfrac{\log 2}{\log 1.06} = t \qquad \text{Divide both sides by 2}$

Chapter 5

$\qquad$ 5.9 ≈ t $\qquad$ Calculator

To the nearest half-year, it would take 6 years.

33. Using $A = Pe^{rt}$, $2000 = 1000 e^{.12t}$

$\qquad$ $2 = e^{.12t}$ $\qquad$ Divide both sides by 1000

$\qquad$ ln 2 = .12t $\qquad$ Write as a logarithm

$\qquad$ $\dfrac{\ln 2}{.12} = t$ $\qquad$ Divide by .12

$\qquad$ 5.7762265 ≈ t $\qquad$ Calculator

It will take 5.7762265 years, or to the nearest day, (5.7762265)(3.65) = 2108 days (5 years, 283 days).

35. Using $A = Pe^{rt}$, $3000 = 1000e^{.12t}$

$\qquad$ $3 = e^{.12t}$ $\qquad$ Divide both sides by 1000

$\qquad$ ln 3 = .12t $\qquad$ Write as a logarithm

$\qquad$ $\dfrac{\ln 3}{.12} = t$

$\qquad$ 9.1551 ≈ t

It will take 9.1551 years or (9.1551)(365) = 3342 days (9 years, 57 days).

37. Using $P = P_0 e^{rt}$, $5 = 4e^{r(11.04)}$

$\qquad$ $1.25 = e^{11.04r}$

$\qquad$ 11.04r = ln 1.25

$\qquad$ $r = \dfrac{\ln 1.25}{11.04}$

$\qquad$ ≈ .0202

Growth rate of 2.02%

39. a. R = 80 - 27ln (3) ≈ 50.3; about 50%

 b. 10 = 80 - 27 ln t

$\qquad$ -70 = -27 ln t $\qquad$ Subtract 80 from both sides

$\qquad$ $\dfrac{70}{27} = \ln t$ $\qquad$ Divide both sides by -27

$\qquad$ $e^{\frac{70}{27}} = t$ $\qquad$ Write in exponential form

$\qquad$ 13.364375 ≈ t $\qquad$ Calculator, using [ex] key

About 13 seconds.

Chapter 5   121

41. Using the given formula,

$$m = \frac{(10,000)(.01)}{1 - (1 + .01)^{-48}}$$ 48 months = 4 years

$$= \frac{100}{1 - .6202604}$$ Calculator

$$= \$263.34$$

43. $$P = (\frac{1}{2})^{\frac{t}{5700}}$$

$$\log_{\frac{1}{2}} P = \frac{t}{5700}$$ Write in logarithmic form

$$5700 \log_{\frac{1}{2}} P = t$$ Multiply both sides by 5700

45. Using the formula: $584,600 = \frac{19,473}{.08}(e^{.08t} - 1)$

$$\frac{(.08)(584,600)}{19,473} = e^{.08t} - 1$$

$$2.401684 = e^{.08t} - 1$$

$$3.401684 = e^{.08t}$$

$$\ln(3.401684) = .08t$$

$$t = \frac{\ln(3.401684)}{.08} \approx 15.3 \text{ years}$$

47. Let $A = b^x$ and $B = b^y$, which means $\log_b A = x$ and $\log_b B = y$.

$$\log_b \frac{A}{B} = \log_b \frac{b^x}{b^y} = \log_b b^{x-y} = x - y = \log_b A - \log_b B$$

5.4 DERIVATIVES OF LOGARITHMIC AND EXPONENTIAL FUNCTIONS, PAGES 168 - 169

1. $y' = e^{2x}(2x)' = e^{2x}(2) = 2e^{2x}$ **3.** $y' = e^{-x}(-x)' = e^{-x}(-1) = -e^{-x}$

5. Use the Product Rule:
$f'(x) = x(e^x)' + (x)'(e^x)$
$= x(e^x) + (1)e^x$
$= xe^x + e^x$

7. Use the Product Rule:
$f'(x) = (2x^3 + 1)(e^{5x^2})' + (2x^3 + 1)'(e^{5x^2})$
$= (2x^3 + 1)(e^{5x^2}(5x^2)') + (6x^2)(e^{5x^2})$
$= (2x^3 + 1)(e^{5x^2}(10x)) + 6x^2 e^{5x^2}$

Chapter 5

$$= 10xe^{5x^2}(2x^3 + 1) + 6x^2 e^{5x^2}$$
$$= 2xe^{5x^2}[5(2x^3 + 1) + 3x]$$
$$= 2xe^{5x^2}[10x^3 + 3x + 5]$$

9. Use the Product Rule:
$y' = -2e^{5x}(3x^2 + 5)' + (-2e^{5x})'(3x^2 + 5)$
$\quad = -2e^{5x}(6x) + (-2e^{5x}(5x)')(3x^2 + 5)$
$\quad = -12xe^{5x} + (-2e^{5x}(5))(3x^2 + 5)$
$\quad = -12xe^{5x} + (-10e^{5x})(3x^2 + 5)$
$\quad = -2e^{5x}(6x + 15x^2 + 25)$

11. $y' = \dfrac{1}{5-x}(5-x)' = \dfrac{1}{5-x}(-1) = \dfrac{1}{x-5}$

13. $y' = 4(2x^3 + e^{5x})^3 (2x^3 + e^{5x})' = 4(2x^3 + e^{5x})^3 (6x^2 + 5e^{5x})$

15. $y' = \dfrac{1}{x^4}(x^4)' = \dfrac{1}{x^4}(4x^3) = \dfrac{4}{x}$

17. It might be easier to first use the power rule for logarithms:

$$y = \ln\sqrt{x} = \ln x^{\frac{1}{2}} = \dfrac{1}{2}\ln x$$

Now, $y' = \dfrac{1}{2}(\ln x)' = \dfrac{1}{2} \cdot \dfrac{1}{x} = \dfrac{1}{2x}$

19. First rewrite $\ln x^3 = 3\ln x$, then use the quotient rule:

$$y = \dfrac{\ln x^3}{e^x} = \dfrac{3\ln x}{e^x} \ ; \quad y' = \dfrac{e^x(3\ln x)' - 3\ln x(e^x)'}{(e^x)^2}$$

$$= \dfrac{e^x\left(\dfrac{3}{x}\right) - 3\ln x(e^x)}{(e^x)^2}$$

$$= \dfrac{\dfrac{3}{x} - 3\ln x}{e^x} \cdot \dfrac{x}{x} \qquad \text{Reduce by dividing out } e^x$$

$$= \dfrac{3 - 3x\ln x}{xe^x} \qquad \text{Multiply numerator and denominator by } x$$

21. $f'(x) = \dfrac{1}{\left(\dfrac{x}{x+2}\right)} \cdot \left(\dfrac{x}{x+2}\right)' = \dfrac{x+2}{x}\left[\dfrac{(x+2)x' - x(x+2)'}{(x+2)^2}\right]$

Chapter 5

$$= \frac{x+2}{x}\left[\frac{(x+2)(1) - x(1)}{(x+2)^2}\right]$$

$$= \frac{x+2}{x}\left[\frac{2}{(x+2)^2}\right]$$

$$= \frac{2}{x(x+2)}$$

<u>Another way</u> would be to rewrite $y = \ln\left|\frac{x}{x+2}\right| = \ln|x| - \ln|x+2|$

So $y' = \frac{1}{x} - \frac{1}{x+2} = \frac{1}{x} \cdot \frac{x+2}{x+2} - \frac{1}{x+2} \cdot \frac{x}{x} = \frac{x+2}{x(x+2)} - \frac{x}{x(x+2)}$

$$= \frac{2}{x(x+2)}$$

23. Use the Product Rule:

$$f'(t) = e^{t^2}\left(1 - e^{-t}\right)' + \left(e^{t^2}\right)'\left(1 - e^{-t}\right)$$

$$= e^{t^2}\left(e^{-t}\right) + \left(e^{t^2}(2t)\right)\left(1 - e^{-t}\right)$$

$$= e^{t^2}e^{-t}\left[1 + 2te^{t}\left(1 - e^{-t}\right)\right]$$

$$= e^{t^2-t}(1 + 2te^t - 2t)$$

25. $y' = 2(e^x - e^{-x})(e^x - e^{-x})' = 2(e^x - e^{-x})(e^x + e^{-x}) = 2(e^{2x} - e^{-2x})$

27. $y' = \dfrac{(5x+2)(\ln|x|)' - (5x+2)'\ln|x|}{(5x+2)^2} = \dfrac{(5x+2)(\frac{1}{x}) - (5)\ln|x|}{(5x+2)^2}$

$$= \dfrac{(5x+2)(\frac{1}{x}) - 5\ln|x|}{(5x+2)^2} \cdot \frac{x}{x}$$

$$= \dfrac{5x + 2 - 5x \ln|x|}{x(5x+2)^2}$$

29. $f'(x) = \dfrac{\ln|3x|(e^x)' - e^x(\ln|3x|)'}{(\ln|3x|)^2}$

$$= \dfrac{\ln|3x|(e^x) - e^x(\frac{1}{3x} \cdot 3)}{\ln^2|3x|}$$

Chapter 5   124

$$= \frac{e^x \ln |3x| - e^x(\frac{1}{x})}{\ln^2 |3x|} \cdot \frac{x}{x}$$

$$= \frac{e^x x \ln |3x| - e^x}{x \ln^2 |3x|} = \frac{e^x(x \ln |3x| - 1)}{x \ln^2 |3x|}$$

31. $y = 2000\left(1 + 6e^{.3x}\right)^{-1}$; $y' = 2000\left[-1\left(1 + 6e^{.3x}\right)^{-2}\right]\left(1 + 6e^{.3x}\right)'$

$$= -2000\left(1 + 6e^{.3x}\right)^{-2}\left(6e^{.3x}(.3)\right)$$

$$= \frac{-2000(6)(.3)e^{.3x}}{\left(1 + 6e^{.3x}\right)^2}$$

$$= \frac{-3600e^{.3x}}{\left(1 + 6e^{.3x}\right)^2}$$

33. $y' = \dfrac{\left(1 + 40e^{-.2x}\right)(100 \ln |x|)' - \left(1 + 40e^{-.2x}\right)'(100 \ln |x|)}{\left(1 + 40e^{-.2x}\right)^2}$

$$= \frac{\left(1 + 40e^{-.2x}\right)\left(\frac{100}{x}\right) - \left(8e^{-.2x}\right)(100 \ln |x|)}{\left(1 + 40e^{-.2x}\right)^2}$$

$$= \frac{\left(1 + 40e^{-.2x}\right)100 - 800xe^{-.2x} \ln |x|}{x\left(1 + 40e^{-.2x}\right)^2}$$

$$= \frac{100e^{.2x} + 400 - 800x \ln |x|}{xe^{.2x}\left(1 + 40e^{-.2x}\right)^2}$$

$$= \frac{100(e^{.2x} + 40 + 8x \ln |x|)}{xe^{.2x}(1 + 40e^{-.2x})^2}$$

35. Revenue = (number of items)(price) $R(x) = (x)\left(\dfrac{500 \ln (x + 10)}{x^2}\right) = \dfrac{500 \ln (x + 10)}{x}$

Chapter 5

$$\text{Marginal revenue} = R'(x) = 500\left[\frac{x\,(\ln(x+10))' - \ln(x+10)(x)'}{x^2}\right]$$

$$= 500\left[\frac{x\left(\dfrac{1}{x+10}\right) - \ln(x+10)}{x^2}\right]$$

$$= 500\left[\frac{x - (x+10)\ln(x+10)}{x^2(x+10)}\right]$$

37. a. $N(0) = \dfrac{5000}{1 + 4999e^0} = \dfrac{5000}{5000} = 1$

   b. $N(t) = 5000(1 + 4999e^{-.1t})^{-1}$

   $N'(t) = -5000(1 + 4999e^{-.1t})^{-2}(1 + 4999e^{-.1t})'$

   $= -5000(1 + 4999e^{-.1t})^{-2}(499.9e^{-.1t})$

   $= 2499500e^{-.1t}(1 + 4999e^{-.1t})^{-2}$

   $= \dfrac{2499500}{e^{.1t}(1 + 4999e^{-.1t})^2}$

   $N'(0) = \dfrac{2499500}{e^0(1 + 4999e^0)^2} = \dfrac{2499500}{5000^2}$

   $\approx .09998$ (people per day)

39. $A' = P(e^{.12t})' = P(.12e^{.12t})$

   After 1 year: $A'(1) = P(.12e^{.12(1)}) \approx .12(1.27)P$

   $= .1353P$ or 13.5% of P

   After 5 years: $A'(5) = P(.12e^{.12(5)}) \approx .12(1.822)P$

   $\approx .2187P$ or 21.9% of P

41. $A'(t) = 5(e^{-.03t})' = 5(-.03e^{-.03t})$

   $= -.15e^{-.03t}$

43. Revenue, $R(x) = (x)(d(x))$

   $= x(500e^{-.1x})$

   $= 500xe^{-.1x}$

   $R'(x) = 500[x(e^{-.1x})' + x'(e^{-.1x})]$

Chapter 5   126

$$= 500[x(-.1e^{-.1x}) + e^{-.1x}]$$
$$= -50xe^{-.1x} + 500e^{-.1x}$$

45. Use the change of base formula:
$$y = \log x = \frac{\ln x}{\ln 10} = \left(\frac{1}{\ln 10}\right) \ln x$$
$$y' = \frac{1}{\ln 10}(\ln x)' = \frac{1}{\ln 10}\left(\frac{1}{x}\right)$$
$$= \frac{1}{\left(\frac{\log 10}{\log e}\right)} \cdot \frac{1}{x}$$
$$= \frac{\log e}{\log 10} \cdot \frac{1}{x}$$
$$= \frac{\log e}{1} \cdot \frac{1}{x}$$
$$= \frac{\log e}{x}$$

47. If $y = \log_b u$ where u is a function of x, then
$$y' = \frac{dy}{dx} = \frac{\log_b e}{u} \cdot \frac{du}{dx}$$

By the chain rule,
$$\frac{dy}{dx} = \frac{dy}{du} \cdot \frac{du}{dx} \quad \text{but} \quad \frac{dx}{du} = \frac{\log_b e}{u}$$
$$y' = \frac{\log_b e}{u} \cdot \frac{du}{dx}$$

5.5 CHAPTER REVIEW, PAGE 170
1. In exponential form the equation is $6^x = 36$, so $x = 2$.

3. $2 \log 2 - \frac{1}{2} \log 2 = \log \sqrt{x}$ Given

$\phantom{2 \log 2 - \frac{1}{2} \log 2 =} \frac{3}{2} \log 2 = \log \sqrt{x}$ Combining like terms

$\phantom{2 \log 2 - \frac{1}{2} \log 2 =} \log 2^{\frac{3}{2}} = \log \sqrt{x}$ Power rule for logarithms

$\phantom{2 \log 2 - \frac{1}{2} \log 2 =} 2^{\frac{3}{2}} = \sqrt{x}$

Chapter 5

$$\sqrt{2^3} = \sqrt{x}$$
$$2^3 = x$$
$$x = 8$$

5. In logarithmic form: $\log 250 = x - 1$

$$\log 250 + 1 = x$$
$$2.39794 + 1 = x$$
$$x = 3.39794$$

7. In logarithmic form: $2x + 3 = \ln 10$

$$x = \frac{\ln(10) - 3}{2} = \frac{2.3026 - 3}{2}$$
$$\approx -.348707$$

9. $y' = 5.5(e^{-.5x^2})' = 5.5(e^{-.5x^2})(-.5x^2)'$

$$= 5.5(e^{-.5x^2})(-1x)$$
$$= -5.5xe^{-.5x^2}$$

11. First, use the power rule for logarithms: $y = 5 \ln x$
So $y' = 5(\ln x)' = 5(\frac{1}{x}) = \frac{5}{x}$.

13. First use the product rule for logarithms:
$$y = \ln |x^2 (4 - x^3)| = \ln |x^2| + \ln(4 - x^3)$$

Now, $y' = \frac{1}{x^2}(x^2)' + \frac{1}{4 - x^3}(4 - x^3)'$

$$= \frac{1}{x^2}(2x) + \frac{1}{4 - x^3}(-3x^2)$$

$$= \frac{2}{x} + \frac{-3x^2}{4 - x^3} = \frac{2(4 - x^3)}{x(4 - x^3)} + \frac{-3x^3}{x(4 - x^3)}$$

$$= \frac{8 - 5x^3}{x(4 - x^3)}$$

15. $y' = \frac{(4 - x)(\ln x^2)' - (\ln x^2)(4 - x)'}{(4 - x)^2}$

$$= \frac{(4-x)(\frac{1}{x^2} \cdot 2x) - \ln x^2 \cdot (-1)}{(4-x)^2} = \frac{(4-x)(2) + x \ln x^2}{x(4-x)^2}$$

$$= \frac{8 - 2x + x \ln x^2}{x(4-x)^2}$$

17. $A = 10(\frac{1}{2})^{\frac{t}{5700}}$

$\frac{A}{10} = (\frac{1}{2})^{\frac{t}{5700}}$

$\frac{t}{5700} = \log_{\frac{1}{2}}(\frac{A}{10})$

$t = 5700 \log_{\frac{1}{2}}(\frac{A}{10})$

19. Find the equilibrium point by solving $d(x) = s(x)$

$$200e^{-0.2x} = 20e^{0.1x}$$

$200e^{-0.2x} - 20e^{0.1x} = 0$ Divide by 20 and factor $e^{0.1x}$

$e^{0.1x}(10e^{-0.3x} - 1) = 0$

$e^{0.1x} = 0$ $10e^{-0.3x} - 1 = 0$
No solution $e^{-0.3x} = 0.1$
                      $-0.3x = \ln 0.1$
                      $x \approx 7.6752836$

This is about 7675 units.

5.6 STUDENT'S TEST REVIEW AND ADDITIONAL PRACTICE
OBJECTIVES

The material of this chapter is reviewed in the following list of objectives. After each objective there are some practice questions. Answers to these problems immediately follow. Detailed solutions are given for every third problem. For a sample test select the first question of each set and check your answers. Additional practice is given by the other questions in each set. If you are having trouble with a particular type of problem, or if you want additional practice, look back at the indicated section in the test.

Chapter 5

[5.1] Objective 1: *Simplify expressions with positive, negative, and fractional exponents.*

1. $125^{\frac{2}{3}}$
2. $2^{\frac{1}{2}} \cdot 3^{\frac{1}{3}}$
3. $\dfrac{27^{\frac{2}{3}}}{27^{\frac{1}{2}}}$
4. $(x^{\frac{1}{2}} - y^{\frac{1}{2}})(x^{\frac{1}{2}} + y^{\frac{1}{2}})$

Objective 2: *Sketch the graph of exponential functions.*

5. $y = (\frac{1}{2})^x$
6. $y = -2^x$
7. $y = 2^{-x}$
8. $y = e^{-\frac{x}{2}}$

Objective 3: *Evaluate expression with the natural base.*

9. e^1
10. e^4
11. $e^{1.05}$
12. $e^{-.005}$

[5.2] Objective 4: *Write an exponential equation in logarithmic form.*

13. $10^{.5} = \sqrt{10}$
14. $e^0 = 1$
15. $9^3 = 729$
16. $(\sqrt{2})^3 = 2\sqrt{2}$

Objective 5: *Write a logarithmic equation in exponential form.*

17. $\log 1 = 0$
18. $\ln \dfrac{1}{e} = -1$
19. $\log_2 64 = 6$
20. $\log_\pi \pi = 1$

Objective 6: *Evaluate common and natural logarithms.*

21. $\ln 3$
22. $\log 3$
23. $\log .0021$
24. $\ln .013$

Objective 7: *Graph logarithmic functions.*

25. $y = \log x$
26. $y = \ln x$
27. $y = \log_5 x$
28. $y = \log_{\frac{1}{5}} x$

[5.3] Objective 8: *Solve logarithmic equations.*

29. $\log_5 25 = x$
30. $\log_x(x + 6) = 2$
31. $3 \log 3 - \dfrac{1}{2} \log 3 = \log \sqrt{x}$
32. $2 \ln \dfrac{e}{\sqrt{7}} = 2 - \ln x$

Objective 9: *Solve exponential equations.*

33. $10^{x+2} = 125$
34. $5^{2x-3} = .5$
35. $e^{4-3x} = 15$
36. $10^{-x^2} = .45$

[5.4] Objective 10: *Find the derivative of exponential functions.*

37. $y = 4.9e^{-.05x^2}$
38. $y = 3500x - 500e^{3x}$
39. $y = x^3 e^{x^4}$
40. $y = \dfrac{1000}{x - 999e^{-.2x}}$

Chapter 5

Objective 11: *Find the derivative of logarithmic functions.*

41. $y = \ln x^2$

42. $y = \ln |x^4(x^2 - 5)|$

43. $y = \dfrac{\ln x^2}{x + 3}$

44. $y = \dfrac{250 \ln |3x^2 - 5|}{1 + 3e^x}$

[5.1-5.3] Objective 12: *Solve applied problems based on the preceeding objectives. For specific examples of the types of applications look at the list of applications in this chapter on page 146.*

45. If a person's present salary is $20,000 per year, use the formula $A = P(1 + i)^N$ to determine the salary necessary to equal this salary in 15 years if you assume the 1984 inflation rate or 3.2%.

46. If $5,500 is invested at 13.5% compounded daily, how long will it take for this to grow to $10,000? (Use a 365–day year.)

47. The half life formula for Carbon-14 dating is

$$A = 10\left(\dfrac{1}{2}\right)^{\frac{t}{5700}}$$

where A is the amount present (in milligrams) after t years. Solve this equation for t.

48. An advertising agency conducted a survey and found that the number of units sold, N, is related to the amount spent on advertising (in dollars) by the formula

$$A = 1500 + 300 \ln a \quad (a \geq 1)$$

What is the rate at which N is changing at the instant that a = $10,000?

49. The atmospheric pressure P in pounds per square inch (psi) is given by

$$P = 14.7e^{-.21a}$$

where a is the altitude above sea level (in miles). As a hot-air balloon is rising the pressure is constantly changing. How fast is the pressure changing when the balloon is one mile above sea level?

50. A satellite has an initial radioisotope power supply of 50 watts. The power output in watts is given by the equation

$$P = 50e^{\frac{-t}{250}}$$

where t is the time in days. Solve for t.

51. Suppose the price-demand and price-supply equations for x-thousands of units of a commodity are given by

DEMAND: $d(x) = 300e^{-.3x}$

SUPPLY: $s(x) = 30e^{.1x}$

Find the equilibrium point.

52. Suppose the President makes a major policy announcement and that the model to predict the percent, N, of the population that will have heard the announcement is a function of the time, t (in days), after the announcement according to the formula $N = 1 - e^{-2.5t}$

How long will it take for 90% of the population (N = .9) to hear of the announcement?

ANSWERS TO STUDENT'S TEST REVIEW AND PRACTICE QUESTIONS

1. 25   2. $2^{\frac{1}{2}} \cdot 3^{\frac{1}{3}} = 2^{\frac{3}{6}} \cdot 3^{\frac{2}{6}} = (2^3 \cdot 3^2)^{\frac{1}{6}} = \sqrt[6]{72}$ 3. $\dfrac{27^{\frac{2}{3}}}{27^{\frac{1}{2}}} = 27^{\frac{2}{3} - \frac{1}{2}} = 27^{\frac{1}{6}} = (3^3)^{\frac{1}{6}} = 3^{\frac{1}{2}} = \sqrt{3}$ 4. x - y

5.

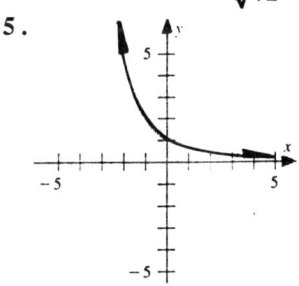

6. Some points are:

Let x = -1, y = -2^{-1} = $-\dfrac{1}{2}$

Let x = 0, y = -2^0 = -1

Let x = 1, y = -2^1 = -2

Let x = 2, y = -2^2 = -4

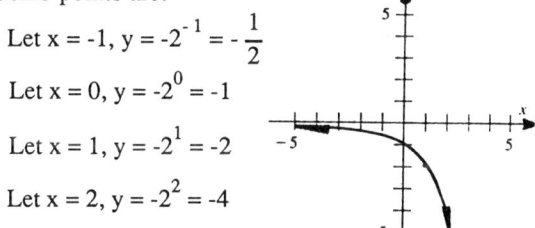

7.

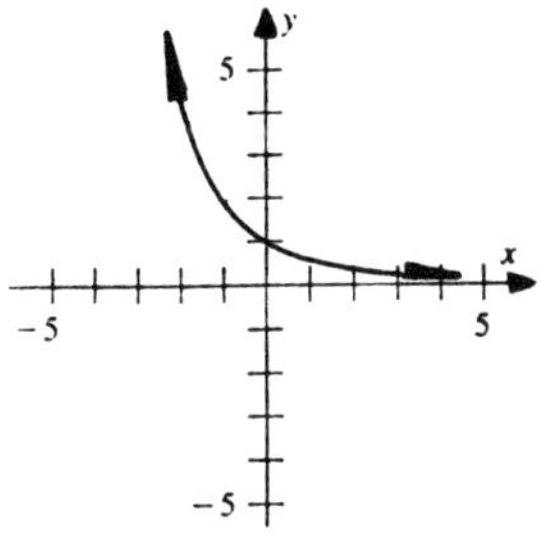

8.

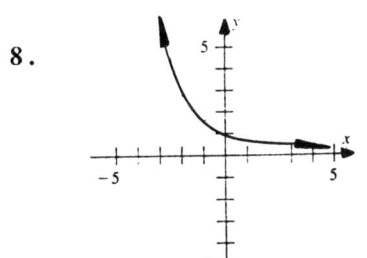

9. e ≈ 2.71828   10. 54.59815   11. 2.8576511   12. .99501248   13. log $\sqrt{10}$ = .5
14. ln 1 = 0   15. $\log_9 729 = 3$ 16. $\log_{\sqrt{2}} 2\sqrt{2} = 3$ 17. $10^0 = 1$
18. $e^{-1} = \dfrac{1}{e}$ 19. $2^6 = 64$ 20. $\pi^1 = \pi$ 21. 1.0986123   22. .47712125

23. -2.6777807   24. -4.3428-59

25.

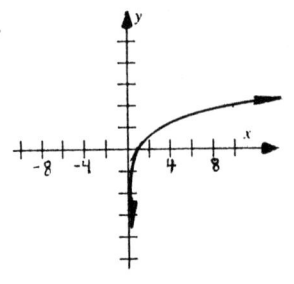

26.

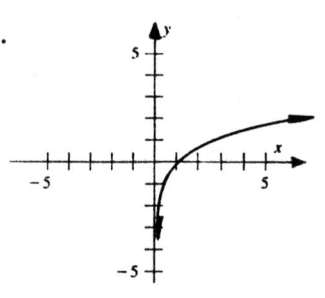

27. Some points are:

Let $x = \frac{1}{5}$, $y = \log_5(\frac{1}{5}) = \log_5 5^{-1} = -1$

Let $v = 1$, $y = \log_5(1) = 0$

Let $x = 5$, $y = \log_5(5) = 1$

Let $x = \sqrt{5} \approx 2.236$, $y = \log_5(\sqrt{5}) = \log_5 5^{\frac{1}{2}} = \frac{1}{2}$

28.

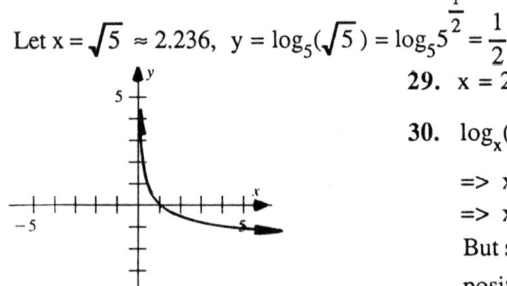

29. $x = 2$

30. $\log_x(x + 6) = 2 \iff x^2 = x + 6$
$\Rightarrow x^2 - x - 6 = 0 \Rightarrow (x - 3)(x + 2) = 0$
$\Rightarrow x = 3$ or -2
But since the base of a logarithm must be positive, $x = 3$.

31. $x = 243$ 32. $x = 7$

33. $10^{x+2} = 125 \iff \log 125 = x + 2 \Rightarrow x = \log 125 - 2 \approx .09691$ 34. $x = 1.28466$

35. $x = .43065$

36. $10^{-x^2} = .45 \iff \log(.45) = -x^2 \Rightarrow -.3467875 = -x^2 \Rightarrow x = \pm .58888665$

37. $y' = -.49xe^{-.05x^2}$ 38. $y' = 3500 - 1500e^{3x}$

39. $y' = x^3(e^{x^4})' + (x^3)'(e^{x^4}) = x^3(4x^3 e^{x^4}) + (3x^2)e^{x^4} = x^2 e^{x^4}(4x^4 + 3)$

40. $y' = \dfrac{-1000(1 + 199.8e^{-.2x})}{(x - 999e^{-.2x})^2}$ 41. $y' = \dfrac{2}{x}$

Chapter 5   133

42. $y' = \dfrac{1}{x^4(x^2-5)} \cdot (x^4(x^2-5))' = \dfrac{1}{x^4(x^2-5)} \cdot (6x^5 - 20x^3) = \dfrac{2x^3(3x^2-10)}{x^4(x^2-5)} = \dfrac{2(3x^2-10)}{x(x^2-5)}$

43. $y' = \dfrac{2x + 6 - x \ln x^2}{x(x+3)^2}$ 44. $y' = \dfrac{750[2x(1+3e^x) - (3x^2-5)e^x \ln |3x^2-5|]}{(3x^2-5)(1+3e^x)^2}$

45. $A = 20{,}000(1+.032)^{15} = \$32{,}079.34$ 46. 1617 days (4 years 157 days)

47. $t = 5700 \log_{\frac{1}{2}} \dfrac{A}{10} = 5700 \dfrac{\log(\frac{A}{10})}{\log(\frac{1}{2})}$

48. $N'(a) = \dfrac{300}{a}$; $N'(10{,}000) = \dfrac{300}{10{,}000} = .03$ units per (advertising) dollar

49. $P'(1) \approx -2.5023$ psi per mile   50. $t = -250 \ln(\dfrac{P}{50})$ or $250 \ln(\dfrac{50}{P})$

51. Equilibrium when $d(x) = s(x)$ => $300e^{-.3x} = 30e^{.1x}$ => $10e^{-.3x} = e^{.1x}$ => $10 = e^{.1x}e^{.3x}$
    => $10 = e^{.4x}$ => $.4x = \ln(10)$ => $x = \dfrac{\ln(10)}{.4} \approx 5.756$; this is 5,756 items.

52. $t \approx .92$ days (about 22 hours)

5.7 MODELING APPLICATION 4: SAMPLE ESSAY
World Running Records
Model 1: Time and Year Linearly Related
  The raw data for this extended application are found in the list of records in Table 1. We begin by carrying out an analysis of the data assuming a linear relationship as discussed in Section 9.7. The trend is illustrated in a scatter diagram as shown in Figure 1. The calculations for finding the least squares line were done by Oakley and Baker in 1977.*

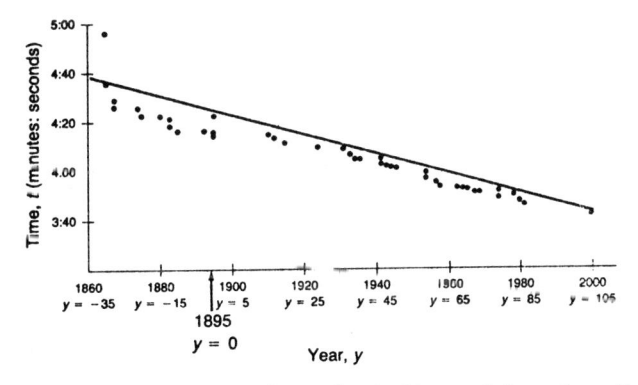

Figure 1 World record times for mile run

---

* Cletus O. Oakley and Justine C. Baker, "Least Squares and the 3:40-Minute Mile," *Mathematics Teacher,* April 1977, pp. 322-324.

Table 1 World Records for the Mile Run (up to date as of Jan., 1985)

Year	Name, Country	Time	Rate	$\ln(1 - \frac{s}{615.38})$
1864	Lawes, U.K.	4:56.0	326.22	-.755
1865	Webster, U.K.	4:36.5	349.22	-.838
1868	Chinnery, U.K.	4:29.0	358.96	-.875
1868	Gibbs, U.K.	4:28.8	359.23	-.876
1874	Slade, U.K.	4:26.0	363.01	-.891
1875	Slade, U.K.	4:24.5	365.07	-.900
1880	George, U.K.	4:23.2	366.87	-.907
1882	George, U.K.	4:21.4	369.40	-.917
1882	George, U.K.	4:19.4	372.25	-.929
1884	George, U.K.	4:18.4	373.69	-.935
1894	Bacon, U.K.	4:18.2	373.98	-.936
1895	Bacon, U.K.	4:17.0	375.72	-.943
1895	Conneff, U.S.	4:15.6	377.78	-.952
1911	Jones, U.S.	4:15.4	378.08	-.953
1913	Jones, U.S.	4:14.6	379.26	-.958
1915	Taber, U.S.	4:12.6	382.27	-.971
1923	Nurmi, Finland	4:10.4	385.63	-.985
1931	Ladoumegue, France	4:09.2	387.48	-.993
1922	Lovelock, N.Z.	4:07.6	389.99	-1.004
1934	Cunningham, U.S.	4:06.8	391.25	-1.010
1937	Wooderson, U.K.	4:06.4	391.89	-1.013
1942	Haegg, Sweden	4:06.2	392.20	-1.014
1942	Andersson, Sweden	4:06.2	392.20	-1.014
1942	Haegg, Sweden	4:04.6	394.77	-1.026
1943	Andersson, Sweden	4:02.6	398.02	-1.041
1944	Andersson, Sweden	4:01.6	399.67	-1.048
1945	Haegg, Sweden	4:01.4	400.00	-1.050
1954	Bannister, U.K.	3:59.4	403.34	-1.065
1954	Landy, Australia	3:58.0	405.72	-1.077
1957	Ibbotson, U.K.	3:57.2	407.09	-1.083
1958	Elliott, Australia	3:54.5	411.77	-1.106
1962	Snell, N.Z.	3:54.4	411.95	-1.107
1964	Snell, N.Z.	3:54.1	412.48	-1.110
1965	Jazy, France	3:53.6	413.36	-1.114
1966	Ryun, U.S.	3:51.3	417.47	-1.134
1967	Ryun, U.S.	3:51.1	417.83	-1.136
1975	Bayi, Tanzania	3:51.0	418.01	-1.137
1975	Walker, N.Z.	3:49.4	420.93	-1.152
1979	Coe, U.K.	3:49.0	421.66	-1.156
1980	Ovett, U.K.	3:48.8	422.03	-1.158
1981	Coe, U.K.	3:48.53	422.80	-1.160
1981	Ovett, U.K.	348.40	422.80	-1.162
1981	Coe, U.K.	3:47.33	424.76	-1.172

Oakley and Baker obtained the relationship

$$t = 4.339 - .006450y$$

where t is the time for race and y is the year (1865 = 0). They then extrapolated to predict that the record would be lowered to 3 minutes 40 seconds in about the year 2001.

Chapter 5

Extrapolation is risky business, no matter how good the fit to historical data. This is especially true when the model is clearly inappropriate for the long run. The above equation "predicts" that in the year 2567 the record time for the mile will be .3 seconds, with negative times thereafter! Indeed, Oakley and Baker comment "it is safe to say that in the future the records will be leveling off and that some curve other than a straight line would give better predictions."

Model 2: Hyperbolic Relationship

So, we seek a more realistic model. Fortunately, at least two are available, which do not use any more sophisticated mathematics. Our second model is suggested by an aritcle titled, "Future Performance in Footracing" by Ryder, Carr, and Herget.* This article, which concludes that current running records are well below human physiological limits and that performance limits are chiefly psychological, discusses runners' accomplishments in terms of running speed rather than time. A graph of year versus world record speed is shown in Figure 2.

Using linear least squares we find that the line of best fit is

$$s = .58076y - 729.08$$

where y is the year (actual) and s is the average running speed in meters per minute. Now, running speed and time are inversely related, so they cannot both be linearly related to year. In this second model, replacing s by 1,609.344/t (there are 6,609.344 meters in 1 mile), we get

$$t = \frac{1,609.344}{.58076y - 729.08}$$

As the years go by, t will asymptotically approach zero, an unlikely asymptote, but at least t will not be equal to zero or negative. Using this model the 3:40 mile would be predicted for the year 2011.

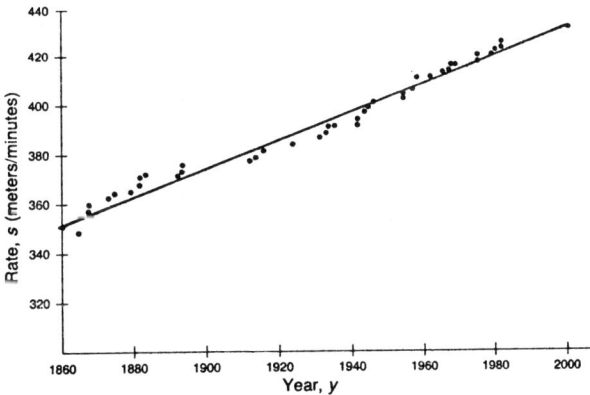

Figure 2 World record running rates for mile run

---

* Henry W. Ryder, Harry Jay Carr, and Paul Herget, "Future Performance in Footracing," *Scientific American* June 1976, pp. 109-119.

Model 3: Exponential Relationship

A more likely model would have running speed asymptotically approaching some maximum possible speed, M, for a human, as in Figure 3. The available data cannot really support the assumption of this specific curve, but this is probably because we have data over too short a span of years. Almost any curve is a fair approximation of a straight line if we look at a small enough portion of it (see Figure 3). The real stumbling block to fitting this kind of model, however, is not lack of data, but lack of a value of M.

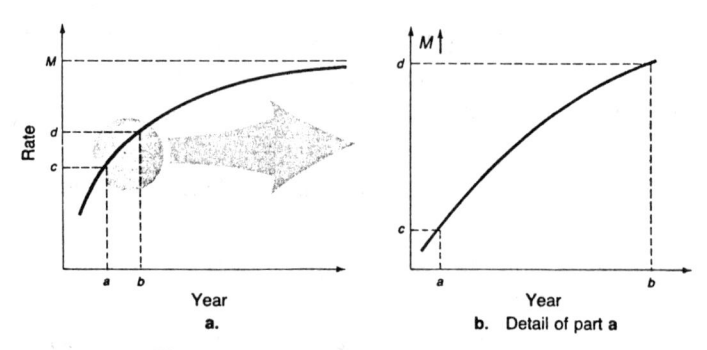

Figure 3 Exponential model for mile run

Ryder, Carr, and Herget (1976) indicate that we are not close to attaining that level yet and make no attempt to estimate it. Scientists have used bone size and estimated body weight to calculate a maximum speed some dinosaurs could run without breaking their own bones from the stresses and shocks. A similar calculation might be made for humans, but a risk would be that future diet improvements and medical advances might alter it. Instead of attempting this approach, let us take the attitude of Ryder, Carr, and Herget that records, in events other than the sprints, are set not by running faster than anyone has before, but rather by increased endurance at a rate previously maintainable only for lesser distances. The most estreme case we can imagine, based on rates actually attained, would be to use the fastest speed ever run by a human for any distance. This would make M = 615.38 meters/minute, based on the world record of 19.5 seconds for 200 meters. (The 100-meter record has a slower average speed because of time lost at the start.)

The curve for this model, known as the *learning curve* after its use by Clark Hull (1844-1952) to describe his learning theories,* has an equation of the form

---

* Ernest Hilgard, *Theories of Learning*, 2nd ed. (New York: Appleton-Century-Crofts), 1956, p. 372.

$$s = M\left[1 - e^{\alpha(y-\beta)}\right]$$

where α and β are constants. The constant α controls the rate of climb of the curve, and β provides the proper translation of y. This relationship between s and y is, of course, not linear, but if we solve for y,

$$y = \frac{1}{\alpha}\ln\left(1 - \frac{s}{M}\right) + \beta$$

we do get a linear relationship between y and $\ln\left(1 - \frac{s}{M}\right)$. Using linear least squares here we get

$$y = -369.02 \ln\left(1 - \frac{s}{615.38}\right) + 1{,}554.5$$

or, if solved for s,

$$s = 615.38\left[1 - e^{-.0027099(h - 1{,}554.5)}\right]$$

In this model, the 3:40 mile would be expected in about 2015 and the "ultimate" mile, the one we would be approaching asymptotically, would have a time of 2 minutes 36.9 seconds.

The curves for all three models are shown in Figure 4. All the models give a respectable fit to the available data, but they differ markedly as we look at years to come. The lesson here is that when we attempt to make models of "real-life" situations, we must not limit our considerations to the past, but incorporate what seem to be reasonable assumptions about how things might change in the future. It is these assumptions, as much as or more than mathematical techniques, which make the model a valuable predictor.

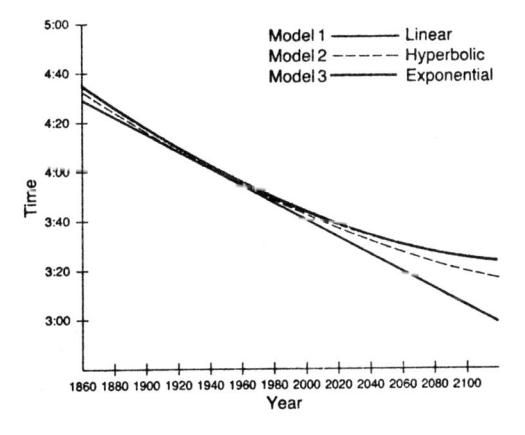

Figure 4 Projected world record times for mile run

CHAPTER 6
INTEGRATION

6.1 THE ANTIDERIVATIVE, PAGES 178 - 179

1. $\int x^7 dx = \dfrac{x^8}{8} + C$ **3.** $\int 4x^3 dx = 4\int x^3 dx = 4\left(\dfrac{x^4}{4}\right) + C = x^4 + C$ **5.** $\int 3dx = 3\int dx = 3x + C$

7. $\int (5x+7)dx = 5\int x\,dx + 7\int dx = 5\left(\dfrac{x^2}{2}\right) + 7x + C = \dfrac{5x^2}{2} + 7x + C$ **9.** $\int dx = x + C$

11. $\int (18x^2 - 6x + 5)dx = 18\int x^2 dx - 6\int x\,dx + 5\int dx = 18\left(\dfrac{x^3}{3}\right) - 6\left(\dfrac{x^2}{2}\right) + 5x + C$

$$= 6x^3 - 3x^2 + 5x + C$$

13. $\int \dfrac{dx}{9} = \dfrac{1}{9}\int dx = \dfrac{1}{9}x + C$ **15.** $\int \dfrac{5dx}{x^2} = 5\int x^{-2}dx = 5\left(\dfrac{x^{-1}}{-1}\right) + C = -5x^{-1} + C$

17. $\int (1 - e^x)dx = \int dx - \int e^x dx = x - e^x + C$

19. $\int (\sqrt[3]{x} + \sqrt{2})dx = \int x^{\frac{1}{3}}dx + \sqrt{2}\int dx = \dfrac{x^{\frac{4}{3}}}{\frac{4}{3}} + \sqrt{2}x + C = \dfrac{3}{4}x^{\frac{4}{3}} + \sqrt{2}x + C$

21. $\int \dfrac{3x-1}{\sqrt{x}}dx = \int \dfrac{3x}{\sqrt{x}}dx - \int \dfrac{1}{\sqrt{x}}dx = 3\int x^{\frac{1}{2}}dx - \int x^{-\frac{1}{2}}dx = 3\left(\dfrac{x^{\frac{3}{2}}}{\frac{3}{2}}\right) - \dfrac{x^{\frac{1}{2}}}{\frac{1}{2}} + C = 2x^{\frac{3}{2}} - 2x^{\frac{1}{2}} + C$

23. $\int \dfrac{x}{\sqrt[3]{x}}dx = \int x^{\frac{2}{3}}dx = \dfrac{x^{\frac{5}{3}}}{\frac{5}{3}} + C = \dfrac{3}{5}x^{\frac{5}{3}} + C$

25. $\int \dfrac{x^2 + x + 1}{x}dx = \int \left(\dfrac{x^2}{x} + \dfrac{x}{x} + \dfrac{1}{x}\right)dx = \int \left(x + 1 + \dfrac{1}{x}\right)dx = \dfrac{x^2}{2} + x + \ln|x| + C$

27. $\int y^3 \sqrt{y}\,dy = \int y^{\frac{7}{2}}dy = \dfrac{y^{\frac{9}{2}}}{\frac{9}{2}} + C = \dfrac{2}{9}y^{\frac{9}{2}} + C$

Chapter 6

29. $\int (1+z^2)^2 dz = \int (1 + 2z^2 + z^4)dz = z + \dfrac{2z^3}{3} + \dfrac{z^5}{5} + C$

31. $\int (3u^2 - u^{-1} + e^u)du = \dfrac{3u^3}{3} - \ln |u| + e^u + C = u^3 - \ln |u| + e^u + C$

33. $F(x) = \int (3x^2 + 4x + 1)dx = x^3 + 2x^2 + x + C$

$10 = F(0) = 0^3 + 2(0)^2 + 0 + C$; hence $C = 10$, and $F(x) = x^3 + 2x^2 + x + 10$

35. $F(x) = \int \dfrac{5x+1}{x} dx = \int (5 + \dfrac{1}{x}) dx = 5x + \ln |x| + C$

$5000 = F(1) = 5(1) + \ln |1| + C = 5 + 0 + C$

Solving for C gives $C = 4995$; hence $F(x) = 5x + \ln |x| + 4995$.

37. $C(x) = \int (4x - 8)dx = 2x^2 - 8x + c$

$5000 = C(0) = 2(0)^2 - 8(0) + c$. Solving for c gives $c = 5000$; hence $C(x) = 2x^2 - 8x + 5000$.

39. $C(x) = \int .009x^2 dx = .009 \dfrac{x^3}{3} + c = .003x^3 + c$

$18,500 = C(0) = .003(0)^3 + c$. Solving for c gives $c = 18,500$; hence $C(x) = .003x^3 + 18,500$.

41. If P(t) is the population at time t, we are given $P'(t) = 450 + 600\sqrt{t}$.

$P(t) = \int P'(t)dt = \int (450 + 600t^{\frac{1}{2}})dt = 450t + \dfrac{600t^{\frac{3}{2}}}{\frac{3}{2}} + C = 450t + 400t^{\frac{3}{2}} + C$

Presently the population is 420,000; which means $420,000 = P(0) = 450(0) + 400(0)^{\frac{3}{2}} + C$.

Solving for C gives $C = 420,000$; hence $P(t) = 450t + 400t^{\frac{3}{2}} + 420,000$.

In 5 years, $P(5) = 450(5) + 400(5)^{\frac{3}{2}} + 420,000 \approx 426,722$.

43. $P(x) = \int (200x - 10,000)dx = \dfrac{200x^2}{2} - 10,000x + C = 100x^2 - 10,000x + C$

We are given $-50,000 = P(0) = 100(0)^2 - 10,000(0) + C$.

Solving for C gives $C = -50,000$; hence $P(x) = 100x^2 - 10,000x - 50,000$.

45. If N(t) is the number of cases at time t (in days), we are given $N'(t) = 240t - 3t^2$.

$N(t) = \int (240t - 3t^2)dt = \dfrac{240t^2}{2} - \dfrac{3t^3}{3} + C = 120t^2 - t^3 + C$

Chapter 6

We are given $50 = N(1) = 120(1)^2 - 1^3 + C$. Solving for C gives $C = -69$; hence

$N(t) = 120t^2 - t^3 - 69$, and on the tenth day $N(10) = 120(10)^2 - 10^3 - 69 = 10{,}931$ (cases).

47. $v(t) = \int a(t)dt = \int -32dt = -32t + C$ Initial velocity is -72: $-72 = v(0) = -32(0) + C$.

Solving for C gives $C = -72$, hence $v(t) = -32t - 72$.

49. The curve is $f(x) = \int f'(x)dx = \int (3x^2 + 5)dx = \dfrac{3x^3}{3} + 5x + C = x^3 + 5x + C$.

Since (1,2) is on the curve, $2 = f(1) = 1^3 + 5(1) + C = 6 + C$. Solve for C:
$C = -4$; hence $f(x) = x^3 + 5x - 4$.

6.2 INTEGRATION BY SUBSTITUTION, PAGES 184 - 185

1. Let $u = 5x + 3$, then $du = 5dx$:

$$\int (5x+3)^3 dx = \int u^3 \frac{du}{5} = \frac{1}{5}\int u^3 du = \frac{1}{5}\cdot\frac{u^4}{4} + C = \frac{(5x+3)^4}{20} + C$$

3. Let $u = 5 - x$, then $du = -dx$:

$$\int \frac{5}{(5-x)^4}dx$$

$$= 5\int (5-x)^{-4}dx = 5\int u^{-4}(-du) = -5\int u^{-4}du = \frac{-5u^{-3}}{-3} + C = \frac{5}{3u^3} + C = \frac{5}{3(5-x)^3} + C$$

5. Let $u = 3x + 5$, then $du = 3dx$:

$$\int \sqrt{3x+5}\,dx = \int \sqrt{u}\,(\frac{du}{3}) = \frac{1}{3}\int u^{\frac{1}{2}}du = \frac{1}{3}\cdot\frac{u^{\frac{3}{2}}}{\frac{3}{2}} + C = \frac{2u^{\frac{3}{2}}}{9} + C = \frac{2(3x+5)^{\frac{3}{2}}}{9} + C$$

7. Let $u = 3x^2 + 1$, then $du = 6xdx$:

$$\int 6x(3x^2+1)dx = \int 6x(u)(\frac{du}{6x}) = \int u\,du = \frac{u^2}{2} + C = \frac{(3x^2+1)^2}{2} + C$$

9. Let $u = x^2 + 5x$, then $du = (2x + 5)dx$:

$$\int \frac{2x+5}{\sqrt{x^2+5x}}dx = \int \frac{2x+5}{\sqrt{u}}(\frac{du}{2x+5}) = \int \frac{1}{\sqrt{u}}du = \int u^{-\frac{1}{2}}du = \frac{u^{\frac{1}{2}}}{\frac{1}{2}} + C = 2\sqrt{u} + C$$

$$= 2\sqrt{x^2+5x} + C$$

11. Let $u = 6 + 5x$, then $du = 5dx$:

Chapter 6   141

$$\int \frac{dx}{6+5x} = \int \frac{(\frac{du}{5})}{u} = \frac{1}{5}\int \frac{du}{u} = \frac{1}{5}\ln|u| + C = \frac{1}{5}\ln|6+5x| + C$$

13. Let $u = 1 - 3x^2$, then $du = -6xdx$:

$$\int \frac{xdx}{1-3x^2} = \int \frac{x(\frac{du}{-6x})}{u} = -\frac{1}{6}\int \frac{du}{u} = -\frac{1}{6}\ln|u| + C = -\frac{1}{6}\ln|1-3x^2| + C$$

15. Let $u = 5x$, then $du = 5dx$:

$$\int e^{5x}dx = \int e^u(\frac{du}{5}) = \frac{1}{5}\int e^u du = \frac{1}{5}e^u + C = \frac{1}{5}e^{5x} + C$$

17. Let $u = 4x^3$, then $du = 12x^2 dx$:

$$\int 5x^2 e^{4x^3}dx = \int 5x^2 e^u (\frac{du}{12x^2}) = \frac{5}{12}\int e^u du = \frac{5}{12}e^u + C = \frac{5}{12}e^{4x^3} + C$$

19. Let $u = 4x^2 - 4x$, then $du = (8x-4)dx = 4(2x-1)dx$

$$\int \frac{2x-1}{(4x^2-4x)^2}dx = \int \frac{2x-1}{u^2}(\frac{du}{4(2x-1)}) = \frac{1}{4}\int u^{-2}du = \frac{1}{4}\frac{u^{-1}}{-1} + C = -\frac{1}{4u} + C$$

$$= -\frac{1}{4(4x^2-4x)} + C$$

21. Let $u = x^4 - 2x^2 + 3$, then $du = (4x^3 - 4x)dx$:

$$\int \frac{4x^3-4x}{x^4-2x^2+3}dx = \int \frac{4x^3-4x}{u}(\frac{du}{4x^3-4x}) = \int \frac{du}{u} = \ln|u| + C = \ln|x^4-2x^2+3| + C$$

23. Let $u = \ln|x|$, then $du = \frac{1}{x}dx$:

$$\int \frac{\ln|x|}{x}dx = \int \frac{u}{x}(xdu) = \int u\,du = \frac{u^2}{2} + C = \frac{\ln^2|x|}{2} + C$$

25. $\ln e^x = x$, so $\int \ln e^x dx = \int x\,dx = \frac{x^2}{2} + C$

27. Let $u = 4x^2 + 1$, then $du = 8xdx$. Also let $v = x^3$, then $dv = 3x^2 dx$:

$$\int \left(\frac{3x}{4x^2+1} + x^2 e^{x^3}\right)dx = \int \frac{3x}{4x^2+1}dx + \int x^2 e^{x^3}dx = \int \frac{3x}{u}(\frac{du}{8x}) + \int x^2 e^v (\frac{dv}{3x^2})$$

$$= \frac{3}{8}\int \frac{du}{u} + \frac{1}{3}\int e^v dv = \frac{3}{8}\ln|u| + \frac{1}{3}e^v + C = \frac{3}{8}\ln(4x^2+1) + \frac{1}{3}e^{x^3} + C$$

(Since $4x^2 + 1 > 0$, the absolute value symbols may be dropped.)

29. Let $u = x^2 + 1$, then $du = 2xdx$:

$$\int x\sqrt[3]{(x^2+1)^2}\,dx = \int x\sqrt[3]{u^2}\,(\frac{du}{2x}) = \frac{1}{2}\int u^{\frac{2}{3}}\,du = \frac{1}{2}\frac{u^{\frac{5}{3}}}{\frac{5}{3}} + C = \frac{3}{10}(x^2+1)^{\frac{5}{3}} + C$$

31. Let $u = \sqrt{x+1}$, then $u^2 = x+1$, or $u^2 - 1 = x$. Differentiate implicitly: $(u^2 - 1)' = x'$

$$2u\frac{du}{dx} = 1$$

$$2u\,du = dx$$

$$\int \frac{x\,dx}{\sqrt{x+1}} = \int \frac{(u^2 - 1)}{u}(2u\,du) = 2\int(u^2 - 1)\,du = 2(\frac{u^3}{3} - u) + C = 2\left(\frac{\sqrt{x+1}^3}{3} - \sqrt{x+1}\right) + C$$

$$= \frac{2}{3}(x+1)^{\frac{3}{2}} - 2\sqrt{x+1} + C$$

33. $P(t) = \int P'(t)\,dt = \int 40t(5 - t^2)^2\,dt$ (Let $u = 5 - t^2$, then $du = -2t\,dt$)

$$= \int 40t(u)^2(\frac{du}{-2t}) = -20\int u^2\,du = -20(\frac{u^3}{3}) + C = -\frac{20}{3}(5 - t^2)^3 + C$$

Now, $10 = P(0) = -\frac{20}{3}(5 - 0^2)^3 + C$

$$10 = -833\frac{1}{3} + C$$

$$\frac{2530}{3} = C$$

So, $P(t) = -\frac{20}{3}(5 - t^2)^3 + \frac{2530}{3} = \frac{20}{3}(t^2 - 5)^3 + \frac{2530}{3}$

35. $P(x) = \int \frac{100x}{\sqrt[3]{(x^2 - 36)^2}}\,dx$ Let $u = x^2 - 36$, then $du = 2x\,dx$

$$= \int \frac{100x}{\sqrt[3]{u^2}}(\frac{du}{2x}) = \int \frac{50\,du}{\sqrt[3]{u^2}} = 50\int u^{-\frac{2}{3}}\,du = 50\frac{u^{\frac{1}{3}}}{\frac{1}{3}} + C = 150(x^2 - 36)^{\frac{1}{3}} + C$$

Now, $0 = P(10) = 150(10^2 - 36)^{\frac{1}{3}} + C$

$$0 = 150(4) + C$$

$$-600 = C$$

So $P(x) = 150(x^2 - 36)^{\frac{1}{3}} - 600$, and specifically $P(100) = 150(100^2 - 36)^{\frac{1}{3}} - 600 \approx \$2{,}627.77$

37. $S(t) = \int 2000e^{-.2t}dt$ Let $u = -.2t$, then $du = -.2dt$

$= 2000 \int e^u (\frac{du}{-.2}) = -10,000 \int e^u du = -10,000 e^u + C = -10,000e^{-.2t} + C$

Now $15,000 = S(0) = -10,000e^0 + C = -10,000 + C$

$25,000 = C$

So $S(t) = -10,000e^{-.2t} + 25,000$. In 3 years:

$S(3) = -10,000e^{-.2(3)} + 25,000 \approx \$19,511.88$

39. $P(t) = \int 14,000e^{.0175t}dt$ Let $u = .0175t$, $du = .0175dt$

$= 14,000 \int e^u (\frac{du}{.0175}) = \frac{14,000}{.0175} \int e^u du = 800,000 (e^u) + C = 800,000e^{.0175t} + C$

We are given in 1984 ($t = 4$) the population is ($P = 842,779$):

$842,779 = P(4) = 800,000e^{.0175(4)} + C = 858,006.54 + C$

$-15,227.54 = C$

So $P(t) = 800,000e^{.0175t} - 15,227.54$.

In the year 2000 ($t = 20$): $P(20) = 800,000e^{.0175(20)} - 15,227.54 \approx 1,120,026$ people

41. $T(t) = \int 32.5e^{.048t}dt$ Let $u = .048t$, $du = .048dt$

$= 32.5 \int e^u (\frac{du}{.048}) = \frac{32.5}{.048} e^u + C = 677.083e^{.048t} + C$

We are given that $1950 = T(0) = 677.083e^0 + C$

$1273 \approx C$

So $T(t) = 677.083e^{.048t} + 1273$. $T(0)$ is the total consumption from 1925 to 1985; $T(15)$ is the total consumption from 1925 to 2000. Thus, from 1985 to 2000 find $T(15) - T(0)$:

$T(15) = 677.083e^{.048(15)} + 1273 \approx 2264$

$T(10) = 677.083 + 1273 \approx 1950$

$T(15) - T(0) \approx 2664 - 1950 \approx 714$

6.3 THE FUNDAMENTAL THEOREM OF CALCULUS, PAGES 194 - 195

1. $\int_2^3 x^2 dx = \frac{x^3}{3}\Big|_2^3 = \frac{27}{3} - \frac{8}{3} = \frac{19}{3}$

3. $\int_1^4 \sqrt{x}\, dx = \int_1^4 x^{\frac{1}{2}} dx = \frac{x^{\frac{3}{2}}}{\frac{3}{2}}\Big|_1^4 = \frac{2}{3}\sqrt{x^3}\Big|_1^4 = \frac{2}{3}(\sqrt{4^3} - \sqrt{1^3}) = \frac{2}{3}(\sqrt{4^3} - \sqrt{1^3}) = \frac{2}{3}(8 - 1) = \frac{14}{3}$

Chapter 6   144

5. $\displaystyle\int_{-1}^{2} x(1+x^4)dx = \int_{-1}^{2}(x+x^5)dx = \frac{x^2}{2}+\frac{x^6}{6}\Big|_{-1}^{2} = (\frac{4}{2}+\frac{64}{6}) - (\frac{1}{2}+\frac{1}{6}) = 12$

7. Let $u = x - 2$, then $du = dx$. Also, if $x = 3$ then $u = 1$, and if $x = 8$ then $u = 6$:

$$\int_{x=3}^{x=8}(x-2)^{-1}dx = \int_{u=1}^{u=6} u^{-1}du = \ln|u|\Big|_{1}^{6} = \ln 6 - \ln 1 = \ln 6$$

9. $\displaystyle\int_{1}^{32} x^{-\frac{2}{5}}dx = \frac{x^{\frac{3}{5}}}{\frac{3}{5}}\Big|_{1}^{32} = \frac{5}{3}(32^{\frac{3}{5}} - 1^{\frac{3}{5}}) = \frac{5}{3}(8-1) = \frac{35}{3}$

11. Let $u = 2x - 1$, then $du = 2dx$. Also, if $x = 0$ then $u = -1$, and if $x = 1$ then $u = 1$:

$$\int_{x=0}^{x=1}\frac{dx}{(2x-1)^2} = \int_{u=-1}^{u=1}\frac{\frac{du}{2}}{u^2} = \frac{1}{2}\int_{-1}^{1}u^{-2}du = -\frac{1}{2}(\frac{1}{u})\Big|_{-1}^{1} = -\frac{1}{2}(1-(-1)) = -1$$

13. $\displaystyle\int_{1}^{2}(x^{-2} - \frac{2}{x} + x^3)dx = \frac{x^{-1}}{-1} - 2\ln|x| + \frac{x^4}{4}\Big|_{1}^{2} = (-\frac{1}{2} - 2\ln 2 + 4) - (-1 - 2\ln 1 + \frac{1}{4}) = \frac{17}{4} - 2\ln 2$

$$= \frac{17}{4} - \ln 4$$

15. Since $\displaystyle\int_{a}^{a} f(x)dx = 0,\ \int_{6}^{6}(x^2 + \frac{\sqrt{5x}}{3} - 5)dx = 0$

17. Let $u = 1 + 3x^2$, then $du = 6xdx$. Also, if $x = -1$ then $u = 4$, and if $x = 1$ then $u = 4$:

$$\int_{-1}^{1}\frac{xdx}{\sqrt{1+3x^2}} = \int_{4}^{4}\frac{\frac{du}{6}}{u^{\frac{1}{2}}} = 0 \quad (\text{Since } \int_{a}^{a} f(u)du = 0)$$

19. Let $u = x^3 + 1$, then $du = 3x^2 dx$. Also, if $x = -1$ then $u = 0$, and if $x = 0$ then $u = 1$:

$$\int_{-1}^{0} 5x^2(x^3+1)^{10}dx = \int_{0}^{1} 5x^2 u^{10}\frac{du}{3x^2} = \frac{5}{3}\int_{0}^{1}u^{10}du = \frac{5}{3}(\frac{u^{11}}{11}\Big|_{0}^{1}) = \frac{5}{3}(\frac{1}{11} - 0) = \frac{5}{33}$$

Chapter 6

21. $\int_0^4 3dx = 3x \Big|_0^4 = 3(4-0) = 12$

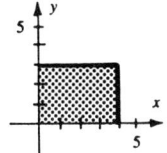

23. $\int_0^4 3xdx = \frac{3^2}{2}\Big|_0^4 = \frac{3}{2}(4^2 - 0^2) = 24$

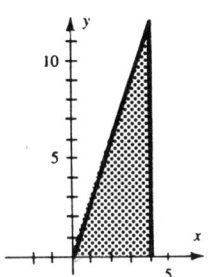

25. $\int_0^4 (2x+1)dx = (x^2 + x)\Big|_0^4$
$= (16 + 4) - (0+0) = 20$

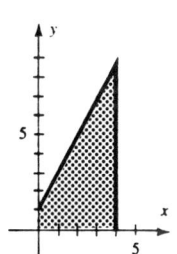

27. $\int_2^5 (x+4)dx = \frac{x^2}{2} + 4x \Big|_2^5$
$= (\frac{25}{2} + 20) - (\frac{4}{2} + 8) = \frac{45}{2}$

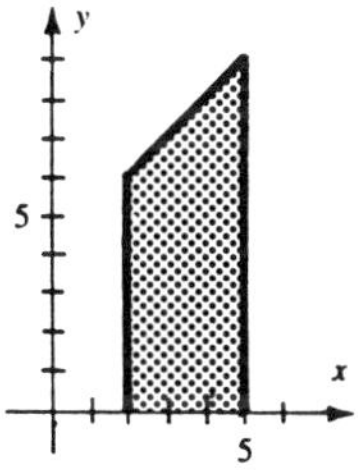

29. $\int_{-2}^1 (x^2 + 1)dx = \frac{x^3}{3} + x \Big|_{-2}^1 = 6$

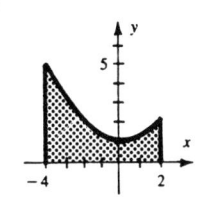

31. Let $u = 2x + 1$, then $du = 2x$. Also, if $x = 1$ and 3 then $u = 3$ and 7, respectively:

$\int_1^3 \frac{1}{(2x+1)^2} dx = \int_3^7 \frac{1}{u^2}(\frac{du}{2}) = \frac{1}{2}\int_3^7 u^{-2}du$

$= -\frac{1}{2}(u^{-1}\Big|_3^7) = -\frac{1}{2}(\frac{1}{7} - \frac{1}{3}) = \frac{2}{21}$

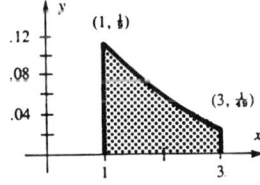

33. Let u = 5x then du = 5dx. Also, if x = 0 and 2 then u = 0 and 10, respectively:

$$\int_0^2 e^{5x}dx = \int_0^{10} e^u \left(\frac{du}{5}\right) = \frac{1}{5}(e^u)\Big|_0^{10} = \frac{1}{5}(e^{10} - 1)$$

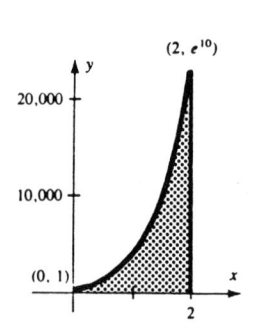

35. a. $\int_{-3}^{2} f(x)dx = \int_{-3}^{0} f(x)dx + \int_{0}^{2} f(x)dx = 4 + 10 = 14$ **b.** $\int_{2}^{-3} f(x)dx = -\int_{-3}^{2} f(x)dx = 5$

    c. $\int_{-3}^{2} [2f(x) - 3g(x)]dx = 2\int_{-3}^{2} f(x)dx - 3\int_{-3}^{2} g(x)dx = 2(14) - 3(-5) = 43$

37. $\int_3^5 f(y)dy = \int_3^1 f(y)dy + \int_1^5 f(y)dy = -\int_1^3 f(y)dy + \int_1^5 f(y)dy = -9 + 15 = 6$

39. $\int_2^{15} 78e^{-.04t}dt = -1950e^{-.04t}\Big|_2^{15} = -1950(e^{-.6} - e^{-.08}) \approx 730$ (billion barrels)

41. $\int_2^{15} 32.5e^{.048t} \approx 677e^{.048t}\Big|_2^{15} = 677(e^{.72} - e^{.096}) \approx 645.6$ (billion)

43. We want to approximate the shaded area from S to D: October + November + December
≈ 7.5 + 6.5 + 7.5 = 21.1 or, to the million, about 21 million.

45. $\int_0^{10} 50e^t dt = 50e^t \Big|_0^{10} = 50(e^{10} - e^0) \approx \$1{,}101{,}273.29$

47. $G = 1 - \int_0^1 2x^3 dx = 1 - \left(\frac{x^4}{2}\Big|_0^1\right) = 1 - \left(\frac{1}{2} - 0\right) = \frac{1}{2}$

49. The curve is a parabola that opens downward, with vertex at (0, 9) and x-intercepts at (-3, 0) and (3, 0).

Chapter 6

$$\text{Area} = \int_{-3}^{3}(9-x^2)dx = 9x - \frac{x^3}{3}\Big|_{-3}^{3} = (27-\frac{27}{3}) - (-27+\frac{27}{3}) = 36$$

6.4 AREA BETWEEN CURVES, PAGES 203 - 204

1. Since $f(x) > g(x)$ on $[a, b]$, the area = $\int_a^b [f(x) - g(x)]dx$

3. Since $g(x) > f(x)$ on $[a, b]$, the area = $\int_a^b [g(x) - f(x)]dx$

5. Same as #1. Since $f(x) > g(x)$ on $[a, b]$, the area = $\int_a^b [f(x) - g(x)]dx$

7. Since $g(x) \geq f(x)$ on $[a, c]$ and $f(x) \geq g(x)$ on $[c, b]$, the area
$$= \int_a^c [g(x) - f(x)]dx + \int_c^b [f(x) - g(x)]dx$$

9. To see where the graph intersects, set $x^2 = 6 - x$ and solve for x:
$x^2 + x - 6 = 0$; $(x+3)(x-2) = 0$, so $x = -3$ or 2.
The area is $\int_{-3}^{2}[(6-x) - x^2]dx = 6x - \frac{x^2}{2} - \frac{x^3}{3}\Big|_{-3}^{2} = 20\frac{5}{6}$

11. To see where $y = x^2$ and $y = 6x$; intersect, set $x^2 = 6x$ and solve:
$x^2 - 6x = 0$; $x(x-6) = 0$; $x = 0$ or $x = 6$.
The area from -2 to 5 is: $\int_{-2}^{0}(x^2 - 6x)dx + \int_0^5 (6x - x^2)dx = \frac{x^3}{3} - 3x^2\Big|_{-2}^{0} + 3x^2 - \frac{x^3}{3}\Big|_0^5$
$$= (\frac{18}{3} + 12) + (75 - \frac{125}{3}) = 48$$

13. To see where the graphs intersect, set $x^2 + 4 = 2x + 4$ and solve:
$x^2 - 2x = 0 \Rightarrow x(x-2) = 0 \Rightarrow x = 0$ or 2. The area is:
$$\int_{-1}^{0}[(x^2 + 4) - (2x+4)]dx + \int_0^1 [(2x+4) - (x^2+4)]dx$$
$$= \int_{-1}^{0}(x^2 - 2x)dx + \int_0^1 (2x - x^2)dx = \frac{x^3}{3} - x^2\Big|_{-1}^{0} + x^2 - \frac{x^3}{3}\Big|_0^1 = \frac{4}{3} + \frac{2}{3} = 2$$

Chapter 6   148

15. The graph of $y = x^2 - 4$ crosses the x-axis at $(-2, 0)$ and $(2, 0)$. Since $0 > x^2 - 4$ for $-2 < x < 2$; the area is:

$$\int_{-2}^{2} (0 - (x^2 - 4))dx = -\frac{x^3}{3} + 4x \Big|_{-2}^{2} = \frac{32}{3}$$

17. The graph of $y = x^2$ is above $y = x^3$ for $0 < x < 1$, so the area is:

$$\int_{0}^{1} (x^2 - x^3)dx = \frac{x^3}{3} - \frac{x^4}{4} \Big|_{0}^{1} = \frac{1}{12}$$

19. To see where the graphs intersect, set $x^2 - x = 6$ and solve: $x^2 - x - 6 = 0$; $(x - 3)(x + 2) = 0$ so $x = -2$ or $x = 3$. Since the graph of $y = x^2 - x$ is below the graph of $y = 6$ for $-2 < x < 3$, the area is:

$$\int_{-2}^{3} (6 - (x^2 - x))dx = 6x - \frac{x^3}{3} + \frac{x^2}{2} \Big|_{-2}^{3} = 20\frac{5}{6}$$

21. The graph of $y = x^2$ and $y = \sqrt{x}$ intersect at $(0, 0)$ and $(1, 1)$. Since the graph of $y = \sqrt{x}$ is above the graph of $y = x^2$ for $0 < x < 1$, the area is:

$$\int_{0}^{1} (\sqrt{x} - x^2)dx = \frac{x^{\frac{3}{2}}}{\frac{3}{2}} - \frac{x^3}{3} \Big|_{0}^{1} = \frac{2}{3}x^{\frac{3}{2}} - \frac{1}{3}x^3 \Big|_{0}^{1} = \frac{1}{3}$$

23. Since the graph $y = \sqrt{x}$ is below $y = 4$ for $0 < x < 16$ and they intersect at $(16, 4)$, the area is:

$$\int_{0}^{16} (4 - \sqrt{x})dx = 4x - \frac{2}{3}x^{\frac{3}{2}} \Big|_{0}^{16} = 64 - \frac{2}{3}(16)^{\frac{3}{2}} = \frac{64}{3}$$

25. The graphs of $y = x^3$ and $y = x$ intersect at $(-1, 1)$ and $(1, 1)$. The graph of $y = x$ is above $y = x^3$ for $0 < x < 1$ and $y = x$ is below $y = x^3$ for $-1 < x < 0$. The area is:

$$\int_{-1}^{0} (x^3 - x)dx + \int_{0}^{1} (x - x^3)dx = \frac{x^4}{4} - \frac{x^2}{2} \Big|_{-1}^{0} + \frac{x^2}{2} - \frac{x^4}{4} \Big|_{0}^{1} = \frac{1}{4} + \frac{1}{4} = \frac{1}{2}$$

27. $\int_{0}^{2000} (\frac{100}{\sqrt{x+1}} + 50)dx = \int_{0}^{2000} [100(x + 1)^{-\frac{1}{2}} + 50]dx$

Let $u = x + 1$, $du = dx$. If $x = 0$ then $u = 1$ and if $x = 2000$ then $u = 2001$:

Chapter 6

$$= \int_0^{2000} 100(x+1)^{-\frac{1}{2}}dx + \int_0^{2000} 50dx = \int_1^{2001} 100u^{-\frac{1}{2}}du + \int_0^{2000} 50dx = 100(2u^{\frac{1}{2}})\Big|_1^{2001} + 50x\Big|_0^{2000}$$

$$= 100(2(2001)^{\frac{1}{2}} - 2) + 50(2000 - 0) = \$108{,}747.$$

29. First find where supply and demand intersect:

$$1000 - 15x - x^2 = x^2 - 3x - 40$$

$$0 = 2x^2 + 12x - 1040$$

$$0 = x^2 + 6x - 520$$

$$0 = (x - 20)(x + 26)$$

$$x = 20 \text{ or } x = -26 \qquad \text{(Reject negative value.)}$$

At $x = 20$, $D(20) = S(20) = 300$. Since the domain is $[10, 30]$, the producer's surplus is:

$$\int_{10}^{20} [300 - (x^2 - 3x - 40)]dx = 300x - \frac{x^3}{3} + \frac{3x^2}{2} + 40x \Big|_{10}^{20} = 340x - \frac{x^3}{3} + \frac{3x^2}{2}\Big|_{10}^{20} = \$1{,}516.67$$

31. The graphs of supply and demand intersect at $x = 2$ (graph both carefully for $0 \leq x < 3$); and $S(2) = D(2) = 22$.

Producer's surplus $= \int_0^2 [22 - (x + 20)]dx = 22x - \frac{x^2}{2} - 20x \Big|_0^2 = 2x - \frac{x^2}{2}\Big|_0^2 = (4 - 2) - (0) = \$2$

Consumer's surplus $= \int_0^2 \left(\frac{236 - 107x}{(x-3)^2} - 22\right)dx = \int_0^2 \left(\frac{236}{(x-3)^2} - \frac{107x}{(x-3)^2} - 22\right)dx$

$$= 236\int_0^2 (x-3)^{-2}dx - 107\int_0^2 \frac{x}{(x-3)^2}dx - 22\int_0^2 dx$$

Let $u = x - 3$, then $x = u + 3$ and $du = dx$. Also, when $x = 0$, $u = -3$ and if $x = 2$ then $u = -1$:

$$236\int_{-3}^{-1} u^{-2}du - 107\int_{-3}^{-1} \frac{u+3}{u^2}du - 22\int_0^2 dx = 236\left(\frac{u^{-1}}{-1}\Big|_{-3}^{-1}\right) - 107\int_{-3}^{-1}\left(\frac{1}{u} + 3u^{-2}\right)du - 22x\Big|_0^2$$

$$= -236\left(\frac{1}{u}\Big|_{-3}^{-1}\right) - 107\left(\ln|u| - 3u^{-1}\Big|_{-3}^{-1}\right) - 22(2)$$

$$= -236\left(-1 + \frac{1}{3}\right) - 107[(\ln 1 + 3) - (\ln 3 + 1)] - 44$$

$$= 157\frac{1}{3} - 107[.9013877] - 44 \approx \$16.88$$

Chapter 6   150

33. $\int_0^{30} \frac{100{,}000}{\sqrt{t+250}} dt = 100{,}000 \int_0^{30} (t+250)^{-\frac{1}{2}} dt = 100{,}000 \frac{(t+250)^{\frac{1}{2}}}{\frac{1}{2}} \Big|_0^{30}$

$= 200{,}000 \,[(30+250)^{\frac{1}{2}} - (0+250)^{\frac{1}{2}}] \approx 200{,}000 \,[.921812] = \$184{,}362$

6.5 CHAPTER REVIEW, PAGES 204 - 205

1. $\int x^5 dx = \frac{x^6}{6} + C$

3. $\int e^u du = e^u + C$

5. $\int (18x^2 - 6x - 3)dx = 18 \left(\frac{x^3}{3}\right) - 6\left(\frac{x^2}{2}\right) - 3x + C = 6x^3 - 3x^2 - 3x + C$

7. $\int (e^{3x+1} + \frac{1}{e})dx = \int e^{3x+1} dx + \frac{1}{e} \int dx = \int e^u \left(\frac{du}{3}\right) + \frac{1}{e} \int dx$

   Let $u = 3x + 1$ then $du = 3dx$: $= \frac{1}{3} e^u + \frac{1}{e} x + C = \frac{1}{3} e^{3x+1} + \frac{1}{e} x + C$

9. $\int_{-1}^{3} \frac{x^3 + 2}{x^2} dx = \int_{-1}^{3} (x + 2x^{-2})dx = \frac{x^2}{2} + \frac{2x^{-1}}{-1} \Big|_{-1}^{3} = \left(\frac{9}{2} - \frac{2}{3}\right) - \left(\frac{1}{2} + 2\right) = \frac{4}{3}$

11. Let $u = 1 + x^{\frac{4}{3}}$ then $du = \frac{4}{3} x^{\frac{1}{3}} dx$. Also, if $x = 1$ and 8, $u = 2$ and 17, respectively:

$\int_1^8 x^{\frac{1}{3}} (1 + x^{\frac{4}{3}})^3 dx = \int_2^{17} x^{\frac{1}{3}} u^3 \frac{3du}{4x^{\frac{1}{3}}} = \frac{3}{4} \int_2^{17} u^3 du = \frac{3}{4} \frac{u^4}{4} \Big|_2^{17} = \frac{3}{16}(17^4 - 2^4) = 15{,}657.1875$

13. Area $= \int_{\frac{5}{3}}^{4} \sqrt{3x+4} \, dx$. Let $u = 3x + 4$, then $du = 3dx$. Also, if $x = \frac{5}{3}$ and 4, then $u = 9$ and 16, respectively.

$\int_{\frac{5}{3}}^{4} \sqrt{3x+4} \, dx = \int_9^{16} u^{\frac{1}{2}} \frac{du}{3} = \frac{1}{3} \frac{u^{\frac{3}{2}}}{\frac{3}{2}} \Big|_9^{16} = \frac{2}{9} u^{\frac{3}{2}} \Big|_9^{16} = \frac{74}{9}$

15. Area $= \int_{-1}^{0} (x^3 - x)dx + \int_0^1 (x - x^3)dx = \frac{x^4}{4} - \frac{x^2}{2} \Big|_{-1}^0 + \frac{x^2}{2} - \frac{x^4}{4} \Big|_0^1 = \frac{1}{4} + \frac{1}{4} = \frac{1}{2}$

Chapter 6

17. The height is $h(t) = \int (1.2 + 5t^{-4})dt = 1.2t - \frac{5}{3}t^{-3} + C$. We are given $1 = h(1) = 1.2 - \frac{5}{3} + C$.

Solving for C gives $\frac{22}{15} = C$, so $h(t) = 1.2t - \frac{5}{3t^3} + \frac{22}{15}$.

$h(6) = 1.2(6) - \frac{5}{3(6)^3} + \frac{22}{15} \approx 8.66$ feet, so it will be economically feasible.

19. Set 670 equal to the total oil consumed, $\int_0^t 32.4e^{.048t}dt$

$670 = \int_0^t 32.4e^{.048t}dt = 675e^{.048t}\Big|_0^t = 675e^{.048t} - 675$

Add 675 to both sides: $\quad 1345 = 675e^{.048t}$

Divide by 675: $\quad 1.9926 \approx e^{.048t}$

$\ln 1.9926 \approx .048t$

$\frac{\ln 1.9926}{.048} = t$

$t \approx 14.4$ years

6.6 STUDENT'S TEST REVIEW AND ADDITIONAL PRACTICE

OBJECTIVES

The material of this chapter is reviewed in the following list of objectives. After each objective there are some practice questions. Answers to these problems immediately follow. Detailed solutions are given for every third problem. For a sample test select the first question of each set and check your answers. Additional practice is given by the other questions in each set. If you are having trouble with a particular type of problem, or if you want additional practice, look back at the indicated section in the test.

[6.1] Objective 1: *Evaluate an indefinite integral using one of the three direct integration formulas.*

1. $\int x^6 dx$ 2. $\int \frac{dx}{x}$ 3. $\int e^u du$ 4. $\int u^{-5} du$

Objective 2: *Evaluate an indefinite integral using one of the integration formulas listed in the box on Page 176.*

5. $\int (6x^2 - 2x - 5)dx$ 6. $\int \frac{du}{12}$

7. $\int \frac{(x-2)dx}{3x^2 - 5x - 2}$ 8. $\int (e^{5x+1} + \frac{1}{e}) dx$

Chapter 6

Objective 3: *Find an antiderivative which satisfies a given initial conditions.*

9. $f(x) = e^x + 9$; $F(0) = 10$
10. $f(x) = 3x^2 - 9$; $F(10) = 0$
11. $f(x) = \dfrac{1}{x}$; $F(1) = 0$
12. $f(x) = \dfrac{x^2 - 9}{x + 3}$; $F(10) = 100$

Objective 4: *Find the cost function, given the marginal cost function.*

13. $C'(x) = 9x^2 + 1$; fixed cost, $25,000
14. $C'(x) = 250 - 8x$; fixed cost, $8,000
15. $C'(x) = 10 - \dfrac{1}{\sqrt{x}}$; fixed cost, $1,500
16. $C'(x) = .005x^3$; fixed cost, $35,000

[6.2] Objective 5: *Evaluate indefinite integrals by substitution.*

17. $\displaystyle\int (e^{-3x} - \dfrac{1}{e})\, dx$
18. $\displaystyle\int (2 + 3x^2)^5 x\, dx$
19. $\displaystyle\int \dfrac{2x - 5}{\sqrt{x^2 - 5x)^3}}\, dx$
20. $\displaystyle\int x^2 \sqrt{10 - x}\, dx$

[6.3] Objective 6: *Evaluate definite integrals.*

21. $\displaystyle\int_1^2 \dfrac{x^3 + 2}{x^2}\, dx$
22. $\displaystyle\int_0^2 x\sqrt{2x^2 + 1}\, dx$
23. $\displaystyle\int_1^8 x^{\frac{1}{3}}(1 + x^{\frac{4}{3}})^3\, dx$
24. $\displaystyle\int_{\sqrt{17}}^5 x\sqrt{x^2 - 16}\, dx$

Objective 7: *Find the area bounded by the x-axis, the curve $y = f(x)$ and two given vertical lines.*

25. $y = \sqrt{5x + 1}$; $x = 3$, $x = \dfrac{8}{5}$
26. $y = 5x^4 - 20$; $x = -1$; $x = 2$
27. $y = \dfrac{1}{(1 - 2x)^3}$; $x = 1$, $x = 5$
28. $y = \dfrac{\ln(3x + 4)}{3x + 4}$; $x = 1$; $x = 3$

[6.4] Objective 8: *Write an integral to express a shaded region.*

29.

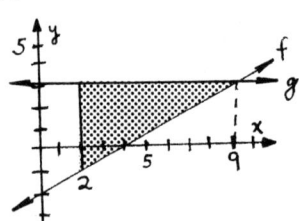

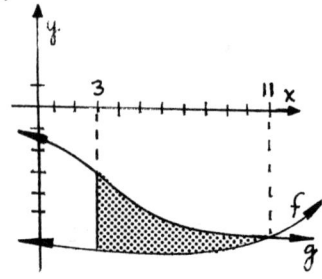

30.

Chapter 6   153

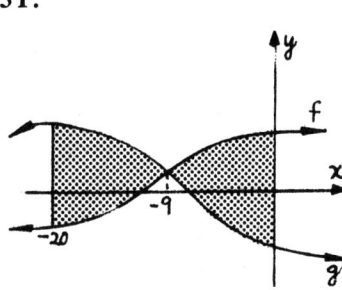

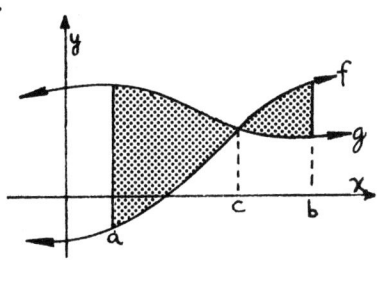

31.   **32.**

Objective 9: *Find the area between two given curves.*

33. $y = \dfrac{1}{2}x^2$ and $y = 4 - x$ from $x = -2$ to $x = 3$
34. $y = x^2 - 16$ and the x-axis
35. $y = x^3$ and $y = x$
36. $y = x^2 - 3x$, $x - 4y = 0$

ANSWERS TO STUDENT'S TEST REVIEW AND PRACTICE QUESTIONS

1. $\dfrac{x^7}{7} + C$ 2. $\ln |x| + C$ 3. $e^u + C$ 4. $-\dfrac{1}{4u} + C$ 5. $2x^3 - x^2 - 5x + C$

6. $\dfrac{u}{12} + C$ 7. $\dfrac{1}{3}\ln |3x + 1| + C$ 8. $\dfrac{e^{5x+1}}{5} + \dfrac{x}{e} + C$

9. $F(x) = \int (e^x + 9)dx = e^x + 9x + C$
   $10 = F(0) = e^0 + 9(0) + C = 1 + C \Rightarrow C = 9$, so $F(x) = e^x + 9x + 9$

10. $F(x) = x^3 - 9x - 910$ 11. $F(x) = \ln |x|$

12. $F(x) = \int \dfrac{x^2 - 9}{x + 3} dx = \int (x - 3) dx = \dfrac{x^2}{2} - 3x + C$
    $100 = F(10) = 20 + C \Rightarrow C = 80$, so $F(x) = \dfrac{x^2}{2} - 3x + 80$

13. $C(x) = 3x^3 + x + 25{,}000$ 14. $C(x) = 250x - 4x^2 + 8000$

15. $C(x) = 10x - 2\sqrt{x} + 15{,}000$ 16. $C(x) = .00125x^4 + 35{,}000$

17. $-\dfrac{e^{-3x}}{3} - \dfrac{x}{e} + C$

18. Let $u = 2 + 3x^2$ so $du = 6x dx$: $\int (2 + 3x^2)^5 x dx = \int u^5 (\dfrac{1}{6}) du = \dfrac{1}{6}(\dfrac{u^6}{6}) + C = \dfrac{(2 + 3x^2)^6}{36} + C$

Chapter 6

19. $\dfrac{-2}{\sqrt{x^2 - 5x}} + C$ **20.** $-\dfrac{2}{21}(10 - x)^{\frac{3}{2}}(3x^2 - 48x + 280) + C$

21. $\displaystyle\int_1^2 \dfrac{x^3 + 2}{x^2}\,dx = \int_1^2 (x + 2x^{-2})\,dx = \dfrac{x^2}{2} - 2x^{-1}\Big|_1^2 = 2\dfrac{1}{2}$ **22.** $\dfrac{13}{3}$

23. $\dfrac{3}{16}[17^4 - 2^4] = \dfrac{250{,}515}{16}$

24. $\displaystyle\int_{\sqrt{17}}^5 x\sqrt{x^2 - 16}\,dx = \dfrac{1}{3}(x^2 - 16)^{\frac{3}{2}}\Big|_{\sqrt{17}}^5 = \dfrac{1}{3}(9^{\frac{3}{2}} - 1^{\frac{3}{2}}) = \dfrac{26}{3}$ **25.** $\dfrac{74}{15}$

26. $32\sqrt{2} + 11$ (Do not forget to notice the graph crosses the x-axis at $x = \sqrt{2}$.)

27. For $1 \le x \le 5$ the graph of $y = \dfrac{1}{(1 - 2x)^3}$ lies below the x axis. The area, then, is

$\displaystyle\int_1^5 \dfrac{-1}{(1 - 2x)^3}\,dx = \int_1^5 -(1 - 2x)^{-3}\,dx = \dfrac{1}{2} \cdot \dfrac{(1 - 2x)^{-2}}{-2}\Big|_1^5 = \dfrac{20}{81}$

28. $\dfrac{\ln 13 - \ln 7}{6}$ **29.** $\displaystyle\int_2^9 [g(x) - f(x)]\,dx$

30. Since $g(x)$ is above $f(x)$ throughout the shaded region, the area is $\displaystyle\int_3^{11} [g(x) - f(x)]\,dx$

31. $\displaystyle\int_{-20}^{-9} [g(x) - f(x)]\,dx + \int_{-9}^0 [f(x) - f(x)]\,dx$ **32.** $\displaystyle\int_a^c [g(x) - f(x)]\,dx + \int_c^b [f(x) - f(x)]\,dx$

33. The two graphs intersect at $x = -4$ and $x = 2$. The line $y = 4 - x$ is above the parabola $y = \dfrac{1}{2}x^2$ between $x = -2$ and $x = 2$. The area is:

$\displaystyle\int_{-2}^2 [(4 - x) - \dfrac{1}{2}x^2]\,dx + \int_2^3 [\dfrac{1}{2}x^2 - (4 - x)]\,dx = 4x - \dfrac{x^2}{2} - \dfrac{x^3}{6}\Big|_{-2}^2 + \dfrac{x^3}{6} - 4x + \dfrac{x^2}{2}\Big|_2^3 = 15$

Chapter 6

34. $\dfrac{256}{3}$ **35.** $\dfrac{1}{2}$

36. The parabola $y = x^2 - 3x$ and the line $x - 4y = 0$ intersect at $x = 0$ and $x = \dfrac{13}{4}$. The area between them is: $\displaystyle\int_0^{\frac{13}{4}} [\tfrac{1}{4}x - (x^2 - 3x)]dx = \dfrac{x^2}{8} - \dfrac{x^3}{3} + \dfrac{3x^2}{2} \Big|_0^{\frac{13}{4}} = \dfrac{2197}{384}$

MODELING APPLICATION 5: SAMPLE ESSAY

Computers in Mathematics; muMath

In the late 1970s, mathematics instruction was forever altered with the advent of powerful, inexpensive hand-held calculators. They allowed both the instructor and the student the ability to consider realistic examples while focusing on concepts and processes rather than on tedious and complicated calculations. While computers can be programmed to carry out calculator-type simplifications there have been several fundamental stumbling blocks to prohibit computers from being used to as great an extent as calculators.

1. Expense and availability
2. Computers offer little advantage (if any) over simplification of arithmetic calculations
3. Computers can do only very limited algebraic simplification and manipulation

The second and third objections have been met with a computer software package called muMATH*. This software package is written in a computer language called muSIMP. MuSIMP is a general-purpose programming language which was designed for implementing computer algebra systems. For example,

$$5x(x + y)^3 - 3x^2(2x - 3y)$$

will be simplified on the computer as**

$$5 x y \wedge 3 + 9 x \wedge 2 y + 15 x \wedge 2 y \wedge 2 + 15 x \wedge 3 y - 6 x \wedge 3 + 5 x \wedge 4$$

Although it is not our intention to make you proficient in using muMATH in this extended application, it is our purpose to let you know the scope and power of available software. In order to use muMATH you will need (1) and IBM PC with 128K RAM memory, or (2) Apple II or IIe with 64K memory, or (3) CP/M-80 computer with 56K memory.

---

* The software called muMATH is a copyrighted program marketed by the Soft Warehouse, Box 11174, Honolulu, Hawaii, 46828-0174. The author has no financial interest in the sale of this software, but is recommended here because of its educational value.

** See Table 2 for an interpretation of muSIMP notation.

Chapter 6

MuMATH is a software package of programs written in a computer language called muSIMP. You can call up specific programs to do arithmetic, algebra, trigonometry, or calculus. The list of available programs is growing, but includes those listed in Table 1.

Table 1 muMATH Source Files

Rational arithmetic
Elementary algebra
Equation simplification
Equation solver
Array operations
Matrix operations
Simultaneous linear algebraic equations
Absolute-value simplification
Trigonometric simplification
Inverse trigonometric simplification
Hyperbolic trignometric simplification
Symbolic differentiation and Taylor series
Symbolic integration
Extended symbolic integration
Limits of Functions
Closed-form summation and products
First-order ordinary differential equations
Higher order ordinary differential equations
Vector algebra
Vector calculus

In this section we will dexcribe the mathematical notation used in muSIMP, as well as some of the capabilities of the arithmetic, algebra, and integration packages.

The first step in using muMATH (or any symbolic software package) is the notation. Usual mathematical notation is used except

1. Use * instead of · or ×. You can use juxtaposition for multiplication except between a variable and a following parenthesized expression. For example, 2*x, 2x, 2 x, 2(x + y), or x*(x + y) are permitted, but x(x + y) is not permitted.
2. Use / for division or fractions.
3. Use ^ for exponentiation.
4. Expressions entered by the user must be followed by a semicolon and the return keys.

Chapter 6

Table 2

Algebraic Notation	muSIMP Notation
$2 \cdot 3$ or 2×3	2 3 or 2 (3) NOTE: 2 3 ≠ 23
$x^2 + 2y^2$	x ^ 2 + 2 y ^ 2
$x^2 + (2y)^2$	x ^ 2 + (2 y) ^ 2
$(x+y)(5x^4 - 5y)^2$	(x + y)*(5 x ^ 4 - 5 y) ^ 2
$\dfrac{x+y}{x-z}$	(x + y)/(x - z)
$x + \dfrac{y}{x} - z$	x + y/x - z

ARITHMETIC PROGRAM (ARITH.MUS)*

This program will, of course, do all the numerical calculations that you can do on a calculator. However, there are some additional features.

Fractions

Unlike a calculator, muMATH will give exact fractional answers. For example,

$$\frac{1}{2} + \frac{1}{3} = \frac{5}{6}$$

On the computer this would look like

   ? 1/2 + 1/3;

   @: 5/6

Notice that the computer prints "@:" before giving the answer. Here is another more complicated example:

   ?19/300 + 55/144 + 25/108;

   @: 7309/10800

Accuracy

The only limitation of accuracy with muMATH is the amount of memory in your computer. The calculation 2^{100} is quite easy in muMATH

   ? 2 ^ 100

   @: 1267650600228229401496703205376

In all of the muMATH programs there are functions that instruct the computer to carry out certain mathematical processes. These functions are followed by parentheses enclosing the expression upon which you wish to operate. A list of some arithmetic functions you will find in muMATH is given in Table 3.

---

* This is the muMATH name for the arithmetic program.

Table 3 muMATH Arithmetic Functions

ABS(x)	Finds the absolute value of x.
GCD(x,y)	Finds the greatest common divisor of x and y
LCM(x,y)	Finds the least common multiple of x and y
MIN(x,y)	Finds the minimum of x and y

ALGEBRA PROGRAM (ALGEBRA.ARI, EQN.ALG, SOLVE.EQN)

In addition to all arithmetic simplifications, this program will also carry out algebraic simplifications. For example, similar terms and factors are combined, powers are evaluated, and expressions are simplified. When using muMATH, the first difference you will note is that the order of terms in the answers in muMATH are not in the order in which you are accustomed. For example, if you expand $(x + y)^2$ in algebra you will write $x^2 + 2xy + y^2$ but in muMATH the result will be printed out as

@: 2 x y + x ^ 2 + y ^ 2

If you ask type

?2 (x + y);

@: 2 x + 2 y

is the output. On the other hand,

?(x + y) ^ 2;

@: (x + y) ^ 2

is the output. What is wrong here? You have not used an algebraic function to tell the computer what you want done. The function, or operator, to carry out this calculation is called EXPAND and is used as shown below:

?EXPAND((x + y) ^ 2);

@: 2 x y + x ^ 2 + y ^ 2

Do you see why the double parentheses are necessary after the word "EXPAND"?

A list of algebraic functions are given in Table 4.

Table 4 Algebraic Functions in muMATH

EXPAND(x)	Evaluates x to give a fully expanded expression
FCTR(x)	Finds common factors in the expression x
RATIONALIZE(x)	Multiplies the numerator and denominator of x by the complex conjugate of the denominator
SOLVE(a,x)	Solves the equation specified in location a for the variable specified in location x

The solve command is one of the most powerful functions in muMATH's albegra package. Equations are denoted by a double equal sign (==) in muMATH.

Some examples are indicated below.

equation to be solved ↓   variable you are solving for ↓

?SOLVE(5 x ^ 2 + 2 x - 2 == 0, x);
@: {x == (-2 + (11) ^ (1/2))/5
    x == (-2 - (11) ^ (1/2))/5}

In solving equations it is very common to have one side of an equation equal to zero, so as a convenience, muMATH will automatically set the right side equal to zero if it is omitted. So, in the above example you could have simply given the command "SOLVE(5 x ^ 2 + 2 x - 2, x);".

? SOLVE(5 x ^ 2 + (1 - w)*x - (w + 4), x);
@: {x == -1
    x == (w + 4)/5}

INTEGRATION PROGRAM (INT.DIF, INTMORE.INT, DEFINT)

This program does both indefinite and definite symbolic integration. In order to find an antiderivative (indefinite integral) you should use the INT command followed by a parentheses which encloses the integrand (expr) and the variable (var) as shown by the following examples:

? INT (3 + a x ^ 2, x);   ? INT (x/5 x ^ 2 + 3), x);
@: 3 x + a x ^ 3/3   @: LN (5 (x ^ 2 + 3) / 10)

Notice that the answers given by the computer (after the "@" symbol) do not show the constant of integration, and you must add this when you are intrepreting the solution. The program works much like an integral table and will search for an appropriate form. Sometimes this means giving back the integral as an unevaluated integral as shown:

? INT (1 + E ^ x/x, x);
@: x + INT(E ^ x/x, x)

This program will, generally speaking, automatically expand or factor in an attempt to find the integral of the expression.

The command DEFINT will give a definite integral. The lower and upper limits of integration are added to the parentheses as shown by the following example.

integrand ⤸   variable ↓
? DEFINT (9 x ^ 2, x, 0, 1);
                     ↑ ↑
                     | upper limit of integration
      lower limit of integration

?DEFINT (a x ^ 2, x, 0, 1);

@: a/3

With these examples you can begin to see some of the many possibilities of using a computer to aid you in mathematics. Just as calculators became inexpensive and readily available by the early 80s, perhaps we can look forward to the first objection, that of availability and expense, being overcome by the early 90s.

CHAPTER 7
ADDITIONAL INTEGRATION TOPICS

7.1 INTEGRATION BY PARTS, PAGES 211 - 212

1.
$$\text{Let } u = x \quad dv = e^{3x}dx$$
$$du = dx \quad v = \int e^{3x}dx = \frac{1}{3}e^{3x}$$

$\int 12xe^{3x}dx = 12[uv - \int vdu] = 12[\frac{1}{3}xe^{3x} - \int \frac{1}{3}e^{3x}dx] = 4xe^{3x} - 4\int e^{3x}dx = 4xe^{3x} - 4[\frac{1}{3}e^{3x}]$

$= 4xe^{3x} - \frac{4}{3}e^{3x} + C$

3.
$$\text{Let } u = x \quad dv = e^{-x}dx$$
$$du = dx \quad v = \int e^{-x}dx = -e^{-x}$$

$\int xe^{-x}dx = uv - \int vdu = -xe^{-x} - \int -e^{-x}dx = -xe^{-x} + \int e^{-x}dx = -xe^{-x} - e^{-x} + C$

Therefore, $\int_0^1 xe^{-x}dx = -xe^{-x} - e^{-x} \Big|_0^1 = (-2e^{-1}) - (-1) = 1 - 2e^{-1}$

5.
$$\text{Let } u = x \quad dv = \sqrt{1-x}\,dx$$
$$du = dx \quad v = \int (1-x)^{\frac{1}{2}}dx = -\frac{2}{3}(1-x)^{\frac{3}{2}}$$

$\int x\sqrt{1-x}\,dx = uv - \int vdu = -\frac{2}{3}x(1-x)^{\frac{3}{2}} - \int -\frac{2}{3}(1-x)^{\frac{3}{2}}dx = -\frac{2}{3}x(1-x)^{\frac{3}{2}} + \frac{2}{3}\int (1-x)^{\frac{3}{2}}dx$

$= -\frac{2}{3}x(1-x)^{\frac{3}{2}} + \frac{2}{3}\left[-\frac{2}{5}(1-x)^{\frac{5}{2}}\right] + C = -\frac{2}{3}x(1-x)^{\frac{3}{2}} - \frac{4}{15}(1-x)^{\frac{5}{2}} + C$

7.
$$\text{Let } u = x \quad dv = (x+2)^3 dx$$
$$du = dx \quad v = \frac{(x+2)^4}{4}$$

$\int x(x+2)^3 dx = uv - \int vdu = \frac{x(x+2)^4}{4} - \int \frac{(x+2)^4}{4}dx = \frac{x(x+2)^4}{4} - \frac{1}{4}\left[\frac{(x+2)^5}{5}\right] + C$

$= \frac{x(x+2)^4}{4} - \frac{(x+2)^5}{20} + C$

Chapter 7

9.
$$\boxed{\begin{array}{ll} u = \ln(x+1) & dv = dx \\ du = \dfrac{1}{x+1}\,dx & v = \int dx = x \end{array}}$$

$\int \ln(x+1)\,dx = uv - \int v\,du = x\ln(x+1) - \int \dfrac{x}{x+1}\,dx$

This last integral can be done without <u>too</u> much work, but an easier way would be to go back to:

$$\boxed{\begin{array}{ll} u = \ln(x+1) & dv = dx \\ du = \dfrac{1}{x+1}\,dx & v = \int dx = x+1 \end{array}}$$

Since $\int dx = x +$ (**any** constant we like).

Now, $\int \ln(x+1)\,dx = uv - \int v\,du = (x+1)\ln(x+1) - \int (x+1)\cdot\dfrac{1}{x+1}\,dx$

$= (x+1)\ln(x+1) - \int dx = (x+1)\ln(x+1) - x + C$

11.
$$\boxed{\begin{array}{ll} u = x & dv = \dfrac{1}{\sqrt{1-x}}\,dx \\ du = dx & v = \int (1-x)^{-\frac{1}{2}}\,dx = -2(1-x)^{\frac{1}{2}} \end{array}}$$

$\int \dfrac{x\,dx}{\sqrt{1-x}} = uv - \int v\,du = -2x(1-x)^{\frac{1}{2}} - \int -2(1-x)^{\frac{1}{2}}\,dx = -2x(1-x)^{\frac{1}{2}} + 2\int (1-x)^{\frac{1}{2}}\,dx$

$= -2x(1-x)^{\frac{1}{2}} + 2\left[-\dfrac{2}{3}(1-x)^{\frac{3}{2}}\right] + C = -2x(1-x)^{\frac{1}{2}} - \dfrac{4}{3}(1-x)^{\frac{3}{2}} + C$

13.
$$\boxed{\begin{array}{ll} u = x^2 & dv = e^{2x}\,dx \\ du = 2x\,dx & v = \int e^{2x}\,dx = \dfrac{1}{2}e^{2x} \end{array}}$$

$\int 6x^2 e^{2x}\,dx = 6\left[\dfrac{1}{2}x^2 e^{2x} - \int xe^{2x}\,dx\right] = 3x^2 e^{2x} - 6\int xe^{2x}\,dx$

This last integral can be done by parts also:

$$\boxed{\begin{array}{ll} u = x & dv = e^{2x}\,dx \\ du = dx & v = \dfrac{1}{2}e^{2x} \end{array}}$$

$3x^2 e^{2x} - 6\left[uv - \int v\,du\right] = 3x^2 e^{2x} - 6\left[\dfrac{1}{2}xe^{2x} - \int \dfrac{1}{2}e^{2x}\,dx\right]$

Chapter 7

$$= 3x^2 e^{2x} - 6\left[\frac{1}{2}xe^{2x} - \frac{1}{4}e^{2x}\right] + C = 3xe^{2x} - 3xe^{2x} + \frac{3}{2}e^{2x} + C$$

15.

Let $u = x^2$	$dv = \sqrt{1-2x}\, dx$
$du = 2x\, dx$	$v = \int (1-2x)^{\frac{1}{2}} dx = -\frac{1}{3}(1-2x)^{\frac{3}{2}}$

$$\int x^2 \sqrt{1-2x}\, dx = uv - \int v\, du = -\frac{x^2(1-2x)^{\frac{3}{2}}}{3} + \frac{2}{3}\int x(1-2x)^{\frac{3}{2}} dx$$

This last integral can be done by parts also:

Let $u = x$	$dv = (1-2x)^{\frac{3}{2}} dx$
$du = dx$	$v = \int (1-2x)^{\frac{3}{2}} dx = -\frac{1}{5}(1-2x)^{\frac{5}{2}}$

$$-\frac{x^2(1-2x)^{\frac{3}{2}}}{3} + \frac{2}{3}[uv - \int v\,du] = -\frac{x^2(1-2x)^{\frac{3}{2}}}{3} + \frac{2}{3}\left[-\frac{x(1-2x)^{\frac{5}{2}}}{5} - \int -\frac{1}{5}(1-2x)^{\frac{5}{2}} dx\right]$$

$$= \frac{-x^2(1-2x)^{\frac{3}{2}}}{3} - \frac{2x(1-2x)^{\frac{5}{2}}}{15} + \frac{2}{3}\left[-\frac{(1-2x)^{\frac{7}{2}}}{35}\right] + C = -\frac{x^2(1-2x)^{\frac{3}{2}}}{3} - \frac{2x(1-2x)^{\frac{5}{2}}}{15} - \frac{2(1-2x)^{\frac{7}{2}}}{105} + C$$

17. Integrate using substitution:

Let $u = 1 - x^2$ $\Rightarrow$ $x^2 = 1 - u$
$du = -2x\, dx$

$$\int x^3 \sqrt[3]{1-x^2}\, dx = \int (1-u)xu^{\frac{1}{3}} \frac{du}{-2x}$$

$$= -\frac{1}{2}\int (1-u)u^{\frac{1}{3}} du = -\frac{1}{2}\int u^{\frac{1}{3}} du + \frac{1}{2}\int u^{\frac{4}{3}} du$$

$$= -\frac{3}{8}(1-x^2)^{\frac{4}{3}} + \frac{3}{14}(1-x^2)^{\frac{7}{3}} + C$$

If you integrate by parts first and then use substitution you will obtain a very different (but equivalent) result: $-\dfrac{3x^2(1-x^2)^{\frac{4}{3}}}{8} - \dfrac{9(1-x^2)^{\frac{7}{3}}}{56} + C$

Chapter 7

19. Let $u = \ln x$, $dv = x^2 dx$
$du = \dfrac{1}{x} dx$, $v = \dfrac{x^3}{3}$

$\int x^2 \ln x\, dx = uv - \int v\, du = \dfrac{x^3 \ln x}{3} - \int \dfrac{x^2}{3}\, dx = \dfrac{x^3 \ln x}{3} - \dfrac{x^3}{9} + C$

21. Let $u = x^2$, $dv = e^{4x} dx$
$du = 2x\, dx$, $v = \dfrac{1}{4} e^{4x}$

$\int 6x^2 e^{4x}\, dx = 6[uv - \int v\, du] = 6\left[\dfrac{1}{4} x^2 e^{4x} - \int \dfrac{1}{2} xe^{4x}\, dx\right] = \dfrac{3}{2} x^2 e^{4x} - 3\int xe^{4x}\, dx.$

This last integral can also be done by parts:

Let $u = x$, $dv = e^{4x} dx$
$du = dx$, $v = \dfrac{1}{4} e^{4x}$

$\dfrac{3}{2} x^2 e^{4x} - 3\int xe^{4x}\, dx = \dfrac{3}{2} x^2 e^{4x} - 3[uv - \int v\, du]$

$= \dfrac{3}{2} x^2 e^{4x} - 3\left[\dfrac{1}{4} xe^{4x} - \int \dfrac{1}{4} e^{4x}\, dx\right] = \dfrac{3}{2} x^2 e^{4x} - \dfrac{3}{4} xe^{4x} + \dfrac{3}{16} e^{4x} + C$

23. By substitution, let $u = 1 - x^2$ then

$\int \dfrac{x^3\, dx}{\sqrt{1 - x^2}} = \int x(1 - u)u^{-\frac{1}{2}} \dfrac{du}{-2x} = -\dfrac{1}{2} \int (u^{-\frac{1}{2}} - u^{\frac{1}{2}})du = -(1 - x^2)^{\frac{1}{2}} + \dfrac{1}{3}(1 - x^2)^{\frac{3}{2}} + C$

If you integrate by parts first, then use substitution you will obtain

$x^2(1 - x^2)^{\frac{1}{2}} - \dfrac{2}{3}(1 - x^2)^{\frac{3}{2}} + C$

25. Rather than doing this by parts, make the substitution $u = \ln x$. Then $du = \dfrac{1}{x} dx$:

$\int \dfrac{dx}{x \ln x} = \int \dfrac{1}{u} du = \ln |u| = \ln |\ln x| + C$

27. Let $u = \ln x$, $dv = \ln x\, dx$
$du = \dfrac{1}{x} dx$, $v = \int \ln x\, dx = x \ln x - x$ (See Example 3.)

164

Chapter 7   165

$\int (\ln x)^2 dx = uv - \int v\,du = (\ln x)(x\ln x - x) - \int (\ln x - 1)dx = x\ln^2 x - x\ln x - (x\ln x - x - x)$

$= x\ln^2 x - 2x\ln x + 2x$

Therefore, $\int_1^e (\ln x)^2 dx = x\ln^2 x - 2x\ln x + 2x \Big|_1^e = e - 2$

29. $\boxed{\begin{array}{ll} \text{Let } u = x^m & dv = e^x dx \\ du = mx^{m-1}dx & v = \int e^x dx = e^x \end{array}}$

$\int x^m e^x dx = uv - \int v\,du = x^m e^x - \int mx^{m-1} e^x dx = x^m e^x - m\int x^{m-1} e^x dx$

31. a. $\boxed{\begin{array}{ll} \text{Let } u = x^2 & dv = \dfrac{x}{x^2-1}\,dx \\ du = 2x\,dx & v = \int \dfrac{x\,dx}{x^2-1} = \dfrac{1}{2}\ln|x^2-1| \end{array}}$

$\int \dfrac{x^3}{x^2-1}\,dx = uv - \int v\,du = \dfrac{1}{2}x^2 \ln|x^2-1| - \int x\ln|x^2-1|dx$

This last integral can be done with a substitution,

$z = x^2 - 1$, then $dz = 2x\,dx$:

$\dfrac{1}{2}x^2\ln|x^2-1| - \int \dfrac{1}{2}\ln|z|\,dz$

$= \dfrac{1}{2}x^2\ln|x^2-1| - \dfrac{1}{2}[z\ln z - z] + C_1$ (See Example 3.)

$= \dfrac{1}{2}x^2\ln|x^2-1| - \dfrac{1}{2}[(x^2-1)\ln|x^2-1| - (x^2-1)] + C_1$

$= \dfrac{1}{2}x^2\ln|x^2-1| - \dfrac{1}{2}x^2\ln|x^2-1| + \dfrac{1}{2}\ln|x^2-1| + \dfrac{1}{2}x^2 - \dfrac{1}{2} + C_1$

$= \dfrac{1}{2}\ln|x^2-1| + \dfrac{1}{2}x^2 + C$ (where $C = C_1 - \dfrac{1}{2}$)

b. By long division, $\dfrac{x^3}{x^2-1} = x + \dfrac{x}{x^2-1}$

$\int \dfrac{x^3}{x^2-1}\,dx = \int (x + \dfrac{x}{x^2-1})dx = \int x\,dx + \int \dfrac{x\,dx}{x^2-1} = \dfrac{x^2}{2} + \dfrac{1}{2}\ln|x^2-1| + C$

Chapter 7

33. $N(t) = \int N'(t) \, dt = \int \dfrac{t}{(1+t)^{\frac{1}{3}}} \, dt$

Let $u = t$ $dv = (1+t)^{-\frac{1}{3}} \, dt$

$du = dt$ $v = \int (1+t)^{-\frac{1}{3}} \, dt = \dfrac{3}{2}(1+t)^{\frac{2}{3}}$

$\int \dfrac{t}{(1+t)^{\frac{1}{3}}} \, dt = uv - \int v \, du = \dfrac{3t(1+t)^{\frac{2}{3}}}{2} - \int \dfrac{3}{2}(1+t)^{\frac{2}{3}} \, dt = \dfrac{3t(1+t)^{\frac{2}{3}}}{2} - \dfrac{9}{10}(1+t)^{\frac{5}{3}} + C$

We are given $.01 = N(0) = \dfrac{3(0)(1+0)^{\frac{2}{3}}}{2} - \dfrac{9}{10}(1+0)^{\frac{5}{3}} + C$

$.01 = -\dfrac{9}{10} + C$

$.91 = C$

So $N(t) = \dfrac{3t(1+t)^{\frac{2}{3}}}{2} - \dfrac{9}{10}(1+t)^{\frac{5}{3}} + .91$

$N(7) = \dfrac{3(7)(8)^{\frac{2}{3}}}{2} - \dfrac{9}{10}(8)^{\frac{5}{3}} + .91$

$= 42 - 28.8 + .91 = 14.11$ (million)

35. The total costs are $\int_{5}^{10} \left(100x^2 + \dfrac{5000x}{x^2+1}\right) dx = \dfrac{100x^3}{3} + 2500 \ln(x^2+1) \Big|_{5}^{10} \approx \$32{,}559$

7.2 INTEGRATION BY TABLE, PAGES 216 - 217

1. Using Formula 9, $\int (1+bx)^{-1} \, dx = \dfrac{1}{b} \ln |1 + bx| + C$

3. Using Formula 23, $\int \dfrac{x \, dx}{\sqrt{x^2 + a^2}} = \sqrt{x^2 + a^2} + C$

5. Using Formula 26, $\int \dfrac{dx}{x^2 \sqrt{x^2 - a^2}} = \dfrac{\sqrt{x^2 - a^2}}{a^2 x} + C$

Chapter 7

7. Using Formula 30, $\int x \ln x \, dx = \dfrac{x^2}{2} \ln x - \dfrac{x^2}{4} + C$

9. Using Formula 32 (m = -1), $\int x^{-1} \ln x \, dx = \dfrac{1}{2} \ln^2 x + C$

11. Using Formula 35, $\int x e^{ax} \, dx = \dfrac{e^{ax}}{a^2}(ax - 1) + C$

13. Using Formula 29 (a = 1),
$\int \dfrac{x^2}{\sqrt{x^2 + 1}} \, dx = \dfrac{x}{2}\sqrt{x^2 + 1} - \dfrac{1}{2} \ln |x + \sqrt{x^2 + 1}\,| + C$

15. Using Formula 26 (a = 4), $\int \dfrac{dx}{x^2 \sqrt{x^2 + 16}} = -\dfrac{\sqrt{x^2 + 16}}{16x} + C$

17. Let $u = 4x^2 + 1$, then $du = 8x\,dx$: $\int \dfrac{x\,dx}{\sqrt{4x^2 + 1}} = \int \dfrac{\frac{du}{8}}{\sqrt{u}} = \dfrac{1}{8}\int u^{-\frac{1}{2}}\,du = \dfrac{1}{4} u^{\frac{1}{2}} + C$
$= \dfrac{1}{4}\sqrt{4x^2 + 1} + C$

19. $\int \dfrac{dx}{x\sqrt{1 - 9x^2}} = \int \dfrac{dx}{3x\sqrt{\frac{1}{9} - x^2}} = \dfrac{1}{3}\int \dfrac{dx}{x\sqrt{\frac{1}{9} - x^2}}$

Using Formula 20 with $a = \dfrac{1}{3}$, $= \dfrac{1}{3}\left[-\dfrac{1}{\left(\frac{1}{3}\right)} \ln \left| \dfrac{\frac{1}{3} + \sqrt{\frac{1}{9} - x^2}}{x} \right| \right] + C$

$= -\ln \left| \dfrac{\frac{1}{3} + \sqrt{\frac{1}{9} - x^2}}{x} \right| + C = -\ln \left| \dfrac{\frac{1}{3} + \sqrt{\frac{1}{9} - x^2}}{x} \cdot \dfrac{3}{3} \right| + C$

$= -\ln \left| \dfrac{1 + 3\sqrt{\frac{1}{9} - x^2}}{3x} \right| + C = -\ln \left| \dfrac{1 + \sqrt{1 - 9x^2}}{3x} \right| + C$

21. Using Formula 23, $\int \dfrac{x\,dx}{\sqrt{x^2 + 4}} = \sqrt{x^2 + 4} + C$

Chapter 7   168

23. Using Formula 9, with a = 1, b = 1, and n = 3

$$\int (1+x)^3 dx = \int u^3 du = \frac{u^4}{4} + C = \frac{(1+x)^{3+1}}{(3+1)\cdot 1} + C = \frac{(1+x)^4}{4} + C$$

25. Using Formula 10 with a = 1, b = 1, and n = 3:

$$\int x(1+x)^3 dx = \frac{1}{3+2}(1+x)^{3+2} - \frac{1}{3+1}(1+x)^{3+1} + C = \frac{1}{5}(1+x)^5 - \frac{1}{4}(1+x)^4 + C$$

27. Using Formula 13 with m = 1, a = 1, b = 1

$$\int x\sqrt{1+x}\,dx = \frac{-2(2-3x)\sqrt{(1+x)^3}}{15} + C = \frac{-2(2-3x)\sqrt{(1+x)^3}}{15} + C$$

29. Using Formula 35 with a = 4: $\int xe^{4x} dx = \frac{e^{4x}}{4^2}(4x-1) + C = \frac{e^{4x}}{16}(4x-1) + C$

31. Let u = 2x, then $x = \frac{u}{2}$ and du = 2dx. Using Formula 30,

$$\int x \ln 2x\, dx = \int \frac{u}{2} \ln u\, \frac{du}{2} = \frac{1}{4}\int u \ln u\, du = \frac{1}{4}\left[\frac{u^2}{2}\ln u - \frac{u^2}{4}\right] + C = \frac{x^2}{2}\ln 2x - \frac{x^2}{4} + C$$

33. Let u = 5x, then du = 5dx: $\int x^2 \ln 5x\, dx = \int \frac{u^2}{25}(\ln u)\frac{du}{5} = \frac{1}{125}\int u^2 \ln u\, du$

Now, by Formula 31 this is $\frac{1}{125}\left[\frac{u^3}{3}\ln u - \frac{u^3}{9}\right] + C = \frac{1}{125}\left[\frac{(5x)^3}{3}\ln 5x - \frac{(5x)^3}{9}\right] + C$

$= \frac{x^3}{3}\ln(5x) - \frac{x^3}{9}\ln(5x) + C$

35. Using Formula 39 with a = 3, b = 5, and m = 1: $\int \frac{dx}{3+5e^x} = \frac{x}{3} - \frac{1}{3}\ln|3+5e^x| + C$

37. Using Formula 39 with a = 1, b = 1, and m = 2: $\int \frac{dx}{1+e^{2x}} = \frac{x}{1} - \frac{1}{2}\ln|1+e^{2x}| + C$

$= x - \frac{1}{2}\ln|1+e^{2x}| + C$

39. Let u = 1 + x then du = dx. Using Formula 2:

$$\int \frac{dx}{\sqrt{1+x}} = \int \frac{du}{\sqrt{u}} = \int u^{-\frac{1}{2}} du = 2u^{\frac{1}{2}} + C = 2(1+x)^{\frac{1}{2}} + C$$

41. Using Formula 11 with a = 1, b = 1, and n = 3:

Chapter 7

$$\int x^2(1+x)^3 dx = 1\left[\frac{(1+x)^6}{6} - 2\frac{(1+x)^5}{5} + \frac{(1+x)^4}{4}\right] + C$$

$$= \frac{(1+x)^6}{6} - \frac{2(1+x)^5}{5} + \frac{(1+x)^4}{4} + C$$

43. Using Formula 11 with a = 1, b = -1, and n = 3:

$$\int 5x^2(1-x)^3 dx = 5\left(\frac{1}{-1}\left[\frac{(1-x)^6}{6} - 2\frac{(1-x)^5}{5} + \frac{(1-x)^4}{4}\right]\right) + C$$

$$= \frac{-5(1-x)^6}{6} + 2(1-x)^5 - \frac{5(1-x)^4}{4} + C$$

45. Using Formula 13 with m = 1, a = 1, and b = -2:

$$\int 2x\sqrt{1-2x}\, dx = 2\left(\frac{-2(2+6x)\sqrt{(1-2x)^3}}{15(4)}\right) + C = \frac{-(2+6x)\sqrt{(1-2x)^3}}{15} + C$$

47. Using Formula 11 with a = 2, b = 3, and n = 3:

$$\int x^2(2+3x)^3 dx = \frac{1}{27}\left[\frac{(2+3x)^6}{6} - \frac{4(2+3x)^5}{5} + (2+3x)^4\right] + C$$

49. Let $u = 4x^3 + 1$, then $du = 12x^2 dx$. Using Formula 2,

$$\int x^2\sqrt{4x^3+1}\, dx = \int x^2\sqrt{u}\,\frac{du}{12x^2} = \frac{1}{12}\int u^{\frac{1}{2}} du = \frac{1}{18} u^{\frac{3}{2}} + C = \frac{1}{18}(4x^3+1)^{\frac{3}{2}} + C$$

51. Using Formula 36 with m = 2 and a = 3: $\int x^2 e^{3x} dx = \frac{x^2 e^{3x}}{3} - \frac{2}{3}\int x e^{3x} dx$

This last integral can be done using Formula 35 with a = 3:

$$= \frac{x^2 e^{3x}}{3} - \frac{2}{3}\left[\frac{e^{3x}}{3}(3x-1)\right] + C = \frac{e^{3x}}{27}\left[9x^2 - 6x + 2\right] + C$$

53. Using Formula 14 with m = 2, a = 2, and b = 9:

$$\int \frac{\sqrt{2+9x}}{x^2} dx = -\frac{1}{2}\left[\frac{\sqrt{(2+9x)^3}}{x} + -\frac{9}{2}\int\frac{\sqrt{2+9x}}{x} dx\right] = \frac{-\sqrt{(2+9x)^3}}{2x} + \frac{9}{4}\int\frac{\sqrt{2+9x}}{x} dx$$

Using Formula 14 again with m = 1:

$$= -\frac{\sqrt{(2+9x)^3}}{2x} + \frac{9}{4}\left[2\sqrt{2+9x} + \frac{1}{\sqrt{2}}\ln\left|\frac{\sqrt{2+9x}-\sqrt{2}}{\sqrt{2+9x}+\sqrt{2}}\right|\right] + C$$

Chapter 7   170

55. Let u = 3x, then du = 3dx: $\int \dfrac{\sqrt{2+9x^2}}{x}\,dx = \int \dfrac{\sqrt{2+u^2}}{\frac{u}{3}}\,\dfrac{du}{3} = \int \dfrac{\sqrt{2+u^2}}{u}\,du.$

Using Formula 21 with $a = \sqrt{2}$: $= \sqrt{u^2+2} - \sqrt{2}\ln\left|\dfrac{\sqrt{2}+\sqrt{u^2+2}}{u}\right| + C$

$= \sqrt{9x^2+2} - \sqrt{2}\ln\left|\dfrac{\sqrt{2}+\sqrt{9x^2+2}}{3x}\right| + C$

57. Let u = 3x, then du = 3dx: $\int \dfrac{1}{\sqrt{2+9x^2}}\,dx = \int \dfrac{1}{\sqrt{2+u^2}}\,\dfrac{du}{3}$

Using Formula 16 with $a = \sqrt{2}$ gives: $\dfrac{1}{3}\ln\left|u + \sqrt{u^2+2}\right| + C = \dfrac{1}{3}\ln\left|3x + \sqrt{9x^2+2}\right| + C$

59. Let $u = 9 - 16x^2$, then du = -32xdx:

$\int 5x\sqrt{9-16x^2}\,dx = \int 5x(\sqrt{u})\dfrac{du}{-32x} = -\dfrac{5}{32}\int u^{\frac{1}{2}}\,du = -\dfrac{5}{32}(\dfrac{2}{3}u^{\frac{3}{2}}) + C = -\dfrac{5}{48}(9-16x^2)^{\frac{3}{2}} + C$

61. Using Formula 28 with a = 1: $\int \dfrac{\sqrt{x^2-1}}{x^2}\,dx = -\dfrac{\sqrt{x^2-1}}{x} + \ln\left|x + \sqrt{x^2-1}\right| + C$

63. Using Formula 11 with $n = \dfrac{1}{2}$, a = 3, and b = 10:

$\int x^2(3+10x)^{\frac{1}{2}}dx = \dfrac{1}{1000}\left[\dfrac{(3+10x)^{\frac{7}{2}}}{\frac{7}{2}} - 6\dfrac{(3+10x)^{\frac{5}{2}}}{\frac{5}{2}} + 9\dfrac{(3+10x)^{\frac{3}{2}}}{\frac{3}{2}}\right] + C$

$= \dfrac{(3+10x)^{\frac{7}{2}}}{3500} - \dfrac{3(3+10x)^{\frac{5}{2}}}{1250} + \dfrac{3(3+10x)^{\frac{3}{2}}}{500} + C$

If instead of Formula 11 you use Formula 13, you will obtain the following (equivalent) form: $\dfrac{6 - 30x + 125x^2}{4375}(3+10x)^{\frac{3}{2}} + C$

65. The total amount will be $\int_1^{12}\dfrac{5000}{t\sqrt{100+t^2}}\,dt$ Using Formula 19 with a = 10:

$\int_1^{12}\dfrac{5000}{t\sqrt{100+t^2}}\,dt = -\dfrac{5000}{10}\ln\left|\dfrac{10+\sqrt{t^2+100}}{t}\right|\Big|_1^{12}$

$= -500\left[\ln\left|\dfrac{10+\sqrt{244}}{12}\right| - \ln\left|\dfrac{10+\sqrt{101}}{1}\right|\right] \approx -500\,[.7585 - 2.9982] \approx 1120 \text{ tons.}$

Chapter 7   171

67. The accumulated sales are $\int_{12}^{24} te^{.1t}dt$. Using Formula 35 with a = .1, this is

$$\frac{e^{.1t}}{.1^2}(.1t-1)\Big|_{12}^{24} = \frac{e^{2.4}}{.01}(2.4-1) - \frac{e^{1.2}}{.01}(1.2-1) \approx 1476.842 \text{ thousands or about } \$1,480,000$$

7.3 NUMERICAL INTEGRATION, PAGES 224 - 225

1. $\Delta x = \dfrac{b-a}{n} = \dfrac{6-1}{1} = 5$. $A_1 = 5[f(1)] = 5\sqrt{1+3} = 10$

3. $\Delta x = \dfrac{b-a}{n} = \dfrac{6-1}{3} = \dfrac{5}{3}$. $A_3 = \dfrac{5}{3}[f(1) + f(\dfrac{8}{3}) + f(\dfrac{13}{3})] = \dfrac{5}{3}\left[\sqrt{1+3} + \sqrt{\dfrac{8}{3}+3} + \sqrt{\dfrac{13}{3}+3}\right]$

≈ 11.81415

5. $\Delta x = \dfrac{b-a}{n} = \dfrac{6-1}{1} = 5$. $T_1 = \dfrac{\Delta x}{2}[f(1) + f(6)] = \dfrac{5}{2}[\sqrt{1+3} + \sqrt{6+3}] = 12.5$

7. $\Delta x = \dfrac{b-a}{n} = \dfrac{6-1}{3} = \dfrac{5}{3}$. $T_3 = \dfrac{\Delta x}{2}[f(1) + 2f(\dfrac{8}{3}) + 2f(\dfrac{13}{3}) + f(6)]$

$= \dfrac{5}{6}\left[\sqrt{1+3} + 2\sqrt{\dfrac{8}{3}+3} + 2\sqrt{\dfrac{13}{3}+3} + \sqrt{6+3}\right] \approx 12.64748$

9. $\Delta x = \dfrac{b-a}{n} = \dfrac{6-1}{2} = \dfrac{5}{2}$. $P_2 = \dfrac{\Delta x}{3}[f(1) + 4f(\dfrac{7}{2}) + f(6)] = \dfrac{5}{6}\left[\sqrt{1+3} + 4\sqrt{\dfrac{7}{2}+3} + \sqrt{6+3}\right]$

≈ 12.66503

11. $\Delta x = \dfrac{b-a}{n} = \dfrac{6-1}{6} = \dfrac{5}{6}$. $P_6 = \dfrac{\Delta x}{3}[f(1) + 4f(\dfrac{11}{6}) + 2f(\dfrac{16}{6}) + 4f(\dfrac{21}{6}) + P_6$

$= \dfrac{\Delta x}{3}[f(1) + 4f(\dfrac{11}{6}) + 2f(\dfrac{16}{6}) + 4f(\dfrac{21}{6}) + 2f(\dfrac{26}{6}) + 4f(\dfrac{31}{6}) + f(6)]$

$= \dfrac{5}{18}\left[\sqrt{1+3} + 4\sqrt{\dfrac{11}{6}+3} + 2\sqrt{\dfrac{16}{6}+3} + 4\sqrt{\dfrac{21}{6}+3} + 2\sqrt{\dfrac{26}{6}+3} + 4\sqrt{\dfrac{31}{6}+3} + \sqrt{6+3}\right]$

≈ 12.66664

13. $\Delta x = \dfrac{b-a}{n} = \dfrac{4-2}{1} = 2$. $A_1 = \Delta x[f(2)] = 2\left[\dfrac{2}{(1+2(2))^2}\right] = 0.16$

15. $\Delta x = \dfrac{b-a}{n} = \dfrac{4-2}{3} = \dfrac{2}{3}$. $A_3 = \Delta x[f(2) + f(\dfrac{8}{3}) + f(\dfrac{10}{3})]$

$= \dfrac{2}{3}\left[\dfrac{2}{(1+2(2))^2} + \dfrac{(\dfrac{8}{3})}{(1+2(\dfrac{8}{3}))^2} + \dfrac{\dfrac{10}{3}}{(1+2(\dfrac{10}{3}))^2}\right] \approx .13546$

Chapter 7

17. $\Delta x = \dfrac{b-a}{n} = \dfrac{4-2}{1} = 2.$ $T_1 = \dfrac{\Delta x}{2}[f(2) + f(4)] = (1)\left[\dfrac{2}{(1+2(2))^2} + \dfrac{4}{(1+2(4))^2}\right] \approx .12938$

19. $\Delta x = \dfrac{b-a}{n} = \dfrac{4-2}{3} = \dfrac{2}{3}.$ $T_3 = \dfrac{\Delta x}{2}\left[f(2) + 2f(\dfrac{8}{3}) + 2f(\dfrac{10}{3}) + f(4)\right]$

$= \dfrac{1}{3}\left[\dfrac{2}{(1+2(2))^2} + \dfrac{2(\dfrac{8}{3})}{(1+2(\dfrac{8}{3}))^2} + \dfrac{2(\dfrac{10}{3})}{(1+2(\dfrac{10}{3}))^2} + \dfrac{4}{(1+2(4))^2}\right] \approx .12526$

21. $\Delta x = \dfrac{b-a}{n} = \dfrac{4-2}{2} = 1.$ $P_2 = \dfrac{\Delta x}{3}[f(2) + 4f(3) + f(4)]$

$= \dfrac{1}{3}\left[\dfrac{2}{(1+2(2))^2} + \dfrac{4(3)}{(1+2(3))^2} + \dfrac{4}{(1+2(4))^2}\right] \approx .12476$

23. $\Delta x = \dfrac{b-a}{n} = \dfrac{4-2}{6} = \dfrac{1}{3}.$ $P_6 = \dfrac{\Delta x}{3}\left[f(2) + 4f(\dfrac{7}{3}) + 2f(\dfrac{8}{3}) + 4f(3) + 2f(\dfrac{10}{3}) + 4f(\dfrac{11}{3}) + f(4)\right]$

$= \dfrac{1}{9}\left[\dfrac{2}{(1+2(2))^2} + \dfrac{4(\dfrac{7}{3})}{(1+2(\dfrac{7}{3}))^2} + \dfrac{2(\dfrac{8}{3})}{(1+2(\dfrac{8}{3}))^2} + \dfrac{4(3)}{(1+2(3))^2} + \dfrac{2(\dfrac{10}{3})}{(1+2(\dfrac{10}{3}))^2} + \dfrac{4(\dfrac{11}{3})}{(1+2(\dfrac{11}{3}))^2}\right.$

$\left. + \dfrac{4}{(1+2(4))^2}\right] \approx .12473$

25. $\Delta x = \dfrac{7-3}{4} = 1.$ $A_4 = \Delta x\,[f(3) + f(4) + f(5) + f(6)]$

$= (1)\left[\dfrac{\sqrt{1+3}}{3^3} + \dfrac{\sqrt{1+4}}{4^3} + \dfrac{\sqrt{1+5}}{5^3} + \dfrac{\sqrt{1+6}}{6^3}\right] \approx .140857 \approx .14$

27. $\Delta x = \dfrac{3-1}{4} = \dfrac{1}{2}.$ $A_4 = \Delta x\left[f(1) + f(\dfrac{3}{2}) + f(2) + f(\dfrac{5}{2})\right] = \dfrac{1}{2}\left[1^1 + 1.5^{1.5} + 2^2 + 2.5^{2.5}\right]$

$\approx 8.3596 \approx 8.36$

29. $\Delta x = \dfrac{7-3}{4} = 1.$ $T_4 = \dfrac{\Delta x}{2}[f(3) + 2f(4) + 2f(5) + 2f(6) + f(7)]$

$= \dfrac{1}{2}\left[\dfrac{\sqrt{1+3}}{3^3} + \dfrac{2\sqrt{1+4}}{4^3} + \dfrac{2\sqrt{1+5}}{5^3} + \dfrac{2\sqrt{1+6}}{6^3} + \dfrac{\sqrt{1+7}}{7^3}\right] \approx .1079 \approx .11$

Chapter 7   173

31. $\Delta x = \dfrac{3-1}{4} = \dfrac{1}{2}$. $T_4 = \dfrac{\Delta x}{2}\left[f(1) + 2f(\dfrac{3}{2}) + 2f(2) + 2f(\dfrac{5}{2}) + f(3)\right]$

$= \dfrac{1}{4}\left[1^1 + 2(1.5)^{1.5} + 2(2)^2 + 2(2.5)^{2.5} + 3^3\right] \approx 14.8596 \approx 14.86$

33. $\Delta x = \dfrac{7-3}{4} = 1$. $P_4 = \dfrac{\Delta x}{3}[f(3) + 4f(4) + 2f(5) + 4f(6) + f(7)]$

$= \dfrac{1}{3}\left[\dfrac{\sqrt{1+3}}{3^3} + \dfrac{4\sqrt{1+4}}{4^3} + \dfrac{2\sqrt{1+5}}{5^3} + \dfrac{4\sqrt{1+6}}{6^3} + \dfrac{\sqrt{1+7}}{7^3}\right] \approx .10$

35. $\Delta x = \dfrac{3-1}{4} = \dfrac{1}{2}$. $P_4 = \dfrac{\Delta x}{3}\left[f(1) + 4f(\dfrac{3}{2}) + 2f(2) + 4f(\dfrac{5}{2}) + f(3)\right]$

$= \dfrac{1}{6}\left[1^1 + 4(1.5)^{1.5} + 2(2)^2 + 4(2.5)^{2.5} + 3^3\right] \approx 13.81$

37. Area $\approx \dfrac{5}{3}[0 + 4(10) + 2(11) + 4(13) + 2(16) + 4(18) + 2(23) + 4(25) + 2(27) + 4(25) + 10]$

= 880 square feet. Since it is 3 feet deep, 880 × 3 = 2640 cu. ft. = (2640 ÷ 27) cu. yd.
≈ 97.8. About 100 cu. yd. of fill will be needed.

7.4 CHAPTER REVIEW, PAGES 226 - 227

1. Let $u = 2x - 1$; $\dfrac{du}{dx} = 2$; $\dfrac{du}{2} = dx$.

$\int(2x-1)^3 dx = \int u^3 \dfrac{du}{2} = \dfrac{1}{2}\int u^3 du = \dfrac{1}{2}\dfrac{u^4}{4} + C = \dfrac{(2x-1)^4}{8} + C$

3. By parts:

$u = x^2$	$dv = (2x-1)^3 dx$
$du = 2xdx$	$v = \dfrac{(2x-1)^4}{8}$ (See Problem 1.)

$\int x^2(2x-1)^3 dx = uv - \int v du = \dfrac{x^2(2x-1)^4}{8} - \int \dfrac{2x(2x-1)^4}{8}dx = \dfrac{x^2(2x-1)^4}{8} - \dfrac{1}{4}\int x(2x-1)^4 dx.$

This last integral can be done by parts:

$u = x$	$dv = (2x-1)^4 dx$
$du = dx$	$v = \dfrac{(2x-1)^5}{10}$

We have $\dfrac{x^2(2x-1)^4}{8} - \dfrac{1}{4}\int x(2x-1)^4 dx = \dfrac{x^2(2x-1)^4}{8} - \dfrac{1}{4}[uv - \int v du]$

$= \dfrac{x^2(2x-1)^4}{8} - \dfrac{1}{4}\left[\dfrac{x(2x-1)^5}{10} - \int \dfrac{(2x-1)^5}{10}dx\right] = \dfrac{x^2(2x-1)^4}{8} - \dfrac{1}{4}\left[\dfrac{x(2x-1)^5}{10} - \dfrac{(2x-1)^6}{120}\right] + C$

Chapter 7

$$= \frac{x^2(2x-1)^4}{8} - \frac{x(2x-1)^5}{40} + \frac{(2x-1)^6}{480} + C$$

If instead, you use Formula 11 you will obtain the following (equivalent) form:

$$\frac{(2x-1)^6}{48} + \frac{(2x-1)^5}{40} + \frac{(2x-1)^4}{32} + C$$

5. By parts:

$u = x^2$	$dv = (1-5x)^{\frac{1}{2}}dx$
$du = 2xdx$	$v = \dfrac{-2(1-5x)^{\frac{3}{2}}}{15}$

$$\int x^2\sqrt{1-5x}\,dx = uv - \int v\,du = \frac{-2x^2(1-5x)^{\frac{3}{2}}}{15} + \frac{4}{15}\int x(1-5x)^{\frac{3}{2}}dx$$

This last integral can also be done by parts:

$u = x$	$dv = (1-5x)^{\frac{3}{2}}dx$
$du = dx$	$v = -\dfrac{2(1-5x)^{\frac{5}{2}}}{25}$

We have $\dfrac{-2x^2(1-5x)^{\frac{3}{2}}}{15} + \dfrac{4}{15}[uv - \int v\,du] = \dfrac{-2x(1-5x)^{\frac{3}{2}}}{15} + \dfrac{4}{15}\left[\dfrac{-2x(1-5x)^{\frac{5}{2}}}{25} + \dfrac{2}{25}\int(1-5x)^{\frac{5}{2}}dx\right]$

$$= \frac{-2x^2(1-5x)^{\frac{3}{2}}}{15} - \frac{8x(1-5x)^{\frac{5}{2}}}{375} + \frac{8}{375}\left(\frac{-2(1-5x)^{\frac{7}{2}}}{35}\right) + C$$

$$= \frac{-2x^2(1-5x)^{\frac{3}{2}}}{15} - \frac{8x(1-5x)^{\frac{5}{2}}}{375} - \frac{16(1-5x)^{\frac{7}{2}}}{13125} + C$$

If instead, you use Formula 13 you will obtain the following (equivalent) form:

$$\frac{-2(8 + 60x + 375x^2)}{13125}(1-5x)^{\frac{3}{2}} + C$$

7. By parts:

$u = \ln^2 x$	$dv = dx$
$du = \dfrac{2\ln x}{x}dx$	$v = x$

$\int \ln^2 x\,dx = uv - \int v\,du = x\ln^2 x - 2\int \ln x\,dx = x\ln^2 x - 2[x\ln x - x] + C$ (See Example 3 in Section 7.1 for this last integral) $= x\ln^2 x - 2x\ln x + 2x + C$

Chapter 7

9. Let $u = 4x^2 + 5x - 3$; $\dfrac{du}{dx} = 8x + 5$; $\dfrac{du}{8x + 5} = dx$.

$$\int \dfrac{8x + 5}{4x^2 + 5x - 3} dx = \int \dfrac{8x + 5}{u} \cdot \dfrac{du}{8x + 5} = \int \dfrac{du}{u} = \ln |u| + C = \ln |4x^2 + 5x - 3| + C$$

11. Formula 14   13. Formula 37

15. $\Delta x = \dfrac{3 - 0}{4} = \dfrac{3}{4}$. $T_4 = \dfrac{\Delta x}{2} [f(0) + 2f(\dfrac{3}{4}) + 2f(\dfrac{3}{2}) + 2f(\dfrac{9}{4}) + f(3)]$

$= \dfrac{3}{8} \left[\sqrt{9 - 0^2} + 2\sqrt{9 - (\dfrac{3}{4})^2} + 2\sqrt{9 - (\dfrac{3}{2})^2} + 2\sqrt{9 - (\dfrac{9}{4})^2} + \sqrt{9 - 3^2} \right] \approx 6.74$

17. $N(t) = \int 5e^{-.1t} dt = \dfrac{5e^{-.1t}}{-.1} + C = -50e^{-.1t} + C$

$0 = N(0) = -50e^0 + C = -50 + C$; solving for C gives $C = 50$.

$N(t) = -50e^{-.1t} + 50 = 50(1 - e^{-.1t})$

19. Answers vary, but one approximation is:
    (1) $[11 + 12.8 + 14.5 + 10.6 + 11.9 + 11.5] = 72.3$

7.5 STUDENT'S TEST REVIEW AND ADDITIONAL PRACTICE
OBJECTIVES

The material of this chapter is reviewed in the following list of objectives. After each objective there are some practice questions. Answers to these problems immediately follow. Detailed solutions are given for every third problem. For a sample test select the first question of each set and check your answers. Additional practice is given by the other questions in each set. If you are having trouble with a particular type of problem, or if you want additional practice, look back at the indicated section in the test.

[7.1] Objective 1: *Evaluate indefinite integrals by parts.*

1. $\int \dfrac{x}{e^{2x}} dx$ 2. $\int \dfrac{x^5}{(x^3 + 1)^2} dx$

3. $\int \ln \sqrt{2x} \, dx$ 4. $\int x^2 \sqrt{10 - x} \, dx$

[7.2] Objective 2: *Evaluate indefinite integrals by using the Brief Integral Table.*

5. $\int x^2 \sqrt{x^2 - 16} \, dx$ 6. $\int (3 - 4x)^6 dx$

Chapter 7

7. $\int \ln^3 3x \, dx$

8. $\int x^2 \sqrt{10-x} \, dx$

Objective 3: *Evaluate indefinite integrals using any appropriate method.*

9. $\int x^2 e^{-2x} \, dx$

10. $\int \dfrac{3dx}{1-e^{3x}}$

11. $\int \dfrac{5dx}{2-e^{-x}}$

12. $\int \dfrac{3x+5}{4x^2+5x-3} \, dx$

[7.3] Objective 4: *Approximate integrals using a rectangular approximation.* Find

$$\int_4^{25} \ln\sqrt{x} \, dx$$

correct to two decimal places where n is specified in Problems 13 - 16.

13. n = 1   14. n = 2   15. n = 4   16. n = 8

Objective 5: *Approximate integrals using a trapezoidal approximation.* Find

$$\int_0^2 \dfrac{dx}{1+x^2}$$

correct to two decimal places where n is specified in Problems 17 - 20.

17. n = 1   18. n = 2   19. n = 4   20. n = 8

Objective 6: *Approximate integrals using Simpson's Rule.* Find

$$\int_0^2 \sqrt{4-x^2} \, dx$$

correct to two decimal places where n is specified in Problems 21 - 24.

21. n = 1   22. n = 2   23. n = 4   24. n = 8

Objective 7: *Solve applied problems based on the preceeding objectives. For specific examples of the types of applications look at the list of applications in this chapter on page 207.*

25. The marginal profit of Campress Press-Ons is

   P'(x) = 250x - 5,000

   where x is the number of items. Find the profit function if there is a $1500 loss when no items are produced.

26. The marginal cost of Campress Press-Ons is

   $C'(x) = \dfrac{\sqrt{x}}{1000} + .0001x$

   If the fixed costs are $1500, what is the cost of producing 5000 items (to the nearest thousand dollars)?

27. Industrial waste is pumped into a 1000 gallon holding tank at the rate of 1 gal/min and

Chapter 7

the well-stirred mixture is removed at the same rate. If the concentration of waste in the tank is changing at a rate of

$$c'(t) = .001e^{-.001t}$$

after t minutes, find the concentration of waste in the tank after one hour.

28. If a disease is spreading at a rate of $40t - 6t^2$ cases per day where t is the number of days measured from the first outbreak, give the number of people affected on the seventh day. Assume that there were 4 cases recorded on the first day.

29. In 1986 the growth rate of Chicago, IL was given by the formula

$$P'(t) = -.035e^{-.011t}$$

where t is measured in years after 1980 and P(t) gives the population in millions. If the population in 1984 was 2,992,472, predict the population at the turn of the century.

ANSWERS TO STUDENT'S TEST REVIEW AND PRACTICE QUESTIONS

1. Let $u = x$, $dv = e^{-2x}dx \Rightarrow -\dfrac{e^{-2x}}{4}(2x + 1) + C$

2. Let $u = x^3$, $dv = (x^3 + 1)^{-2}x^2 dx \Rightarrow -\dfrac{x^3}{3(x^3 + 1)} + \dfrac{1}{3}\ln|x^3 + 1| + C$

3. Let $u = \ln\sqrt{2x} = \dfrac{1}{2}\ln(2x)$; $dv = dx$
   $du = \dfrac{1}{2x}dx$; $v = x$

   $\int \ln\sqrt{2x}\, dx = uv - \int v\,du = x\ln\sqrt{2x} - \int x(\dfrac{1}{2x})\,dx = x\ln\sqrt{2x} - \dfrac{1}{2}x + C$

4. $-\dfrac{2}{3}x^2(10-x)^{\frac{3}{2}} - \dfrac{8}{15}x(10-x)^{\frac{5}{2}} - \dfrac{16}{105}(10-x)^{\frac{7}{2}} + C$

5. Use Formula 25 with $a^2 = 16$: $\dfrac{x}{4}\sqrt{(x-16)^3} - 2x^2 - 32\ln\left|x + \sqrt{x^2 - 16}\right| + C$

6. Let $u = 3 - 4x$, then $du = -4dx$; use Formula 2:

   $\int (3-4x)^6 dx = \int u^6(\dfrac{du}{-4}) = -\dfrac{1}{4}\int u^6 du = -\dfrac{1}{4}\dfrac{u^7}{7} = -\dfrac{(3-4x)^7}{28} + C = \dfrac{(4x-3)^7}{28} + C$

7. Let $u = 3x$, then $du = 3dx$. Use Formula 34 then 33:

   $\dfrac{1}{3}\ln^3(3x) - 3x\ln^2(3x) + 6x\ln(3x) - 6x + C$

8. Use Formula 11 with $a = 10$, $b = -1$, $n = \dfrac{1}{2}$:

Chapter 7

$-\dfrac{2}{3}x^2(10-x)^{\frac{3}{2}} - \dfrac{8}{15}x(10-x)^{\frac{5}{2}} - \dfrac{16}{105}(10-x)^{\frac{7}{2}} + C$

9. $-\dfrac{1}{2}x^2 e^{-2x} + \dfrac{1}{2}xe^{-2x} - \dfrac{1}{4}e^{-2x} + C$

10. Use Formula 39 with a = 1, b = -1, and m = 3:

$\displaystyle\int \dfrac{3dx}{1-e^{3x}} = 3\int \dfrac{dx}{1-e^{3x}} = 3\left[\dfrac{x}{1} - \dfrac{1}{1^3}\ln\left|1-e^{3x}\right|\right] + C = 3x - 3\ln\left|1-e^{3x}\right| + C$

11. $\dfrac{5}{2}\left[x + \ln\left|2-e^{-x}\right|\right] + C$

12. Let u = $4x^2$ + 5x - 3, then du = 3x + 5dx:

$\displaystyle\int \dfrac{8x+5}{4x^2+5x-3}dx = \int \dfrac{du}{u} = \ln|u| + C = \ln\left|4x^2 + 5x - 3\right| + C$

13. 24.18   **14.** 21.32

15. $\Delta x = \dfrac{25-4}{4} = 5.25$; $A = \Delta x\,[f(x_0) + f(x_1) + f(x_2) + f(x_3)]$

   $= 5.25\,[f(4) + f(9.25) + f(14.5) + f(19.75)] \approx 24.33$

16. 25.70   **17.** 1.20   **18.** $T = \dfrac{\Delta x}{2}[f(x_0) + 2f(x_1) + f(x_2)] = \dfrac{1}{2}[f(0) + 2f(1) + f(2)] = 1.1$

19. 1.10   **20.** 1.11   **21.** n must be <u>even</u> for Simpson's Rule.   **22.** 2.98   **23.** 3.08

24. $P = \dfrac{\Delta x}{3}[f(x_0) + 4f(x_1) + 2f(x_2) + 4f(x_3) + 2f(x_4) + 4f(x_5) + 2f(x_6) + 4f(x_7) + f(x_8)]$

$= \dfrac{1}{2}[f(0) + 4f(\tfrac{1}{4}) + 2f(\tfrac{1}{2}) + 4f(\tfrac{3}{4}) + 2f(1) + 4f(1\tfrac{1}{4}) + 2f(1\tfrac{1}{2}) + 4f(1\tfrac{3}{4}) + f(2)] = 3.12$

25. $P(x) = 125x^2 - 5000x - 1500$

26. $C(5000) = \dfrac{5000^{\frac{3}{2}}}{1500} + .00005(5000)^2 + 1500 = 2986$. To the nearest thousand dollars, \$3,000.

27. $c(t) = \displaystyle\int .001e^{-.001t}dt = -e^{-.001t} + C$. At t = 0 there is no waste in the holding tank, so

$0 = c(0) = -e^0 + C \Rightarrow C = 1$. We have $c(t) = -e^{-.001t} + 1$.

$c(60) = -e^{-.001(60)} + 1 \approx .058$ or 5.8%.

28. 280   29. 2,501,087

MODELING APPLICATION 6: SAMPLE ESSAY
Modeling the Nervous System

    The nervous system has the ability to respond to a variety of stimuli by generating electrical impulses and transmitting them through nerve cells. Nerve cells vary in length from a few micrometers to more than a meter. Nerve cells that transmit impulses from sense organs to the spinal cord and brain are called <u>sensory neurons</u> or <u>afferent neurons</u>. The neurons in the spinal cord and brain make up the central nervous system (CNS), whereas the neurons outside these structures make up the peripheral nervous system (PNS). The purpose of this extended application is to build a mathematical model to duplicate the length of time a stimulus has to be applied to elicit an action. That is, a mathematical model which describes reaction time in the central nervous system.*

    If you inadvertently stick your finger into an electric outlet you will get a shock and will immediately withdraw it very quickly (although, alas, not quickly enough). The short (but definitely non-zero) interval between the time you first touch the outlet and the time you begin to withdraw your finger is your <u>reaction time</u>. Our intuition, and probably also our experience, suggest that, all other things being equal, the reaction time is a decreasing function of the strength of the stimulus. That is, a more intense stimulus results in a faster reaction. Experimental results** confirm that it is reasonable to assume that the strength S of a stimulus is inversely proportional to its duration. That is, $S = k/d$ where S is measured in milliamperes or in millivolts, t is measured in milliseconds, and k is a constant of proportionality. Thus, there is no single threshold stimulus, but a theoretical range of threshold stimuli of different duration. The purpose of this extended application is to develop a mathematical model which will predict how the reaction time depends upon the stimulus intensity.

    When an above-threshold stimulus of constant intensity S is applied to a portion of the body, its effect is transmitted to the brain along afferent fibers.*** The <u>intensity of excitation</u> $E = E(S)$ in the afferent pathway is an increasing function of S. It does not change with time. It can be thought of as representing the number of neurons in the

---

    * This extended application is adapted from <u>Modeling The Nervous System: Reaction Time and the Central Nervous System</u> by Brindell Horelick, Sinan Koont, and Sheldon F. Gottlieb. © 1982 COMAP, Inc. COMAP is a nonprofit corporation engaged in research and development in mathematics education. This material is reprinted with permission.
    ** <u>Electrical Signs of Nervous Activity</u> by J. Erlanger and H. S. Gasser (Philadelphia: University of Pennsylvania Press, 1937)
    *** This material is based on the work of Dr. Nicholas Rashevsky of the University of Chicago who is a twentieth century pioneer in the development of mathematical biology.

afferent pathway excited by the stimulus.

This excitation in the afferent pathway prompts a response within the brain. The exact physiological nature of this response is not clear, but it is known to build up with the passage of time (assuming the stimulus continues). We may think of it, for example, as the creation of a central excitatory state, with increasing frequency of excitation of these neurons within the brain. Whatever the nature of the response, we shall use M(t), to denote the magnitude of this excitatory state at time t. We call M(t) the <u>excitatory factor</u>.

Whatever the physiological description of M(t), when (and if) it builds up to a certain threshold level h_E it stimulates the efferent neuron, leading from the brain to the effector organ. This results in a response to the stimulus.

The reaction time T is the sum of three times (see Figure 1):
$$T = t_A + t_B + t_E$$
where t_A = the time required for the excitation in the afferent pathway to be transmitted from the point of application of the stimulus to the brain;

t_B = the time required for M(t) to grow to the threshold level h_E;

t_E = the time required for the response to be transmitted along the efferent pathway.

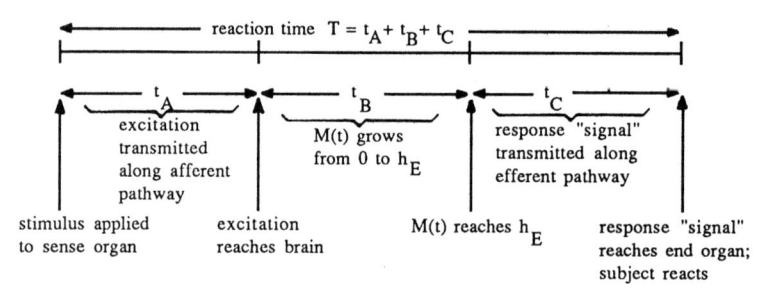

Figure 1 How the Central Nervous System Responds to a Stimulus

Of these, only t_B is dependent upon the stimulus intensity S. Remember that the excitation intensity E depends on S, but its transmission time t_A does not.

We need to determine t_B as a function of S. But all we know about t_B is that it is a certain time interval; namely, the interval from when M(t) starts growing to when it is equal to h_E. Call the beginning of this interval t = 0. Then the end of the interval occurs at t = t_B, since its length is t_B. In other words, t_B is the solution of
$$M(t_B) = h_E$$
From Figure 2 you can see that the stimulus was applied at time t = $-t_A$, since the afferent pathway excitation reaches the brain t_A seconds after the initial application of the stimulus.

Chapter 7

We will make two critical assumptions about M(t). Both are well supported by biological theory. If each excited nerve fiber in the afferent pathway does its part in exciting neuron cycles, producing the central excitatory neuron cycles. producing the central excitatory state, or whatever, the rate at which this response is excited is proportional to E. The second assumption is that there is nothing contradicting the tendency for M(t) to increase other than natural "decay" (for example, accidental failure of a neuron cycle, somewhat like a light bulb's buring out). This decay occurs randomly and the amount of decay per second is proportional to the amount of excited neuron cycles, chemical, or whatever present.

Mathematically, these two assujptions say

$$M'(t) = k_1 E - k_2 M(t)$$

where k_1 and k_2 are positive constants. Now, divide both sides by $k_1 E - k_2 M(t)$:

$$\frac{M'(t)}{k_1 E - k_2 M(t)} = 1$$

Next, antidifferentiate:

$$\int \frac{M'(t) dt}{k_1 E - k_2 M(t)} = \int 1 dt$$

Let $u(t) = k_1 E - k_2 M(t)$, so $u'(t) = -k_2 M'(t)$ or $du = -k_2 M'(t) dt$.

Now, by substitution, this equation becomes

$$\frac{-1}{k_2} \int \frac{u'(t)}{u(t)} dt = \int 1 dt$$

$$\frac{-1}{k_2} \ln u(t) = t + C \qquad \text{There is no need for an absolute value sign because } u(t) > 0.$$

Set $t = 0$ and use the fact that $u(0) = k_1 E - k_2 M(0) = k_1 E$, to obtain

$$C = -\frac{1}{k_2} \ln(k_1 E)$$

Substituting in this value of C the equation for t becomes

$$t = \frac{-1}{k_2} \ln u(t) + \frac{1}{k_2} \ln(k_1 E)$$

$$= \frac{1}{k_2} [\ln(k_1 E) - \ln u(t)]$$

$$= \frac{1}{k_2} \ln \frac{k_1 E}{u(t)}$$

$$= \frac{-1}{k_2} \ln \frac{k_1 E}{k_1 E - k_2 M(t)} \qquad \text{Since } u(t) = k_1 E - k_2 M(t)$$

Now, let $t = t_B$ to find

Chapter 7

$$t_B = \frac{-1}{k_2} \ln \frac{k_1 E}{k_1 E - k_2 M(t_B)}$$

$$= \frac{-1}{k_2} \ln \frac{k_1 E}{k_1 E - k_2 h_E} \qquad \text{since } M(t_B) = h_E$$

Finally,
$$T = t_A + t_B + t_E$$

$$= t_A + t_E + t_B$$

$$= T_0 + t_B$$

$$= T_0 + \frac{-1}{k_2} \ln \frac{k_1 E}{k_1 E - k_2 h_E}$$

CHAPTER 8
APPLICATIONS AND INTEGRATION

8.1 BUSINESS MODELS USING INTEGRATION, PAGES 235 - 236

1. $\int_{4}^{9}(\sqrt{x} + 100)dx = \dfrac{2x^{\frac{3}{2}}}{3} + 100x \Big|_{4}^{9} = 918 - \dfrac{1216}{3} = 512\dfrac{2}{3}$

3. To evaluate $\int_{0}^{2} \dfrac{x - 1}{x + 1} dx$ try letting $u = x + 1$, $du = dx$, $u - 2 = x - 1$

$\int_{x=0}^{x=2} \dfrac{x - 1}{x + 1} dx = \int_{u=1}^{u=3} \dfrac{u - 2}{u} du = \int_{1}^{3} (1 - \dfrac{2}{u})du = u - 2\ln|u| \Big|_{1}^{3} = (3 - 2\ln 3) - 1 = 2 - 2\ln 3 \approx -.197$

5. $\int_{0}^{3} 65e^{-5t} dt = \dfrac{65e^{-5t}}{-5} \Big|_{0}^{3} = -13e^{-5t} \Big|_{0}^{3} = (-13e^{-15}) - (-13) = 13(1 - e^{-15}) \approx 13$

7. $\int_{0}^{6} 8000e^{.08t} dt = \dfrac{8000e^{.08t}}{.08} \Big|_{0}^{6} = 100{,}000 e^{.08t} \Big|_{0}^{6} = 100{,}000(e^{.48} - 1) \approx \$61{,}607.44$

9. $\int_{0}^{1} 2500e^{.05t} dt = \dfrac{2500}{.05} e^{.05t} \Big|_{0}^{1} = 50{,}000 e^{.05t} \Big|_{0}^{1} = 500{,}000(e^{.05} - 1) \approx \$2{,}563.55$

11. $\int_{0}^{5} 6500e^{.07t} dt = \dfrac{6500}{.07} e^{.07t} \Big|_{0}^{5} \approx \$38{,}913.42$

13. $\dfrac{\int_{-1}^{1}(6x^2 + 3x - 10)dx}{1 - (-1)} = \dfrac{2x^3 + \dfrac{3x^2}{2} - 10x \Big|_{-1}^{1}}{2} = \dfrac{-16}{2} = -8$

15. $\dfrac{\int_{1}^{5} \dfrac{x}{x + 1} dx}{5 - 1} = \dfrac{x - \ln(x + 1) \Big|_{1}^{5}}{4} = \dfrac{4 - \ln 6 + \ln 2}{4} = \dfrac{4 - (\ln 6 - \ln 2)}{4} = \dfrac{4 - \ln(\dfrac{6}{2})}{4}$

$= 1 - \dfrac{1}{4}\ln(3) \approx .7253$

Chapter 8

17. $\dfrac{\int_0^3 e^{-5t}dt}{3-0} = \dfrac{\left.\dfrac{e^{-5t}}{-5}\right|_0^3}{3} = \dfrac{e^{-15} - e^0}{-15} = \dfrac{1 - e^{-15}}{15} \approx .0667$

19. $\int_1^2 (60 + 24t^2)dt = 60t + 8t^3 \Big|_1^2 = 116$

21. $\int_0^3 (60 + 24t^2)dt = 60t + 8t^3 \Big|_0^3 = 396$

23. $\int_6^{12} (300x^{\frac{1}{3}} + 50)dx = 225x^{\frac{4}{3}} + 50x \Big|_6^{12} \approx \$4{,}028.35$

25. $\int_0^{10} x^{-.5}dx = 2x^{.5} \Big|_0^{10} = 2\sqrt{10} \approx 6.32$

27. The total amount charged $= \int_0^4 150(1 + x^{\frac{2}{3}})dx = 150(x + .6x^{\frac{5}{3}}) \Big|_0^4 \approx \$1507.14$

The additional monthly charge should be $\dfrac{1507.14}{48 \text{ months}} = \$31.40$

29. Solve $15{,}000 = \int_0^x (2400t + 10{,}500)dt = 1200t^2 + 10{,}500t \Big|_0^x = 1200x^2 + 10{,}500x$.

Subtracting 15,000 from both sides gives $0 = 1200x^2 + 10{,}500x - 15{,}000$

$0 = 4x^2 + 35x - 50$ Divide by 300

$0 = (4x - 5)(x + 10)$

$x = \dfrac{5}{4}$ or $x = -10$ Reject -10 since $x > 0$.

It will take $1\dfrac{1}{4}$ years.

31. $\dfrac{\int_0^{24} (6.44t - .23t^2 + 30)dt}{24 - 0} = \dfrac{3.22t^2 - .07\overline{6}t^3 + 30t \Big|_0^{24}}{24} = \dfrac{1514.88}{24} \approx 63°$

184

Chapter 8

33. $\dfrac{\int_{10}^{50} \dfrac{480}{x} dx}{50-10} = \dfrac{480\ln x \Big|_{10}^{50}}{40} = 12\ln 5 \approx \$19.31$

35. $\dfrac{\int_0^5 (500 + t - .25t^2)dt}{5-0} = \dfrac{500t + \dfrac{t^2}{2} - \dfrac{1}{12}t^3 \Big|_0^5}{5} \approx 500.4$

37. $\int_0^3 3800\,dx = 3800x \Big|_0^3 = \$11{,}400$

39. $E(t) = C + \int_0^t f'(x)dx = 500 + \int_0^t 1000x^{\frac{1}{2}}dx = 500 + \dfrac{2000}{3}x^{\frac{3}{2}}\Big|_0^t = 500 + \dfrac{2000}{3}t^{\frac{3}{2}}$

According to the discussion, we want to set $E'(t) = \dfrac{E(t)}{t}$.

$1000t^{\frac{1}{2}} = \dfrac{500 + \dfrac{2000}{3}t^{\frac{3}{2}}}{t}$

$1000t^{\frac{3}{2}} = 500 + \dfrac{2000}{3}t^{\frac{3}{2}}$ (Multiply by t)

$\dfrac{1000t^{\frac{3}{2}}}{3} = 500$ (Subtract $\dfrac{2000}{3}t^{\frac{3}{2}}$)

$t^{\frac{3}{2}} = \dfrac{3}{2}$ (Multiply by $\dfrac{3}{1000}$)

$t = \left(\dfrac{3}{2}\right)^{\frac{2}{3}} \approx 1.31$ years

41. $\int_a^b f(t)dt = \int_0^5 \sqrt{t}\,dt = \dfrac{2}{3}t^{\frac{3}{2}}\Big|_0^5 \approx 7.45356$ million $= 7{,}453{,}560$

8.2 PROBABILITY DENSITY FUNCTIONS, PAGES 248 - 249

1. Let x = number of people who responded "yes", which is a discrete random variable. The sample space is {0, 1, 2, ..., 10}.

3. Let x = number of words with an error, which is a discrete random variable. The sample space is {0, 1, 2, ..., 499, 500}.

5. Let x = number of boys (or the number of girls), which is a discrete random variable. The

Chapter 8

sample space is {bbb, bbg, bgb, bgg, gbb, gbg, ggb, ggg}.

7. Let x = height in inches.

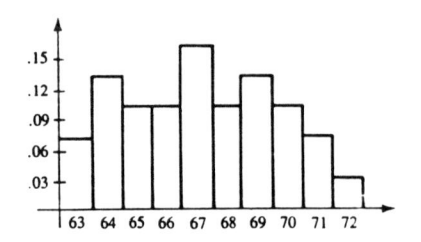

[Histogram with heights: .15, .12, .09, .06, .03 on y-axis; values 63, 64, 65, 66, 67, 68, 69, 70, 71, 72 on x-axis]

9. Let x = time (in minutes) spent on each transaction.

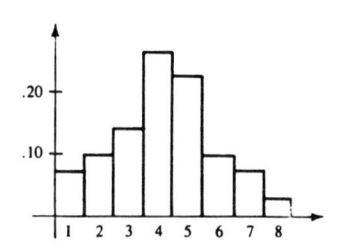

[Histogram with heights .20, .10 on y-axis; values 1, 2, 3, 4, 5, 6, 7, 8 on x-axis]

11. $\int_{1}^{5} \frac{x^2}{72} dx = \frac{x^3}{216}\Big|_{1}^{5} = \frac{124}{216}$. Since this is not 1, f(x) is not a probability density function.

13. $\int_{1}^{5} \frac{1}{4} dx = \frac{1}{4} x \Big|_{1}^{5} = 1$. This is a probability density function.

15. $\int_{1}^{4} 4 dx = 4x \Big|_{1}^{4} = 12$. Since this is not 1, f(x) is not a probability density function.

17. $\int_{1}^{4} \frac{3}{64} x^2 dx = \frac{x^3}{64}\Big|_{1}^{4} = \frac{63}{64}$. Since this is not 1, f(x) is not a probability density function.

19. $\int_{-2}^{3} kx\, dx = \frac{kx^2}{2}\Big|_{-2}^{3} = \frac{9k}{2} - \frac{4k}{2} = \frac{5k}{2}$. Set this equal to 1 and solve for k, $\frac{5k}{2} = 1$, $k = \frac{2}{5}$.

But since $f(x) = \frac{2}{5} x < 0$ for $x < 0$, no value of k will make f(x) a probability density function.

21. $\int_{0}^{5} kx^2 dx = \frac{kx^3}{3}\Big|_{0}^{5} = \frac{125k}{3}$. Set this equal to 1 and solve for k, $\frac{125k}{3} = 1$, $k = \frac{3}{125}$.

23. $\int_1^7 \frac{k}{4} dx = \frac{k}{4} x \Big|_1^7 = \frac{3k}{2}$. Setting this equal to 1 gives $k = \frac{2}{3}$.

25. $\int_1^4 kx^{\frac{1}{2}} dx = \frac{2}{3} kx^{\frac{3}{2}} \Big|_1^4 = \frac{2}{3} k(4^{\frac{3}{2}} - 1^{\frac{3}{2}}) = \frac{14k}{3}$. Setting this equal to 1 gives $k = \frac{3}{14}$.

27. $\int_1^4 (\frac{k}{x^2}) dx = \frac{kx^{-1}}{-1} \Big|_1^4 = \frac{-k}{x} \Big|_1^4 = \frac{3k}{4}$. Setting this equal to 1 gives $k = \frac{4}{3}$.

29. On Table 8.3, locate 1.2 in the left column (x), then 4 columns over to the column headed .03. The entry is 0.8907.

31. On Table 8.3, locate 1.6 in the left column (x), then 10 columns over to the entry 0.9545. This is the area for x < 1.69. The area for x > 1.69 is 1 - 0.9545 = 0.0455.

33. The area under the curve for $-2.8 \le x \le -.46$ is (the area for $x \le -.46$) - (the area for $x \le -2.8$). Using Table 8.3 this is 0.3228 - 0.0026 = 0.3202.

35. $\int_{35}^{75} .01 dx = .01 \, dx \Big|_{35}^{75} = .75 - .35 = .40$

37. $\int_5^{10} .5e^{-.5x} dx = -e^{-.5x} \Big|_5^{10} = -e^{-5} + e^{-2.5} \approx .075$

39. The z value is $z = \frac{x - \mu}{\sigma} = \frac{.91 - .9}{.01} = 1$. $P(x > .91) = P(z > 1) = 1 - P(z \le 1)$
$= 1 - P(z < 1.005) = 1 - \left[\frac{.8413 + .8438}{2} \right] = 1 - .84255 = .1575$

41. The z value is $z = \frac{x - \mu}{\sigma} = \frac{210 - 250}{25} = -1.6$. $P(x < 210) = P(z < -1.6) = .0548$ (By Table 8.3)

43. The z scores for x = 240 and 280 are -.4 and 1.2, respectively. $P(240 < x < 280)$
$= P(-.4 < z < 1.2) = P(z < 1.2) - P(z < -.4) = .8849 - .3446 = .5403$.

8.3 IMPROPER INTEGRALS, PAGES 252 - 253

1. $\int_0^\infty e^{-x} dx = \lim_{b \to \infty} \int_0^b e^{-x} dx = \lim_{b \to \infty} (1 - e^{-b}) = 1 - 0 = 1$

Chapter 8

3. $\int_0^\infty e^{-\frac{x}{2}} dx = \lim_{b \to \infty} \int_0^b e^{-\frac{x}{2}} dx = \lim_{b \to \infty} (-2e^{-\frac{b}{2}} + 2) = 0 + 2 = 2$

5. $\int_1^\infty \frac{dx}{x^3} = \lim_{b \to \infty} \int_1^b x^{-3} dx = \lim_{b \to \infty} (\frac{b^{-2}}{-2} + \frac{1}{2}) = 0 + \frac{1}{2} = \frac{1}{2}$

7. $\int_1^\infty \frac{dx}{\sqrt[3]{x}} = \lim_{b \to \infty} \int_1^b x^{-\frac{1}{3}} dx = \lim_{b \to \infty} \frac{3}{2}(b^{\frac{2}{3}} - 1)$. This limit does not exist, since

$b^{\frac{2}{3}} \to \infty$ as $b \to \infty$.

9. $\int_1^\infty \frac{dx}{x^{1.1}} = \lim_{b \to \infty} \int_1^b x^{-1.1} dx = \lim_{b \to \infty} \frac{x^{-.1}}{-.1} \Big|_1^b = \lim_{b \to \infty} (\frac{-10}{b^{.1}} + 10) = 0 + 10 = 10$

11. $\int_{-\infty}^0 \frac{2x\,dx}{x^2 + 1} = \lim_{a \to -\infty} \int_a^0 \frac{2x}{x^2 + 1} dx = \lim_{a \to -\infty} [\ln(x^2 + 1) \Big|_a^0] = \lim_{a \to -\infty} -\ln(a^2 + 1)$.

This limit does not exist -- the integral diverges.

13. $\int_{-\infty}^\infty x^2 dx = \lim_{a \to -\infty} \int_a^0 x^2 dx + \lim_{b \to \infty} \int_0^b x^2 dx = \lim_{a \to -\infty} (-\frac{a^3}{3}) + \lim_{b \to \infty} (\frac{b^3}{3})$.

These limits do not exist.

15. $\int_{-\infty}^\infty \frac{3x\,dx}{(3x^2 + 2)^3} = \lim_{a \to -\infty} \int_a^0 \frac{3x\,dx}{(3x^2 + 2)^3} + \lim_{b \to \infty} \int_0^b \frac{3x\,dx}{(3x^2 + 2)^3}$

$= \lim_{a \to -\infty} \frac{(3x^2 + 2)^{-2}}{-4} \Big|_a^0 + \lim_{b \to \infty} \frac{(3x + 2)^{-2}}{-4} \Big|_0^b$

$= \lim_{a \to -\infty} (-\frac{1}{16} + \frac{1}{4(3a^2 + 2)^2}) + \lim_{b \to \infty} (\frac{1}{4(3b^2 + 2)} + \frac{1}{16})$

$= -\frac{1}{16} + 0 + 0 + \frac{1}{16} = 0$

17. $\int_e^\infty \frac{dx}{x(\ln x)^2} = \lim_{b \to \infty} \int_e^b \frac{dx}{x(\ln x)^2} = \lim_{b \to \infty} \frac{-1}{\ln x} \Big|_e^b = \lim_{b \to \infty} (-\frac{1}{\ln b} + \frac{1}{\ln e}) = 0 + 1 = 1$

19. $\int_0^\infty 20{,}200 e^{-.06t} dt = \lim_{b \to \infty} \int_0^b 20{,}200 e^{-.06t} dt = \lim_{b \to \infty} \frac{20{,}200}{-.06} e^{-.06t} \Big|_0^b$

Chapter 8

$= \lim_{b \to \infty} (-336{,}666.67)(e^{-.06b} - e^0) = \$336{,}666.67$

21. See Example 2. $V = \int_0^\infty Re^{-rt}dt = \lim_{b \to \infty} \int_0^b Re^{-rt}dt = \lim_{b \to \infty} \left. \frac{Re^{-rt}}{-r} \right|_{t=0}^{t=b} = \lim_{b \to \infty} \left(\frac{Re^{-rb}}{-r} - \frac{Re^0}{-4} \right)$

$= \lim_{b \to \infty} \left(\frac{R}{r} - \frac{R}{re^{rb}} \right) = \frac{R}{r} - 0 = \frac{R}{r}.$

23. $A = \int_0^\infty 200e^{-.05t}dt = \lim_{b \to \infty} \int_0^b 200e^{-.05t}dt = \lim_{b \to \infty} \left. \frac{200e^{-.05t}}{-.05} \right|_0^b = \lim_{b \to \infty} \left(\frac{-4000}{e^{.05b}} + 4000 \right)$

$= 0 + 4000 = 4000$ millirems

25. Area $= \int_6^\infty \frac{2}{(x-4)^3}dx = \lim_{b \to \infty} \int_6^b \frac{2}{(x-4)^3}dx = \lim_{b \to \infty} \left. \frac{-1}{(x-4)^2} \right|_6^b$

$= \lim_{b \to \infty} \left[\frac{-1}{(b-4)^2} + \frac{1}{4} \right] = 0 + \frac{1}{4} = \frac{1}{4}$

27. $\int_{50}^\infty .01e^{-.01t}dt = \lim_{b \to \infty} \int_{50}^b .01e^{-.01t}dt = \lim_{b \to \infty} \left. -e^{-.01t} \right|_{50}^b = \lim_{B \to \infty} \left(\frac{-1}{e^{(.01)b}} + \frac{1}{e^{.5}} \right) = 0 + \frac{1}{e^{.5}}$

≈ 0.6065

29. $\int_{12}^\infty .05e^{-\frac{x}{20}}dx = \lim_{b \to \infty} \int_{12}^b .05e^{-\frac{x}{20}}dx = \lim_{b \to \infty} \left. -e^{-\frac{x}{20}} \right|_{12}^b = \lim_{b \to \infty} \left(\frac{-1}{e^{\frac{b}{20}}} + \frac{1}{e^{.6}} \right)$

$= 0 + \frac{1}{e^{.6}} \approx 0.5488$

8.4 CHAPTER REVIEW, PAGES 253 - 254

1. a. $\int_1^{20} \sqrt[3]{x+7}\, dx = \left. \frac{3}{8}(x+7)^{\frac{4}{3}} \right|_1^{20} = \frac{3}{8}\left(47^{\frac{4}{3}} - 8^{\frac{4}{3}} \right) \approx 49$

   b. $\int_0^{10} e^{-.02x}dx = \left. -50e^{-.02x} \right|_0^{10} = -50(e^{-.2} - 1) \approx 9$

3. $\dfrac{\int_0^3 x\sqrt{25-x^2}\, dx}{3-0} = \dfrac{\left. -\frac{1}{3}(25-x^2)^{\frac{3}{2}} \right|_0^3}{3} = -\frac{1}{9}[16^{\frac{3}{2}} - 25^{\frac{3}{2}}] = -\frac{1}{9}[64 - 125] = \frac{61}{9}$

Chapter 8

5. Answers vary.

7. Refer to Table 8.3 in Section 8.2 for b and c. **a.** Since the standard normal curve is symmetric about zero and has a total area of 1, $x > 0$ is half the area or $\frac{1}{2}$.
b. Find -0.1 in the left-hand column. The first number to the right is .4602, so the area is .4602. **c.** Find -.1 and .2 in the left-hand column. The area for $-.1 \leq x \leq .2$ = (area for $x \leq .2$) - (area for $x \leq -.1$) = .5793 - .4602 = .1191.

9. $\int_{1}^{\infty} .5e^{-.5x}dx = \lim_{b \to \infty} \int_{1}^{b} .5e^{-.5x}dx = \lim_{b \to \infty}(-e^{-.5b} + e^{-.5}) = 0 + e^{-.5} = e^{-.5}$

11. $\int_{-\infty}^{\infty} xe^{-x^2}dx = \int_{-\infty}^{0} xe^{-x^2}dx + \int_{0}^{\infty} xe^{-x^2}dx = \lim_{a \to -\infty} \int_{a}^{0} xe^{-x^2}dx + \lim_{b \to \infty} \int_{0}^{b} xe^{-x^2}dx$

$= \lim_{a \to -\infty} -\frac{1}{2}e^{-x^2}\Big|_{a}^{0} + \lim_{b \to \infty} -\frac{1}{2}e^{-x^2}\Big|_{0}^{b} = \lim_{a \to -\infty}(-\frac{1}{2} + \frac{1}{2e^{-a^2}}) + \lim_{b \to \infty}(-\frac{1}{2e^{-b^2}} + \frac{1}{2})$

$= (-\frac{1}{2} + 0) + (0 + \frac{1}{2}) = 0$

13. $\int_{0}^{5} 200\sqrt{9-t}\, dt = \frac{-400(9-t)^{\frac{3}{2}}}{3}\Big|_{0}^{5} = \frac{-400}{3}(4^{\frac{3}{2}} - 9^{\frac{3}{2}}) = \frac{-400}{3}(8-27) \approx \$2{,}533.33$

15. $\int_{0}^{\infty} 84{,}000e^{-.10t}dt = \lim_{b \to \infty} \int_{0}^{b} 84{,}000e^{-.10t}dt = \lim_{b \to \infty}(-840{,}000e^{-.10t}\Big|_{0}^{b})$

$= \lim_{b \to \infty}(\frac{-840{,}000}{e^{.10b}} + 840{,}000) = 0 + 840{,}000 = \$840{,}000$

8.5 CUMULATIVE REVIEW CHAPTERS 5-8, PAGES 254 - 256

1. Using the $\boxed{e^x}$ key on a calculator, a few values are:
$x = -2, y = e^{-2} \approx .14$
$x = -1, y = e^{-1} \approx .37$
$x = 0, y = e^{0} = 1$
$x = 1, y = e^{1} \approx 2.72$
$x = 2, y = e^{2} \approx 7.39$
The graph is shown on the right.

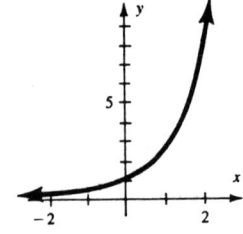

3. $\log(\frac{x-5}{6}) = 2$ is equivalent to $10^2 = \frac{x-5}{6}$ => $600 = x - 5$ => $605 = x$

Chapter 8

5. $10^{-x} = .5$ is equivalent to $\log(.5) = -x$. Using a calculator with a $\boxed{\log}$ key gives $-.301 \approx -x$ so $x \approx .301$.

7. **a.** 0 **b.** $b - a$ **c.** $-\int_b^a f(x)dx$ **d.** $\int_a^b f(x)dx$ **e.** $k\int_a^b f(x)dx$ **f.** $\int_a^b f(x)dx \pm \int_a^b g(x)dx$

   g. $u(x)v(x) \Big|_{x=a}^{x=b} - \int_{x=a}^{x=b} vdu = u(b)v(b) - u(a)v(a) - \int_{x=a}^{x=b} vdu$ where u and v are functions of x.

9. **a.** $\int (5x^4 + 3x^2 + 5)dx = 5\int x^4 dx + 3\int x^2 dx + 5\int dx = 5\frac{x^5}{5} + 3\frac{x^3}{3} + 5x + C = x^5 + x^3 + 5x + C$

   b. Let $u = 2x$, then $\frac{du}{2} = dx$. $\int e^{2x}dx = \int e^u \frac{du}{2} = \frac{1}{2}\int e^u du = \frac{1}{2}e^u + C = \frac{1}{2}e^{2x} + C$

11. Let $u = 5x$, then $\frac{du}{5} = dx$. $\int \ln^4 5x\, dx = \frac{1}{5}\int \ln^4 u\, du$. Using Formula 34 in the Brief Table of Integrals, $\frac{1}{5}\int \ln^4 u\, du = \frac{1}{5}[u\ln^4 u - 4\int \ln^3 u\, du] = \frac{u\ln^4 u}{5} - \frac{4}{5}\int \ln^3 u\, du$

    $= \frac{u\ln^4 u}{5} - \frac{4}{5}[u\ln^3 u - 3\int \ln^2 u\, du] = \frac{u\ln^4 u}{5} - \frac{4u\ln^3 u}{5} + \frac{12}{5}\int \ln^2 u\, du$.

    Using Formula 33 $= \frac{u\ln^4 u}{5} - \frac{4u\ln^3 u}{5} + \frac{12}{5}[u\ln^2 u - 2u\ln u + 2u] + C$

    Substitute $u = 5x$ $= x\ln^4 5x - 4x\ln^3 5x + 12x\ln^2 5x - 24x\ln 5x + 24x + C$

13. Let $u = 4x^2 - 3x + 2$, then $\frac{du}{dx} = 8x - 3$.

    $\int \frac{8x - 3}{4x^2 - 3x + 2} dx = \int \frac{du}{u} = \ln |u| + C = \ln |4x^2 - 3x + 2| + C$

15. $\int_{-\infty}^{\infty} \frac{xdx}{(x^2 + 1)^2} = \int_{-\infty}^{0} \frac{xdx}{(x^2 + 1)^2} + \int_{0}^{\infty} \frac{xdx}{(x^2 + 1)^2} = \lim_{a \to -\infty} \int_{a}^{0} \frac{xdx}{(x^2 + 1)^2} + \lim_{b \to \infty} \int_{0}^{b} \frac{xdx}{(x^2 + 1)^2}$

    $= \lim_{a \to -\infty} \frac{-1}{2(x^2 + 1)}\Big|_a^0 + \lim_{b \to \infty} \frac{-1}{2(x^2 + 1)}\Big|_0^b$

    $= \lim_{a \to -\infty} (-\frac{1}{2} + \frac{1}{2(a^2 + 1)}) + \lim_{b \to \infty}(-\frac{1}{2(b^2 + 1)} + \frac{1}{2}) = -\frac{1}{2} + \frac{1}{2} = 0$

17. First, the graphs intersect when $x^2 = 32 - x^2$. $2x^2 = 32 \Rightarrow x^2 = 16 \Rightarrow x = \pm 4$.

Chapter 8   192

Since the graph of $y = 32 - x^2$ is above the graph of $y - x^2$ between -4 and 4, the area is

$$\int_{-4}^{4} (32 - x^2 - x^2)dx = 32x - \frac{2x^3}{3} \Big|_{-4}^{4} = \frac{512}{3}$$

19. Set $48{,}000 = \int_{0}^{t} (5000x + 2000)dt = 2500x^2 + 2000x \Big|_{0}^{t}$

$\qquad = 2500t^2 + 2000t = 48{,}000$ (Divide by 500)

$\qquad 5t^2 + 4t = 96$

$\qquad 5t^2 + 4t - 96 = 0$

$\qquad (5t + 24)(t - 4) = 0$

$\qquad t = -\frac{24}{5}$ or $t = 4$.

Reject $-\frac{24}{5}$ since t can't be negative, giving us that $t = 4$ years.

21. We are given $A = 5000$, $P = 1000$, $i = .08$:

$5000 = 1000(1 + \frac{.08}{365})^{365t}$ Divide by 1000

$5 = (1.000219178)^{365t}$ Express with a logarithm

$\log_{1.000219178}(5) = 365t$

$\dfrac{\log 5}{\log 1.000219178} = 365t$

$7343.868 \approx 365t$

Dividing by 365 gives $t \approx 20.12$ years.

23. $\dfrac{\int_{0}^{30} (2450 - 2x^2)dx}{30 - 0} = \dfrac{2450x - \dfrac{2x^3}{3}\Big|_{0}^{30}}{30} = 1850$ trees

25. $\int_{60}^{\infty} .01e^{-\frac{x}{100}} dx = \lim_{b \to \infty} \int_{60}^{b} .01e^{-\frac{x}{100}} dx = \lim_{b \to \infty} (-e^{-\frac{x}{100}}\Big|_{60}^{b}) = \lim_{b \to \infty} (\dfrac{-1}{e^{\frac{b}{100}}} + e^{-\frac{60}{200}})$

$= 0 + e^{-\frac{60}{100}} \approx .5488$; about $\frac{1}{2}$ second response time.

8.6 CUMULATIVE REVIEW CHAPTERS 1-8, PAGES 256 - 257

1. Answers vary. The limit of a function f(x), as x approaches a, is the number L if the value of f(x) approaches L as x gets closer to a. The values of x get closer and closer (but never reach) a, the values of f(x) get closer and closer (and may or may not) reach L.

Chapter 8

3. Answers vary. Integration is the inverse process of differentiation; it is anti-differentiation. $\int f(x)dx = F(x)$ means that $F'(x) = f(x)$. That is when looking for the integral of $f(x)$ we are seeking a function, $F(x)$, whose derivative is $f(x)$. The definite integral is used to find areas, as well as to sum or total functional values. Finding areas under a curve can be used in calculating consumer and producer surplus, as well as certain probabilities if we are given a probability density function.

5. **a.** $10^{x-1} = 25$ is equivalent to $\log 25 = x - 1$. Adding 1 to both sides gives $x = \log(25) + 1$
   ≈ 2.3979.
   b. $e^{x-1} = 11$ is equivalent to $\ln 11 = x - 1$. Adding 1 to both sides gives $x = \ln(11) + 1$
   ≈ 3.3979.
   c. $8^{x-1} = 5$ is equivalent to $\log_8 5 = x - 1$. Adding 1 to both sides gives
   $x = \log_8(5) + 1 = \dfrac{\log 5}{\log 8} + 1 \approx 1.774$

7. $y = -3(1 - 2x)^{-1}$. $y' = 3(1 - 2x)^{-2} \cdot (1 - 2x) = 3(1 - 2x)^{-2}(-2) = \dfrac{-6}{(1-2x)^2}$

9. $\int (3x^2 - 2x)dx = 3\int x^2 dx - 2\int x dx = 3(\dfrac{x^3}{3}) - 2(\dfrac{x^2}{2}) + C = x^3 - x^2 + C$

11. $\int_1^3 5x^{-1}dx = \int_1^3 \dfrac{5}{x}dx = 5\ln x \Big|_1^3 - 5(\ln 3 - \ln 1) = 5\ln 3 \approx 5.493$

13. **a.** Break even when Revenue = Cost. Revenue = $R(x) = 130x =$ Cost = $C(x) = 5x + 8000$
   $\Rightarrow 130x = 5x + 8000 \Rightarrow 125x = 8000 \Rightarrow x = 64$ units.
   b. $R(64) = 130(64) = \$8320$

15. $P'(x) = 3x^2 + 10x - 8$. $P'(x) = 0 \Rightarrow x = \dfrac{2}{3}, -4$, but domain is $x \geq 0$.
   Set $P'(x) > 0$, $3x^2 + 10x - 8 > 0 \Rightarrow (3x - 2)(x + 4) > 0 \Rightarrow P'(x) > 0$
   for $x < -4$ or $x > \dfrac{2}{3}$. Since $P'(x) > 0$ for $x > \dfrac{2}{3}$, $P(x)$ is increasing for $x > \dfrac{2}{3}$. This means there can not be a maximum.

17. To find where the curves intersect solve $\begin{cases} y = 2x - 4 \\ y^2 = 4x \end{cases}$
   $y^2 = 4x$ is equivalent to $x = \dfrac{y^2}{4}$. Substitute into $y = 2x - 4$: $y = 2x \cdot 1 = 2\left(\dfrac{y^2}{4}\right) - 4 = \dfrac{y^2}{2} - 4$.
   Solve $y = \dfrac{y^2}{2} - 4$
   $2y = y^2 - 8$ \quad Multiply by 2
   $0 = y^2 - 2y - 8$

$0 = (y - 4)(y + 2)$

$y = 4$ or $y = -2$.   Now substitute into $x = \dfrac{y^2}{4}$, $x = 4$ or 1.

The graph of $y^2 = 4x$, or $y = \pm 2\sqrt{x}$, is above $y = 2x - 4$ for $1 \le x \le 4$.

The area is $\displaystyle\int_1^4 [2\sqrt{x} - (2x - 4)]dx = \int_1^4 (2x^{\frac{1}{2}} - 2x + 4)dx = \dfrac{4}{3}x^{\frac{3}{2}} - x^2 + 4x \Big|_1^4 = \dfrac{19}{3}$

19. $f'(x) = x^3 - 3x$. Setting this equal to zero and solving gives $x(x^2 - 3) = 0 \Rightarrow x = 0$ or $x = \pm\sqrt{3}$.

$f''(x) = 3x^2 - 3$. Using the Second Derivative Test, $f''(0) = -3$, $f''(\sqrt{3}) = 6$, and

$f''(-\sqrt{3}) = 6$, so the graph has local minimums at $x = \pm\sqrt{3}$ and a local maximum at $x = 0$.

Set $f''(x) = 3x^2 - 3 = 0 \Rightarrow x = \pm 1$. For $x < -1$ or $x > 1$, $f''(x) > 0$ so the graph is concave

up for $x < -1$ or $x > 1$. Similarly, the graph is concave down for $-1 < x < 1$ since $f''(x) < 0$ in

that interval. Thus, $x = \pm 1$ are where points of inflection occur.

Some key values are: $x = \pm\sqrt{3}$, $f(\pm\sqrt{3}) = -\dfrac{9}{4}$

$x = \pm 1$, $f(\pm 1) = -\dfrac{5}{4}$

$x = 0$, $f(0) = 0$

You should calculate more ordered pairs. The graph is shown on the right.

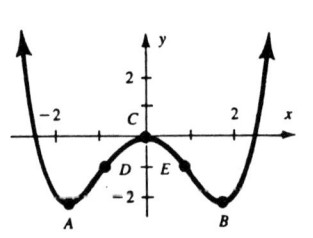

8.7 STUDENT'S TEST REVIEW AND ADDITIONAL PRACTICE
OBJECTIVES

The material of this chapter is reviewed in the following list of objectives. After each objective there are some practice questions. Answers to these problems immediately follow. Detailed solutions are given for every third problem. For a sample test select the first question of each set and check your answers. Additional practice is given by the other questions in each set. If you are having trouble with a particular type of problem, or if you want additional practice, look back at the indicated section in the test.

[8.1] Objective 1: *Find the total value of a given function.*

1. $f(x) = \sqrt[3]{2x + 5}$ on $[1, 40]$ 2. $f(x) = .25e^{-.12x}$ on $[0, 4]$

3. $f(x) = \dfrac{x}{2x - 3}$ on $[2, 4]$ 4. $f(x) = \ln x$ on $[1, 10]$

Objective 2: *Determine the total money flow.*

5. \$150,000 at 6% for 7 years   6. \$85,000 at 15% for 10 years

7. \$1.5 million at 20% for 5 years   8. \$3,900 at 8% for 30 years

Objective 3: *Find the average value for a given function.*

Chapter 8

9. $f(x) = 6x^2 - 2x + 5$ on $[-2, 1]$ 10. $f(x) = \dfrac{3x - 2}{x^2}$ on $[1, 4]$

11. $f(x) = xe^{x^2}$ on $[1, 3]$ 12. $f(x) = x\sqrt{9 - 2x^2}$ on $[0, 2]$

[8.2] Objective 4: *Describe the sample space for a given experiment. Define a random variable for the experiment and be able to tell if it is a discrete or a continuous random variable.*

13. A pair of dice is rolled and the sum of the top faces is recorded.
14. A coin is tossed three times and the sequence of heads and tails is recorded.
15. The number of minutes it takes a rat to make its way through a maze is recorded; compare with Problem 16.
16. The number of rats that can make their way through a maze is recorded; compare with Problem 15.

Objective 5: *Given a set of data, prepare a frequency distribution and draw a histogram.*

17. A pair of dice are rolled and the sum of the spots on the tops of the dice recorded as follows: 3, 2, 6, 5, 3, 8, 8, 7, 10, 9, 7, 5, 12, 9, 6, 8, 11, 11, 8, 7, 7, 7, 10, 7, 9, 7, 9, 6, 6, 9, 4, 4, 6, 3, 4, 10, 6, 9, 6, 11

18. Blane, Inc. a consulting firm was employed to perform an efficiency study at National City Bank. As part of the study, they found the number of times per day that people were waiting in line to be the following: 2 were waiting 20 times; 3 were waiting 15 times; 4, 7 times; 5, 5 times; 6, 2 times; 7, 1 time, and there were never more than 7 people in line at any time during the day.

19. Blane, Inc., the consulting firm described in Problem 18, also noted the transaction times (rounded to the nearest minute) for customers at National City Bank, as follows: 1 minute, 10 times; 2 minutes, 12 times; 3 minutes, 18 times; 4 minutes, 25 times; 5 minutes, 16 times; 6 minutes, 10 times; 7 minutes, 6 times; 8 minutes, 1 time; 9 minutes, none; and 10 minutes, 2 times. No transaction took more than 10 minutes.

20. Three coins are tossed onto a table and the following frequencies are noted:
0 heads, 17 times; 1 head, 59 times; 2 heads 56 times; and 3 heads 18 times.

Objective 6: *Determine whether a given function is a probability density function.*

21. $f(x) = \dfrac{3}{63} x^2$ on $[1, 4]$ 22. $f(x) = \dfrac{x^2}{3}$ on $[-1, 1]$

23. $f(x) = \dfrac{2}{3}$ on $[2, 5]$ 24. $f(x) = .1$ on $[0, 10]$

Objective 7: *Find a constant k in order to define a probability density function over a*

Chapter 8

given interval.

25. $f(x) = kx^{\frac{1}{4}}$ on [0, 16] **26.** $f(x) = k$ on [-4, 0]

27. $f(x) = 5kx$ on [1, 10] **28.** $f(x) = k\sqrt{2x}$ on [2, 8]

Objective 8: *Find the area under the standard normal curve.*

29. $x < 0$ **30.** $x > -.11$

31. $-1.03 < x < 1.59$ **32.** $-.5 \le x \le 1.5$

[8.3] Objective 6: *Evaluate improper integrals that converge.*

33. $\displaystyle\int_0^\infty .5e^{-.5x}\,dx$ **34.** $\displaystyle\int_{-\infty}^1 \frac{x\,dx}{x^2 + 1}$

35. $\displaystyle\int_{-\infty}^\infty e^{-x^2}\,dx$ **36.** $\displaystyle\int_e^\infty \ln x\,dx$

[8.1 - 8.3] Objective 7: *Solve applied problems based on the preceding objectives. For specific examples of the types of applications look at the list of applications in this chapter on page 229.*

37. Suppose that the demand for oil (in billions of barrels) conforms to the formula

$$R(t) = 32.4e^{.048t}$$

for t years after 1985. If the total oil still left in the earth is estimated to be 670 billion barrels, estimate the length of time before all available oil is consumed if the rate does not change.

38. The graph shows the 10-year treasury bond rate for the years 1980-1985 (from the <u>Wall Street Journal</u>, 2/26/86). What is the average bond rate over the years 1980-1985?

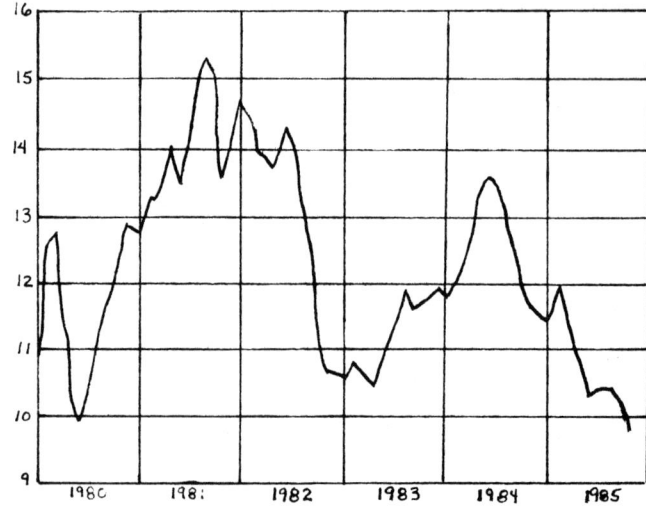

Chapter 8

39. If the total depreciation at the end of t years is given f(t) and the depreciation rate is $f'(t) = 200\sqrt{10 - x}$, find the total depreciation for the first five years.

40. If a light bulb is normally distributed with a mean life of 250 hours and a standard deviation of 25 hours, find the following probabilities.
    a. P(x > 250)
    b. P(x < 220)
    c. P(200 < x < 300)
    d. P(220 ≤ x ≤ 320)

41. Suppose that, for a certain exam, a teacher grades on a curve. It is known that the mean is 50 and the standard deviation is 5. There are 45 students in the class.
    a. How many students should receive an C?
    b. How many students should receive an A?
    c. What score would be necessary to obtain an A?
    d. If an exam paper is selected at random, what is the probability that it will be a failing paper?

42. Suppose that the current interest rate is 12% and the British Embassy in Washington, D. C. has an inderminant lease paying $72,000 per year. What is the capital value of this lease?

ANSWERS TO STUDENT'S TEST REVIEW AND PRACTICE QUESTIONS

1. $\frac{3}{8}(85^{\frac{4}{3}} - 7^{\frac{4}{3}}) \approx 135.13$ 2. $\frac{25}{12}(1 - e^{-.48}) \approx .7942$

3. $\int_{2}^{4} \frac{x}{2x - 3} dx = \frac{1}{2}x + \frac{3}{4}\ln|2x - 3| \Big|_{2}^{4} = 1 + \frac{3}{4}\ln 5 \approx 2.207$

4. $10 \ln 10 - 9 \approx 14.026$ 5. $1,304,903.89

6. $\int_{0}^{10} 85{,}000 e^{.15t} dt \approx 566{,}667 e^{.15t} \Big|_{0}^{10} = \$1{,}972{,}957$ 7. $12,887,114   8. $488,630

9. $\frac{1}{1 - -2}\int_{-2}^{1}(6x^2 - 2x + 5)dx = \frac{1}{3}(2x^3 - x^2 + 5x)\Big|_{-2}^{1} = \frac{1}{3}(36) = 12$

10. $\ln 4 - \frac{1}{2} \approx .8863$ 11. $\frac{1}{4}(e^9 - e)$

12. $\frac{1}{2 - 0}\int_{0}^{2} x\sqrt{9 - 2x^2}\, dx = \frac{1}{2}\left[-\frac{1}{6}(9 - 2x^2)^{\frac{3}{2}}\right]\Big|_{0}^{2} = -\frac{1}{12}(1^{\frac{3}{2}} - 9^{\frac{3}{2}}) = \frac{13}{6}$

13. {1, 2, 3, 5, ..., 12}; x = sum of the top faces; discrete.
14. {HHH, HHT, HTH, HTT, THH, THT, TTH, TTT}; x = number of heads; discrete.

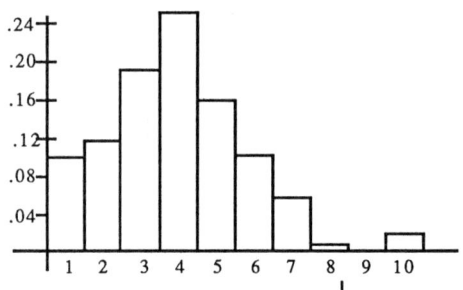

Chapter 8   198

15. $\{x \mid x > 0\}$; x = number of minutes a rat takes to go through a maze; continuous.

16. $\{0, 1, 2, 3, 4, \ldots\}$; x = number of rats that make their way through a maze; discrete.

17.
Roll	2	3	4	5	6	7	8	9	10	11	12
Frequency	1	3	3	2	7	7	4	6	3	3	1
Rel. Frequency	$\frac{1}{40}$	$\frac{3}{40}$	$\frac{3}{40}$	$\frac{2}{40}$	$\frac{7}{40}$	$\frac{7}{40}$	$\frac{4}{40}$	$\frac{6}{40}$	$\frac{3}{40}$	$\frac{3}{40}$	$\frac{1}{40}$

18.
Number of People	2	3	4	5	6	7
Frequency	20	15	7	5	2	1
Rel. Frequency	$\frac{20}{50}$	$\frac{15}{50}$	$\frac{7}{50}$	$\frac{5}{50}$	$\frac{2}{50}$	$\frac{1}{50}$

19.
Number of Minutes	1	2	3	4	5	6	7	8	9	10
Frequency	10	12	18	25	16	10	6	1	0	2
Rel. Frequency	$\frac{10}{100}$	$\frac{12}{100}$	$\frac{18}{100}$	$\frac{25}{100}$	$\frac{16}{100}$	$\frac{10}{100}$	$\frac{6}{100}$	$\frac{1}{100}$	$\frac{0}{100}$	$\frac{2}{100}$

20.
Number of Minutes	0	1	2	3
Frequency	17	59	56	18
Rel. Frequency	$\frac{17}{150}$	$\frac{59}{150}$	$\frac{56}{150}$	$\frac{18}{150}$

21. $\int_1^4 \frac{3}{63}x^2 = \frac{x^3}{63}\Big|_1^4 = \frac{64}{63} - \frac{1}{63} = 1.$ This is a probability density function.

Chapter 8

22. $\int_{-1}^{1} \frac{x^2}{3} dx \neq 1$. Not a probability density function.

23. $\int_{2}^{5} \frac{2}{3} dx \neq 1$. Not a probability density function.

24. $\int_{0}^{10} (.1) dx = 1$. This is a probability density function.

25. $k = \frac{5}{128}$ **26.** $k = \frac{1}{4}$ **27.** $k = \frac{2}{495}$ **28.** $k = \frac{3}{56}$ **29.** 0.5

30. The area for $x < -.11$ is .4562, so the area for $x > -.11$ is $1 - .4562 = .5438$

31. .7926 **32.** .6247

33. $\int_{0}^{\infty} .5e^{-.5x} dx = \lim_{b \to \infty} \int_{0}^{b} .5e^{-.5x} dx = \lim_{b \to \infty} (-3^{-.5x} \Big|_{0}^{b}) = \lim_{b \to \infty} (-e^{-.5b} + 1) = 1$

34. Does not exist **35.** 0

36. $\int_{3}^{\infty} \ln x \, dx = \lim_{b \to \infty} \int_{e}^{b} \ln x \, dx = \lim_{b \to \infty} (x \ln x - x \Big|_{e}^{b}) = \lim_{b \to \infty} [(b \ln b - b) - (e \ln e - e)]$.

This limit does not exist.

37. 14.36 years **38.** Answers vary. Approximation should be between 11.7 and 12.2.

39. $\int_{0}^{5} 200\sqrt{10 - x} \, dx = -\frac{400}{3} (10 - x)^{\frac{3}{2}} \Big|_{0}^{5} \approx \$2,725.66$

40. a. .5 **b.** .1151 **c.** .9544 **d.** .8823

41. a. 31 **b.** 1 **c.** 60 **d.** 2.3%

42. (See Example 2, Section 8.3.)

Capital Value $= \int_{0}^{\infty} 72,000e^{-.12t} dt = \lim_{b \to \infty} \int_{0}^{b} 72,000e^{-.12t} dt = 72,000 \lim_{b \to \infty} \frac{25}{3} (e^{-.12b} - 1)$

$= \$600,000$

CHAPTER 9
FUNCTIONS OF SEVERAL VARIABLES

9.1 THREE DIMENSIONAL COORDINATE SYSTEM, PAGES 265 - 266

1. The area of a rectangle with dimensions 15 by 35 is k(15, 35) = (15)(35) = 525 square units.

3. The volume of a box 3 by 5 by 8 is V(3, 5, 8) = (3)(5)(8) = 120 cubic units.

5. The simple interest from a $500 investment at 5% for 3 years is I(500, .05, 3)
   = 500(1 + (.05)(3)) = 500(1.15) = $575.

7. The future value of a $2500 investment at 12% compounded quarterly for 6 years is
   $A(2500, .12, 6, 4) = 2500(1 + \frac{.12}{4})^{(4)(6)} = 2500(1.03)^{24} \approx 2500(2.0327941) = \$5081.99$

9. The future value of a $110,000 investment at 9% compounded monthly for 30 years is
   $A(110000, .09, 30, 12) = 110{,}000(1 + \frac{.09}{12})^{(30)(12)} = 110{,}000(1.0075)^{360}$
   $\approx 110{,}000(14.730577) = \$1{,}620{,}363$

11. $f(2, 3) = 2^2 - 2(2)(3) + 3^2 = 1$ 13. $f(-2, 5) = (-2)^2 - 2(-2)(5) + 5^2 = 49$

15. $g(2, 1) = \frac{2(2) - 4(1)}{2^2 + 1^2} = 0$ 17. $g(5, -3) = \frac{2(5) - 4(-3)}{5^2 + (-3)^2} = \frac{11}{17}$

19. $h(0, 5) = \frac{e^{(0)(5)}}{\sqrt{0^2 + 5^2}} = \frac{1}{5}$ or 0.2

21. $h(-2, -3) = \frac{e^{(-2)(-3)}}{\sqrt{(-2)^2 + (-3)^2}} = \frac{e^6}{\sqrt{13}} \approx \frac{403.42879}{3.6055513} = 111.89$

23. **a.** Starting at the origin, move (out) 1 unit along the x-axis, 2 units (right) in the direction of the y-axis, then (up) 3 units in the direction of the z-axis. **b.** Starting at the origin, move (in) 3 units along the negative x-axis, 2 units (right) in the direction of the y-axis, then (up) 4 units in the direction of the z-axis. **c.** Starting at the origin, move (out) 1 unit along the x-axis, 4 units (left) in the direction of the negative y-axis, then (up) 3 units in the direction of the z-axis. **d.** Starting at the origin, move (in) 5 units along the negative x-axis, 9 units (left) in the direction of the negative y-axis, then 8 units (down) in the direction of the negative z-axis.

Chapter 9   201

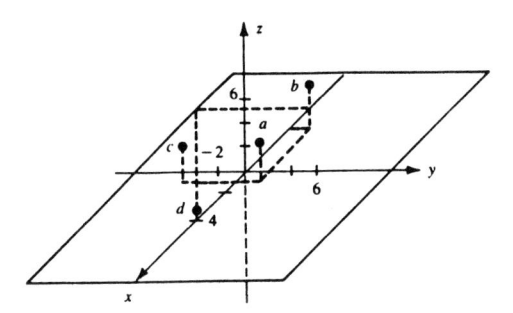

25. a. Starting at the origin, move (out) 10 units along the x-axis, 5 units (right) in the direction of the y-axis, then 20 units (up) in the direction of the z-axis.   **b.** Starting at the origin, move (out) 5 units along the x-axis, 15 units (left) in the direction of the negative y-axis, then 5 units (down) in the direction of the negative z-axis.   **c.** Starting at the origin, move (out) 3 units along the x-axis, 2 units (right) in the direction of the y-axis, then 4 units (down) in the direction of the negative z-axis.   **d.** Starting at the origin, move (in) 5 units along the negative x-axis, 1 unit (left) in the direction of the negative y-axis, then 3 units up.

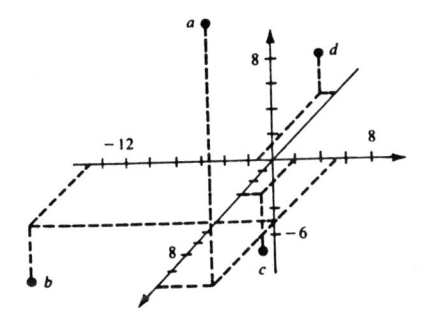

Chapter 9

27. Since all variables are first degree, this is a plane. To graph part of this surface, set y and z equal to zero and solve => x = 10. The x-intercept is 10. Set x and z equal to zero and solve => 2y = 10, so y = 5. The y-intercept is 5. Similarly, set x and y equal to zero to get the z-intercept, z = 2. Connecting these intercepts with line segments gives a (triangular) section of the plane in the first octant (pictured).

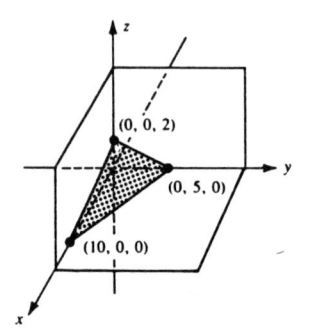

29. Since all variables are first degree, this is a plane. To graph part of this surface, set y and z equal to zero and solve => 3x = 12 => x = 4. The x-intercept is 4. Set x and z equal to zero and solve => -2y = 12 => y = -6. The y-intercept is -6. Similarly, set x and y equal to zero to get the z-intercept, z = -12. Connecting these intercepts with line segments gives a (triangular) section of the plane (pictured).

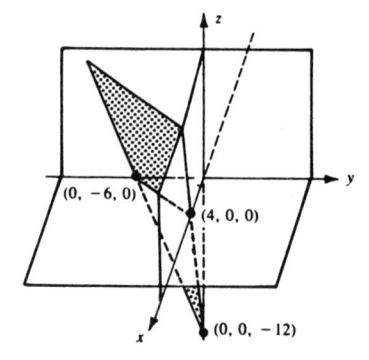

Chapter 9

31. This has the form $z = \dfrac{x^2}{a^2} + \dfrac{y^2}{b^2}$, where $a = 2$ and $b = 3$, which is an elliptic paraboloid. The trace in the xy plane is a point $(0, 0, 0)$, and in planes parallel to the xy-plane the traces are ellipses. The trace in the yz-plane is the parabola $z = \dfrac{y^2}{9}$, and the trace in the xz-plane is the parabola $z = \dfrac{x^2}{4}$.

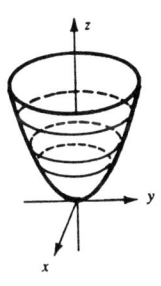

33. This has the form $\dfrac{x^2}{a^2} + \dfrac{y^2}{b^2} + \dfrac{z^2}{c^2} = 1$ (where $a = 3, b = 2, c = 5$) which is an ellipsoid. The trace in the xy-plane is the ellipse $\dfrac{x^2}{9} + \dfrac{y^2}{4} = 1$. The trace in the yz-plane is the ellipse $\dfrac{y^2}{4} + \dfrac{z^2}{25} = 1$. The trace in the xz-plane is the ellipse $\dfrac{x^2}{9} + \dfrac{z^2}{25} = 1$.

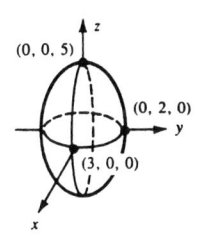

35. This has the form $z = \dfrac{x^2}{a^2} + \dfrac{y^2}{b^2}$ (where a and b are 1), which is an elliptic paraboloid. However, since a and b are equal, it's a circular paraboloid. The traces in planes parallel to the xy-plane are circles. The trace in the yz-plane is the parabola $z = y^2$. The trace in the xz-plane is the parabola $z = x^2$.

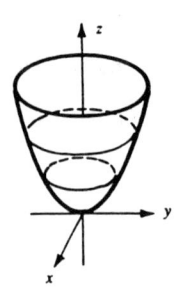

Chapter 9   204

37. This has the form $\dfrac{x^2}{a^2} + \dfrac{y^2}{b^2} - \dfrac{z^2}{c^2} = 1$ which is a hyperboloid of two sheets. If we set x and y equal to zero, we see the z-intercepts are ± 2. The trace in the xz-plane is the hyperbola $\dfrac{x^2}{9} - \dfrac{z^2}{4} = 1$. The trace in the yz-plane is the hyperbola $\dfrac{y^2}{1} - \dfrac{z^2}{4} = 1$. The trace of plane parallel to the xy-plane ($|z| > 2$) are ellipses.

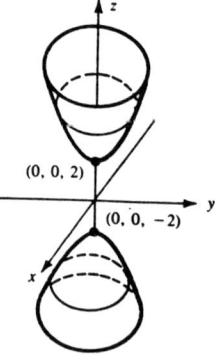

39. The trace in all planes parallel to the xy-plane is the circle $x^2 + y^2 = 36$. The surface is a circular cylinder.

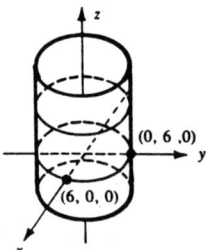

Chapter 9

41. The trace in all planes parallel to the yz-plane is the circle $y^2 + z^2 = 20$. The surface is a circular cylinder.

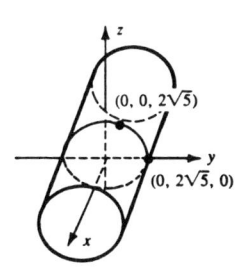

43. The cost for the first type is $2500 + 800x$. The cost for the second type is $1200 + 550y$. The total cost is the sum of these two, $C(x, y) = (2500 + 800x) + (1200 + 550y)$
$= 3700 + 800x + 550y$.
   a. $C(10, 15) = 3700 + 800(10) + 550(15) = \$19{,}950$
   b. $C(5, 25) = 3700 + 800(5) + 550(25) = \$21{,}450$
   c. $C(15, 10) = 3700 + 800(15) + 550(10) = \$21{,}200$
   d. $C(0, 30) = 3700 + 800(0) + 550(30) = \$20{,}200$

45. The ceiling material will cost $(xy)(2)$, the floor material will cost $(xy)(1.25)$, and the four walls will cost $2(xz)(.75) + 2(yz)(.75)$. The total cost is $C(x, y, z)$
$= 2xy + 1.25xy + 1.5xz + 1.5yz$.

47. a. $F(3.1, .002) = \dfrac{.002(3.1)}{.002^4} \approx 3.875 \times 10^8 \text{ ml}$

   b. $F(15.3, .001) = \dfrac{.002(15.3)}{.001^4} \approx 3.06 \times 10^{10} \text{ ml}$

   c. $F(6, .005) = \dfrac{.002(6)}{.005^4} \approx 1.92 \times 10^7 \text{ ml}$

9.2 PARTIAL DERIVATIVES, PAGES 270 - 271

1. $f_x = (5x^2)' - y^4(3x^3)' + \dfrac{\partial}{\partial x}(2y^3 - 15) = 10x - y^4(9x^2) + 0 = 10x - 9x^2y^4$

3. $g_y = 5(4x - 3y)^4 \cdot \dfrac{\partial}{\partial y}(4x - 3y) = 5(4x - 3y)^4(-3) = -15(4x - 3y)^4$

5. See Problem 1. $f_x = 10x - 9x^2y^4 \Rightarrow f_x(1, 2) = 10(1) - 9(1)^2(2)^4 = -134$

7. See Problem 3. $g_y = -15(4x - 3y)^4 \Rightarrow g_y(3, -1) = -15(4(3) - 3(-1))^4 = -759{,}375$

9. $\dfrac{\partial^2 z}{\partial x \partial y} = \dfrac{\partial}{\partial x}(\dfrac{\partial z}{\partial y}) = \dfrac{\partial}{\partial x}(f_y) = \dfrac{\partial}{\partial x}(-12x^3y^3 + 6y^2) = -36x^2y^3 + 0 = -36x^2y^3$

11. $\dfrac{\partial^2 w}{\partial y \partial x} = \dfrac{\partial}{\partial y}(\dfrac{\partial w}{\partial x}) = \dfrac{\partial}{\partial y}(g_x) = \dfrac{\partial}{\partial y}(5(4x-3y)^4 \cdot (4)) = \dfrac{\partial}{\partial y}(20(4x-3y)^4)$

$= 80(4x - 3y)^3 \cdot (0 - 3) = -240(4x - 3y)^3$

13. $f_{xx} = \dfrac{\partial}{\partial x}(f_x) = \dfrac{\partial}{\partial x}(10x - 9x^2y^4) = 10 - 18xy^4 \qquad f_{xx}(0, 2) = 10 - 180(0)(2)^4 = 10$

15. First, $f_y = -12x^3y^3 + 6y^2$, so $f_{yx} = -36x^2y^3 \qquad f_{yx}(-1, 0) = -36(-1)^2(0)^3 = 0$

17. First, $g_x = 5(4x - 3y)^4 \cdot (4) = 20(4x - 3y)^4$, so $g_{xx} = 80(4x - 3y)^3 \cdot (4) = 320(4x - 3y)^3$

$g_{xx}(0, 2) = 320(4(0) - 3(2))^3 = -69{,}120$

19. First, $g_y = -15(4x - 3y)^4$ (See Problem 1.) $g_{yx} = -60(4x - 3y)^3 \cdot \dfrac{\partial}{\partial x}(4x - 3y)$

$= -60(4x - 3y)^3(4) = -240(4x - 3y)^3$ So $g_{yx}(-1, 0) = -240(4(-1) - 3(0))^3 = 15{,}360$

21. $f_x = e^{3x + 2y} \cdot \dfrac{\partial}{\partial x}(3x + 2y) = 3e^{3x + 2y}$

23. First write $g(x, y)$ as $g(x, y) = (x^2 - 3y^2)^{\frac{1}{2}}$

$g_y = \dfrac{1}{2}(x^2 - 3y^2)^{-\frac{1}{2}} \cdot \dfrac{\partial}{\partial y}(x^2 - 3y^2) = \dfrac{1}{2}(x^2 - 3y^2)^{-\frac{1}{2}}(-6y) = \dfrac{-3y}{\sqrt{x^2 - 3y^2}}$

25. $f_x = 3e^{3x + 2y}$ (See Problem 21), so $f_x(1, 2) = 3e^{3(1) + 2(2)} = 3e^7$

27. $g_y = \dfrac{-3y}{\sqrt{x^2 - 3y^2}}$ (See Problem 23), so $g_y(3, -2) = \dfrac{-3(-2)}{\sqrt{3^2 - 3(-2)^2}} = \dfrac{6}{\sqrt{-3}}$ which is not a

real number. $g_y(3, -2)$ does not exist.

29. $\dfrac{\partial^2 z}{\partial x \partial y} = \dfrac{\partial}{\partial x}(\dfrac{\partial z}{\partial y}) = \dfrac{\partial}{\partial x}(2e^{3x + 2y}) = 2\dfrac{\partial}{\partial x}(e^{3x + 2y}) = 2(3e^{3x + 2y}) = 6e^{3x + 2y}$

31. First, $w = g(x, y) = (x^2 - 3y^2)^{\frac{1}{2}}$, so $\dfrac{\partial w}{\partial x} = g_x = \dfrac{1}{2}(x^2 - 3y^2)^{-\frac{1}{2}}\dfrac{\partial}{\partial x}(x^2 - 3y^2)$

$= \dfrac{1}{2}(x^2 - 3y^2)^{-\frac{1}{2}}(2x) = x(x^2 - 3y^2)^{-\frac{1}{2}}. \quad \dfrac{\partial^2 w}{\partial y \partial x} = \dfrac{\partial}{\partial y}(\dfrac{\partial w}{\partial x}) = \dfrac{\partial}{\partial y}(x(x^2 - 3y^2)^{-\frac{1}{2}}).$

$= x\dfrac{\partial}{\partial y}[(x^2 - 3y^2)^{-\frac{1}{2}}] = x[-\dfrac{1}{2}(x^2 - 3y^2)^{-\frac{3}{2}} \cdot \dfrac{\partial}{\partial y}(x^2 - 3y^2)] = x[3y(x^2 - 3y^2)^{-\frac{3}{2}}]$

Chapter 9

$$= \frac{3xy}{(\sqrt{x^2 - 3y^2})^3}$$

33. $f_{xx} = \frac{\partial}{\partial x}(f_x) = \frac{\partial}{\partial x}(3e^{3x + 2y})$ (See Problem 21) $= 3e^{3x + 2y} \cdot \frac{\partial}{\partial x}(3x + 2y) = 9e^{3x + 2y}$

    So $f_{xx}(0, 2) = 9e^{3(0) + 2(2)} = 9e^4$

35. $f_{yx} = \frac{\partial^2 z}{\partial x \partial y} = 6e^{3x + 2y}$ (See Problem 29) So $f_{yx}(-1, 0) = 6e^{3(-1) + 2(0)} = 6e^{-3} = \frac{6}{e^3}$

37. $g_{xx} = 2x\left[-\frac{1}{2}(x^2 - 3y^2)^{-\frac{3}{2}}(2x)\right] + 2(x^2 - 3y^2)^{-\frac{1}{2}}$ So $g_{xx}(2, 0) = -8 \cdot \frac{1}{8} + 2 \cdot \frac{1}{2} = 0$

39. $g_{yx} = g_{xy} = \frac{\partial^2 w}{\partial y \partial x} = \frac{3xy}{(\sqrt{x^2 - 3y^2})^3}$ (See Problem 31) So $g_{yx}(-1, 0)$

    $= \frac{3(-1)(0)}{(\sqrt{(-1)^2 - 3(0)^2})^3} = 0$

41. For f_x, treat y and λ as constants: $f_x = 1 + 2y + \lambda(y - 0) = 1 + 2y + \lambda y$. For f_y, treat x and λ as constants: $f_y = 0 + 2x + \lambda(x - 0) = 2x + \lambda x$. For f_λ treat x and y constants: $f_\lambda = 0 + 0 + (xy - 10) = xy - 10$. For f_x, treat y and λ as constants: f_x

43. For f_x, treat y and λ as constants: $f_x = 2x + 0 - \lambda(3 + 0 - 0) = 2x - 3\lambda$. For f_y, treat x and y as constants: $f_y = 2y - 2\lambda$. Finally; $f_\lambda = 0 + 0 - (3x + 2y - 6) = -3x - 2y + 6$.

45. For $\frac{\partial f}{\partial b}$, treat m as a constant: $\frac{\partial f}{\partial b} = 2(10m + 5b) \cdot \frac{\partial}{\partial b}(10m + 5b) + 0 + 1$

    $= 10(10m + 5b) + 1 = 100m + 50b + 1$. For $\frac{\partial f}{\partial m}$, treat b as a constant:

    $\frac{\partial f}{\partial m} = 2(10m + 5b) \cdot \frac{\partial}{\partial m}(10m + 5b) + 2 + 0 = 20(10m + 5b) + 2 = 200m + 100b + 2$.

47. It might be easier to re-write f(b, m):

    $f(b, m) = (m + b + 1)^2 + 4(m + b + 1)^2 + 9(m + b + 1)^2 = 14(m + b + 1)^2$

    $\frac{\partial f}{\partial b} = 28(m + b + 1) \cdot \frac{\partial}{\partial b}(m + b + 1) = 28(m + b + 1)$.

    $\frac{\partial f}{\partial m} = 28(m + b + 1) \cdot \frac{\partial}{\partial m}(m + b + 1) = 28(m + b + 1)$

49. $\frac{\partial P}{\partial a} = 2p + 0 - 0 - .2ap - 0 = 2p - .2ap$. Since we hold p constant this means at a fixed

price the rate of change of profit per unit change in advertising spent is 2p - .2ap.
$\frac{\partial P}{\partial p} = 2a + 50 - 20p - .1a^2$. Since we hold a constant, this means at a fixed level of advertising the rate of change of profit per unit change in price is $2a + 50 - 20p - .1a^2$.

51. P(x, y) = R(x, y) - C(x, y) = (1500x + 900y) - (3700 + 2500x + 550y)
    = -1000x + 350y - 3700.
    $P_x = -1000$, $P_x(5, 10) = -1000$. This means the rate of decrease in profit is $1000 per unit increase in x, assuming y stays constant. $P_y = 350$, $P_y(5, 10) = 350$. This means the rate of increase in profit is $350 per unit increase in y, assuming x stays constant.

53. For most people answers will vary between 900 and 1700 square feet.

55. $A_r = \frac{\partial A}{\partial r} = 100t(1+r)^{t-1}$. $A_r(r, 5) = 100(5)(1+r)^{5-1} = 500(1+r)^4$.

57. $P = 25(12)\left[\dfrac{1 - (1+\frac{r}{12})^{-12t}}{r}\right]$. Using the Quotient Rule,

$P_r = 300\left[\dfrac{r \cdot \frac{\partial}{\partial r}(1 - (1+\frac{r}{12})^{-12t}) - [1 - (1+\frac{r}{12})^{-12t}]\frac{\partial}{\partial r}(r)}{r^2}\right]$

$= 300\left[\dfrac{r[12t(1+\frac{r}{12})^{-12t-1}(\frac{1}{12})] - [1 - (1+\frac{r}{12})^{-12t}]}{r^2}\right]$

$= 300\left[\dfrac{rt(1+\frac{r}{12})^{-12t-1} - 1 + (1+\frac{r}{12})^{-12t}}{r^2}\right]$

At t = 5: $P_r = 300\left[\dfrac{5r(1+\frac{r}{12})^{-61} - 1 + (1+\frac{r}{12})^{-60}}{r^2}\right]$

$= -\dfrac{300}{r^2}\left[1 - 5r(1+\frac{r}{12})^{-61} - (1+\frac{r}{12})^{-60}\right]$

9.3 MAXIMUM-MINIMUM APPLICATIONS, PAGES 276 - 277

1. $f_x = 6x + 5y$, $f_y = 5x + 2y$. To find critical values we need to solve the system
$\begin{cases} 6x + 5y = 0 \\ 5x + 2y = 0 \end{cases}$. Isolating y in the second equation gives $y = \left(-\frac{5}{2}\right)x$. Substituting into the

Chapter 9   209

first equation gives $6x + 5y = 6x + 5\left(-\frac{5}{2}\right)x = 0 \Rightarrow \left(-\frac{13}{2}\right)x = 0 \Rightarrow x = 0$.

Since $y = \left(-\frac{5}{2}\right)x$, $y = 0$; giving the critical value (0, 0). $A = f_{xx} = 6$, $B = f_{xy} = 5$,
$C = f_{yy} = 2$, $D = B^2 - AC = 5^2 - (6)(2) = 13$. Since $D > 0$, we have a saddle point at
(0, 0, f(0, 0)) = (0, 0, 0).

3. $f_x = 2x + y$, $f_y = x + 2y - 3$. To find critical values we need to solve the system
$\begin{cases} 2x + y = 0 \\ x + 2y - 3 = 0 \end{cases}$. Solving for y in the first equation gives $y = -2x$. Substituting into

the second equation gives $x + 2y - 3 = x + 2(-2x) - 3 = 0 \Rightarrow -3x - 3 = 0 \Rightarrow x = -1$.

Since $y = -2x$, $y = 2$; giving the critical point (-1, 2). $A = f_{xx} = 2$, $B = f_{xy} = 1$,
$C = f_{yy} = 2$, $D = B^2 - AC = 1^2 - (2)(2) = -3$. Since $D < 0$ and $A > 0$ there is a relative
minimum at (-1, 2, f(-1, 2)) = (-1, 2, -3).

5. $f_x = 3x^2 - 3y$, $f_y = -3x - 3y^2$. To find critical values we need to solve the system
$\begin{cases} 3x^2 - 3y = 0 \\ -3x - 3y^2 = 0 \end{cases}$. Solving for y in the first equation gives $y = x^2$. Substituting into the

second equation gives $-3x - 3y^2 = -3x - 3(x^2)^2 = 0 \Rightarrow -3x(1 + x^3) = 0 \Rightarrow x = 0$ or -1.
Since $y = x^2$, the critical points are (0, 0) and (-1, 1). $A = f_{xx} = 6x$, $B = f_{xy} = -3$,

$C = f_{yy} = -6y$, $D = B^2 - AC = (-3)^2 - (6x)(-6y) = 9 + 36xy$. At (0, 0) $D = 9 > 0$, so
we have a saddle point at (0, 0, 0). At (-1, 1), $D = 9 + 36(-1)(1) = -27 < 0$, and
$A = 6(-1) < 0$ which means we have a relative maximum at (-1, 1 f(-1, 1)) = (-1, 1, 1).

7. $f_x = 3y - 10x + 3$, $f_y = 3x - 2y - 5$. To find critical values we need to solve
$\begin{cases} 3y - 10x + 3 = 0 \\ 3x - 2y \ 5 - 0 \end{cases}$. Solving for y in the second equation gives $y = \frac{3x - 5}{2}$. Substituting

this into the first equation gives $3y - 10x + 3 = 3\left(\frac{3x - 5}{2}\right) - 10x + 3 = 0$

$\Rightarrow \frac{9x - 15}{2} - 10x + 3 = 0 \Rightarrow 9x - 15 - 20x + 6 = 0 \Rightarrow -11x - 9 = 0 \Rightarrow x = -\frac{9}{11}$.

Substituting, $y = \frac{3x - 5}{2} = \frac{3(-\frac{9}{11}) - 5}{2} = -\frac{41}{11}$, so our critical point is $(-\frac{9}{11}, -\frac{41}{11})$.

Chapter 9   210

$A = f_{xx} = -10$, $B = f_{xy} = 3$, $C = f_{yy} = -2$, $D = B^2 - AC = 3^2 - (-10)(-2) = -11 < 0$. Since $A < 0$ we have a relative maximum at $(-\frac{9}{11}, -\frac{41}{11}, \frac{45}{11})$.

9. $f_x = y + 2$, $f_y = x - 3$. To find critical values we need to solve the system
$\begin{cases} y + 2 = 0 \\ x - 3 = 0 \end{cases}$. The solution is $x = 3$, $y = -2$. $A = f_{xx} = 0$, $B = f_{xy} = 1$, $C = f_{yy} = 0$, $D = B^2 - AC = 1$. Since $D > 0$ we have a saddle point at $(3, -2, 2)$.

11. $f_x = 2x + e^x$, $f_y = -1$. Since $f_y \neq 0$ there are no critical values.

13. $f_x = e^y$, $f_y = xe^y$. Since $e^y \neq 0$ there are no relative maxima, minima, or saddle points.

15. $f_x = 4y - 4x^3$, $f_y = 4x - 4y^3$. To find critical points we need to solve
$\begin{cases} 4y - 4x^3 = 0 \\ 4x - 4y^3 = 0 \end{cases}$. Solving for y in the first equation gives $y = x^3$. Substitute this into the second equation: $4x - 4y^3 = 4x - 4(x^3)^3 = 0 \Rightarrow 4x(1 - x^8) = 0 \Rightarrow x = 0, 1,$ or -1.
Since $y = x^3$ our critical values are $(0, 0)$, $(1, 1)$, and $(-1, -1)$. $A = f_{xx} = -12x^2$, $B = f_{xy} = 4$, $C = f_{yy} = -12y^2$, $D = B^2 - AC = 16 - 144x^2y^2$. At $(0, 0)$ $D = 16 > 0 \Rightarrow$ saddle point at $(0, 0, 0)$. At $(-1, -1)$ $D = -128 < 0$ and since $A = -12(-1)^2 = -12 < 0 \Rightarrow$ relative maximum at $(-1, -1, 2)$. At $(1, 1)$, $D = -128 < 0$ and since $A = -12(1)^2 < 0 \Rightarrow$ relative maximum at $(1, 1, 2)$.

17. $f_x = 3x^2 + 3y$, $f_y = 3x + 3y^2$. To find critical values we must solve the system
$\begin{cases} 3x^2 + 3y = 0 \\ 3x + 3y^2 = 0 \end{cases}$. Solving for y in the first equation gives $y = -x^2$. Substitute this into the second equation which gives $3x + 3y^2 = 3x + 3(-x^2)^2 = 0 \Rightarrow 3x(1 + x^3) = 0 \Rightarrow x = 0$ or -1.
Since $y = -x^2$ the critical values are $(0, 0)$ and $(-1, -1)$. $A = f_{xx} = 6x$, $B = f_{xy} = 3$, $C = f_{yy} = 6y$, $D = B^2 - AC = 9 - 36xy$. At $(0, 0)$, $D = 9 > 0 \Rightarrow$ saddle point at $(0, 0, 0)$. At $(-1, -1)$, $D = -27 < 0$. Since $A = 6(-1) < 0$ we have a relative maximum at $(-1, -1, 1)$.

19. $P_a = -6a + 34 + 2n$, $P_n = -10n - 2 + 2a$. To find critical values we must solve
$\begin{cases} -6a + 34 + 2n = 0 \\ -10n - 2 + 2a = 0 \end{cases}$. Solving for a in the second equation gives $a = 5n + 1$. Substituting this into the first equation gives $-6a + 34 + 2n = -6(5n + 1) + 34 + 2n = 0 \Rightarrow n = 1$.

Chapter 9   211

Since a = 5n + 1 = 5(1) + 1, our critical point is (a, n) = (6, 1). $A = P_{aa} = -6$, $B = P_{an} = 2$,

$C = P_{nn} = -10$, $D = B^2 - AC = -56 < 0$. Since A < 0 we have a relative maximum at (6, 1).

The maximum is $P(6, 1) = -3(6)^2 - 5(1)^2 + 34(6) - 2(1) + 2(6)(1) + 40 = 141$ (thousand dollars).

21. P(x, y) = Profit = Revenue - Cost
   = [4000 + 50(2y - 3x)]x + [50(x - y)]y - (50x + 110y)
   = $4000x + 100xy - 150x^2 + 50xy - 50y^2 - 50x - 110y$
   = $3950 x + 150xy - 150x^2 - 50y^2 - 110y$

   $P_x = 3950 + 150y - 300x$, $P_y = 150x - 100y - 110$. To find critical points we need to solve the system

   $\begin{cases} 3950 + 150y - 300x = 0 \\ 150x - 100y - 110 = 0 \end{cases}$. Solving the first equation for y gives $y = \dfrac{300x - 3950}{150} = \dfrac{6x - 79}{3}$.

   Substituting into the second equation gives $150x - 100(\dfrac{6x - 79}{3}) - 110 = 0$. Multiply through by $\dfrac{3}{10}$ => $45x - 10(6x - 79) - 33 = 0$ => $-15x + 790 - 33 = 0$ => $x \approx 50.47$. Substituting into $y = \dfrac{6x - 79}{3} = 74.60$. $A = P_{xx} = -300$, $B = P_{xy} = 150$, $C = P_{yy} = -100$,

   $D = B^2 - AC = 150^2 - (-300)(-100) = 22{,}500 - 30{,}000 = -7500$. Since D < 0 and A < 0, this is a relative maximum. Profit will be maximized when about 50 standard and 75 deluxe rings are sold.

23. Let the dimensions be x (width) by y (length) by z (height). Then 64 = xyz, or

   $z = \dfrac{64}{xy} = 64x^{-1}y^{-1}$. The amount of material is

   $xy + 2xz + 2yz = xy + 2x(64x^{-1}y^{-1}) + 2y(64x^{-1}y^{-1}) = xy + 128y^{-1} + 128x^{-1} = f(x, y)$.

   $f_x = y - 128x^{-2}$, $f_y = x - 128y^{-2}$. We need to solve the system

   $\begin{cases} y - 128x^{-2} = 0 \\ x - 128y^{-2} = 0 \end{cases}$. Solving for y in the first equation gives $y = \dfrac{128}{x^2}$. Substituting into the

   second equation gives $x - \dfrac{128}{y^2} = x - \dfrac{128}{\left(\dfrac{128}{x^2}\right)^2} = x - \dfrac{x^4}{128} = 0$. Solving for x we get

   $x(1 - \dfrac{x^3}{128}) = 0$ => x = 0 or $x = \sqrt[3]{128} \approx 5.04$. Since x = 0 does not make sense, we will

Chapter 9

test only at (5.04, 5.04). $A = f_{xx} = 256x^{-3}$, $B = f_{xy} = 1$, $C = f_{yy} = 256y^{-3}$, $D = B^2 - AC$ $= 1 - \dfrac{65536}{x^3 y^3}$. At (5.04, 5.04), $D < 0$ and $A > 0$ which means this is a minimum. The box should therefore be 5.04 by 5.04 by $\dfrac{64}{5.04^2} \approx 2.52$ feet.

25. If $A = 0$, then $D = B^2 - AC = B^2 \geq 0$. Hence, if $A = 0$, then D can't be negative.

9.4 LAGRANGE MULTIPLIERS, PAGES 282 - 283

1. $F(x, y) = f(x, y) + \lambda\, g(x, y) = xy - \lambda(x + y - 20) = xy - \lambda x - \lambda y + 20\lambda$.

$F_x = y - \lambda$, $F_y = x - \lambda$, $F_\lambda = -x - y + 20$. Now solve the system

$\begin{cases} y - \lambda = 0 \\ x - \lambda = 0 \\ -x - y + 20 = 0 \end{cases}$. The solution is $x = 10$, $y = 10$, $\lambda = 10$.

$f(10, 10) = 100$. By checking other points for which $x + y = 20$, we see this is a maximum.

3. $F(x, y) = f(x, y) + \lambda g(x, y) = -2x^2 - 3y^2 + \lambda(x + 2y - 24) = -2x^2 - 3y^2 + \lambda x + 2\lambda y - 24\lambda$.

$F_x = -4x + \lambda$, $F_y = -6y + 2\lambda$, $F_\lambda = x + 2y - 24$. Now solve the system

$\begin{cases} -4x + \lambda = 0 \\ -6y + 2\lambda = 0 \\ x + 2y - 24 = 0 \end{cases}$. The solution is $x = \dfrac{72}{11}$, $y = \dfrac{96}{11}$, $\lambda = \dfrac{288}{11}$.

$f\left(\dfrac{72}{11}, \dfrac{96}{11}\right) = -\dfrac{3456}{11}$. By checking other points for which $x + 2y = 24$ we see this is a maximum.

5. $F(x, y) = f(x, y) + \lambda g(x, y) = x^2 y + \lambda(x + 2y - 14) = x^2 y + \lambda x + 2\lambda y - 14\lambda$.

$F_x = 2xy + \lambda$, $F_y = x^2 + 2\lambda$, $F_\lambda = x + 2y - 14$. Now solve the system

$\begin{cases} 2xy + \lambda = 0 \\ x^2 + 2\lambda = 0 \\ x + 2y - 14 = 0 \end{cases}$. The solutions are $x = \dfrac{28}{3}$, $y = \dfrac{7}{3}$, $\lambda = -\dfrac{392}{9}$.

$f\left(\dfrac{28}{3}, \dfrac{7}{3}\right) = \dfrac{5488}{27}$. By checking other points for which $x + 2y = 14$, we see that this is a maximum.

7. $F(x, y) = f(x, y) + \lambda g(x, y) = x^2 + y^2 + \lambda(x + y - 24) = x^2 + y^2 + \lambda x + \lambda y - 24\lambda$.

Chapter 9   213

$F_x = 2x + \lambda$, $F_y = 2y + \lambda$, $F_\lambda = x + y - 24$. We need to solve

$\begin{cases} 2x + \lambda = 0 \\ 2y + \lambda = 0 \\ x + y - 24 = 0 \end{cases}$. The solution is $x = 12$, $y = 12$, $\lambda = -24$.

$f(12, 12) = 288$ which, by checking other points on $x + y - 24 = 0$, is a relative minimum.

9. $F(x, y) = f(x, y) + \lambda g(x, y) = x^2 + y^2 - xy - 4 + \lambda(x + y - 6)$

$F_x = 2x - 6 + \lambda$, $F_y = 2y - x + \lambda$, $F_\lambda = x + y - 6$. We need to solve

$\begin{cases} 2x - y + \lambda = 0 \\ 2y - x + \lambda = 0 \\ x + y - 6 = 0 \end{cases}$. The solution is $x = 3$, $y = 3$, $\lambda = -3$.

$f(3, 3) = 5$. By checking other points on $x + y - 6 = 0$ this is a relative minimum.

11. $F(x, y) = f(x, y) + \lambda g(x, y) = x^2 + y^2 + \lambda(x + y - 16)$

$F_x = 2x + \lambda$, $F_y = 2y + \lambda$, $F_\lambda = x + y - 16$. We need to solve

$\begin{cases} 2x + \lambda = 0 \\ 2y + \lambda = 0 \\ x + y - 16 = 0 \end{cases}$. The solution is $x = 8$, $y = 8$, $\lambda = -16$.

$f(8, 8) = 128$. By checking other points on $x + y - 16 = 0$, this is a relative minimum.

13. If the two numbers are x and y, we want to maximize $f(x, y) = xy$ subject to $x + y - 10 = 0$.

$F(x, y) = f(x, y) + \lambda g(x, y) = xy + \lambda(x + y - 10)$.

$F_x = y + \lambda$, $F_y = x + \lambda$, $F_\lambda = x + y - 10$. We need to solve

$\begin{cases} y + \lambda = 0 \\ x + \lambda = 0 \\ x + y - 10 = 0 \end{cases}$. The solution is $x = 5$, $y = 5$, $\lambda = -5$.

$f(5, 5) = 25$ which is a maximum because other points on $x + y - 10 = 0$ yield lesser values for $f(x, y)$. The two numbers are therefore both 5.

15. Maximize $f(x, y) = -x^2 - 2y^2 + 140$ subject to $20x + 12y = 100$.

$F(x, y, \lambda) = -x^2 - 2y^2 + 140 + \lambda(20x + 12y - 100)$

$F_x = -2x + 20\lambda$, $F_y = -4y + 12\lambda$, $F_\lambda = 20x + 12y - 100$.

Solve the system $\begin{cases} -2x + 20\lambda = 0 \\ -4y + 12\lambda = 0 \\ 20x + 12y - 100 = 0 \end{cases}$. The solution is $x = \dfrac{250}{59}$, $y = \dfrac{75}{59}$, $\lambda = \dfrac{25}{59}$.

$f\left(\dfrac{250}{59}, \dfrac{75}{59}\right) \approx 119$. Checking points for which $20x + 12y = 100$, such as $(5, 0)$, gives

Chapter 9							214

f(5, 0) = 115 < 119. Hence this is a relative maximum. He should apply $\frac{250}{59}$ acre-feet of water and $\frac{75}{59}$ pounds of fertilizer.

17. Let x be one width of the fence (costing $3 per foot) and y be the other width (costing $9 per foot). We want to maximize area = f(x, y) = xy subject to 3(2x) + 9(2y) = 4000 => 6x + 18y = 4000.

F(x, y, λ) = F(x, y) + λg(x, y) = xy + λ(6x + 18y - 4000)

F_x = y + 6λ, F_y = x + 12λ, $F_λ$ = 6x + 18y - 4000.

Solve the system $\begin{cases} y + 6λ = 0 \\ x + 12λ = 0 \\ 6x + 18y - 4000 = 0 \end{cases}$. The solution is x = $166\frac{2}{3}$, y = $166\frac{2}{3}$. The farmer should build his fence $166\frac{2}{3}$ by $166\frac{2}{3}$ for a maximum area of 27,778 square feet.

19. We want to minimize cost = f(x, y) = .15x + .03y subject to 20xy = 1000.

F(x, y, λ) = f(x, y) + λg(x, y) = .15x + .03y + λ(20xy - 1000)

F_x = .15 + 20λy, F_y = .03 + 20λx, $F_λ$ = 20xy - 1000.

We need to solve $\begin{cases} .15 + 20λy = 0 \\ .03 + 20λx = 0 \\ 20xy - 1000 = 0 \end{cases}$. The solution is x = $\sqrt{10}$, y = $5\sqrt{10}$. Checking other points for which 20xy = 1000 we see that f($\sqrt{10}$, $5\sqrt{10}$) ≈ $.95 is a maximum. The optimum mixture is $\sqrt{10}$ ≈ 3.16 ounces of meat and $5\sqrt{10}$ ≈ 15.81 ounces of vegetables.

21. F(x, y, z, λ) = f(x, y, z) + λg(x, y, z) = x - y + z + λ($x^2 + y^2 + z^2$ - 100).

F_x = 1 + 2λx, F_y = -1 + 2λy, F_z = 1 + 2λz, $F_λ$ = $x^2 + y^2 + z^2$ - 100.

We need to solve $\begin{cases} 1 + 2λx = 0 \\ -1 + 2λy = 0 \\ 1 + 2λz = 0 \\ x^2 + y^2 + z^2 - 100 = 0 \end{cases}$

The two solutions are x = $\pm\frac{10\sqrt{3}}{3}$, y = $\mp\frac{10\sqrt{3}}{3}$, z = $\pm\frac{10\sqrt{3}}{3}$, λ = $-\frac{\sqrt{3}}{20}$.

$f\left(\frac{10\sqrt{3}}{3}, -\frac{10\sqrt{3}}{3}, \frac{10\sqrt{3}}{3}\right)$ = $10\sqrt{3}$. This is a maximum by checking other points on $x^2 + y^2 + z^2$ = 100, like f (0, 0, 10) = 10 < $10\sqrt{3}$.

Chapter 9   215

23. $F(x, y, z, \lambda) = 100 - xy - xz - yz + \lambda(x + y + z - 10)$

$F_x = -y - z + \lambda$, $F_y = -x - z + \lambda$, $F_z = -x - y + \lambda$, $F_\lambda = x + y + z - 10$.

We need to solve $\begin{cases} -y - z + \lambda = 0 \\ -x - z + \lambda = 0 \\ -x - y + \lambda = 0 \\ x + y + z - 10 = 0 \end{cases}$. The solution is $x = \frac{10}{3}$, $y = \frac{10}{3}$, $z = \frac{10}{3}$, $\lambda = \frac{20}{3}$.

The lowest temperature is $T\left(\frac{10}{3}, \frac{10}{3}, \frac{10}{3}\right) = \frac{200}{3}$.

9.5 LINEAR PROGRAMMING, PAGE 289

1. The graph of the feasibility region is shown at right. The corner points and functional values are:

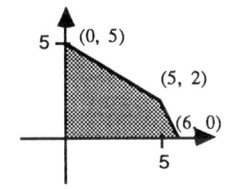

Corner Points	Objective Function
(0, 5)	30(0) + 20(5) = 100
(6, 0)	30(6) + 20(0) = 180
(5, 2)	30(5) + 20(2) = 190
(0, 0)	30(0) + 20(0) = 0

   The maximum is 190, occurs at (5, 2).

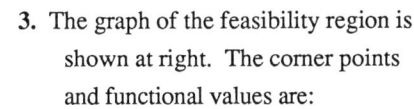

3. The graph of the feasibility region is shown at right. The corner points and functional values are:

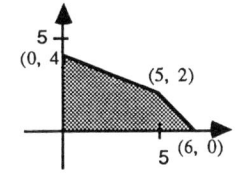

Corner Points	Objective Function
(0, 0)	100(0) + 10(0) = 0
(6, 0)	100(6) + 10(0) = 600
(5, 2)	100(5) + 10(2) = 520
(0, 4)	100(0) + 10(4) = 40

   The maximum is 600, occurs at (6, 0).

5. The graph of the feasibility region is shown at right.

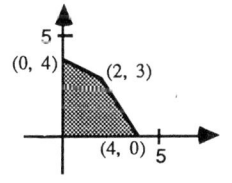

Corner Points	Objective Function
(0, 0)	100(0) + 100(0) = 0
(0, 4)	100(0) + 100(4) = 400
(4, 0)	10(4) + 100(0) = 400
(2, 3)	100(2) + 100(3) = 500

   The maximum is 500, occurs at (2, 3).

Chapter 9   216

7. The feasibility region is shown at right.

Corner Points	Objective Function
(0, 6)	24(0) + 12(6) = 72
(0, 8)	24(0) + 12(8) = 96
(10, 8)	24(10) + 12(8) = 336
(10, 0)	24(10) + 12(0) = 240
(4, 0)	24(4) + 12(0) = 96

The minimum is 72, occurs at (0, 6).

9. The feasibility region is shown at right.

Corner Points	Objective Function
(6, 0)	2(6) - 3(0) = 12
(6, 2)	2(6) - 3(2) = 6
(4, 4)	2(4) - 3(4) = -4
(0, 4)	2(0) - 3(4) = -12
(0, 0)	2(0) - 3(0) = 0

The maximum is 12, occurs at (6, 0).

11. The feasibility region is shown at right.

Corner Points	Objective Function
(0, 6)	90(0) + 20(6) = 120
(0, 9)	90(0) + 20(9) = 180
$(\frac{25}{2}, 9)$	$90(\frac{25}{2}) + 20(9) = 1305$
(8, 0)	90(8) + 20(0) = 720
(6, 0)	90(6) + 20(0) = 540

The minimum is 120, occurs at (0, 6).

13. The feasibility region is the same as Problem 11.

Corner Points	Objective Function
(0, 6)	23(0) + 46(6) = 276
(0, 9)	23(0) + 46(9) = 414
$(\frac{25}{2}, 9)$	$23(\frac{25}{2}) + 46(9) = 701.5$
(8, 0)	23(8) + 46(0) = 184
(6, 0)	23(6) + 46(0) = 138

The maximum is 701.5, occurs at $(\frac{25}{2}, 9)$.

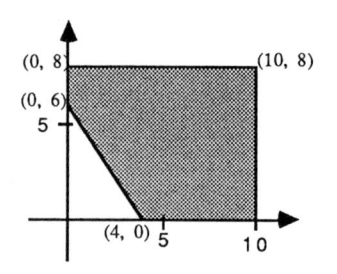

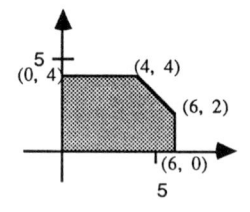

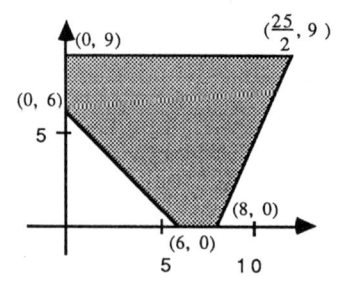

Chapter 9

15. The feasibility region is shown at right.

Corner Points	Objective Function
(0, 0)	6(0) + 3(0) = 0
(0, 2)	6(0) + 3(2) = 6
(6, 3)	6(6) + 3(3) = 45
(3, 0)	6(3) + 3(0) = 18
$(\frac{21}{5}, \frac{24}{5})$	$6(\frac{21}{5}) + 3(\frac{24}{5}) = 39.6$

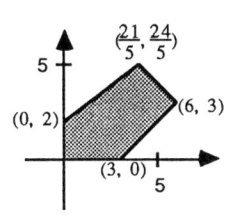

The maximum is 45, occurs at (6, 3).

17. The feasibility region is shown at right.

Corner Points	Objective Function
(3, 2)	5(3) + 3(2) = 21
(5, 5)	5(5) + 3(5) = 40
(7, 5)	5(7) + 3(5) = 50
$(\frac{10}{3}, \frac{4}{3})$	$5(\frac{10}{3}) + 3(\frac{4}{3}) = 20\frac{2}{3}$

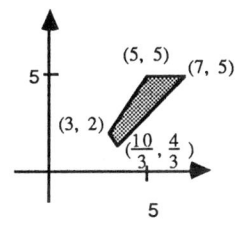

The minimum is $20\frac{2}{3}$, occurs at $(\frac{10}{3}, \frac{4}{3})$.

19. The feasibility region is shown at right.

Corner Points	Objective Function
(9, 12)	140(9) + 250(12) = 4260
(9, 1)	140(9) + 250(1) = 1510
(7, 0)	140(7) + 250(0) = 980
(4, 0)	140(4) + 250(0) = 560
$(\frac{5}{3}, \frac{14}{3})$	$140(\frac{5}{3}) + 250(\frac{14}{3}) = 1400$

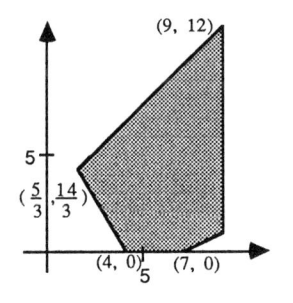

The minimum is 560, occurs at (4, 0).

21. The feasibility region and corner points do not change. The objective function is
P(x, y) = 2.00(110)x + 1.50(30)y = 220x + 45y.

Corner Points	Objective Function
(0, 0)	220(0) + 45(0) = 0
$(0, \frac{500}{7})$	$220(0) + 45(\frac{500}{7}) \approx 3,214$
$(\frac{400}{11}, 0)$	$220(\frac{400}{11}) + 45(0) = 8,000$
(20, 60)	220(20) + 45(60) = 7,100

The maximum is $8,000, occurs at $(\frac{400}{11}, 0)$.

Chapter 9

23. If the farmer plants x acres of corn and y acres of wheat: Maximize P = 120x + 100y.

Subject to $\begin{cases} x \geq 0, y \geq 0 \\ 120x + 60y \leq 24{,}000 \\ 100x + 40y \geq 18{,}000 \\ x + y \leq 500 \end{cases}$

Corner Points	Objective Function
(180, 0)	120(180) + 100(0) = 21,600
(200, 0)	120(200) + 100(0) = 24,000
(100, 200)	120(100) + 100(200) = 32,000

The farmer should plant 100 acres of corn and 200 acres of wheat (leaving 200 acres unplanted).

25. If x standard models and y economy models are produced: Maximize profit,
P = 45x + 30y.

Subject to $\begin{cases} x \geq 0, y \geq 0 \\ 3x + 3y \leq 1500 \\ 2x \leq 800 \end{cases}$

Corner Points	Objective Function
(0, 0)	45(0) + 30(0) = 0
(0, 500)	45(0) + 30(500) = 15,000
(400, 100)	45(400) + 30(100) = 21,000

They should manufacture 400 standard and 100 economy models.

27. If x ounces of Corn Flakes and y ounces of Honeycomb are eaten: Minimize cost,
C = .07x + .19y.

Subject to $\begin{cases} x \geq 0, y \geq 0 \\ 23x + 14y \geq 322 \\ 7x + 17y \geq 119 \end{cases}$

Corner Points	Objective Function
(0, 23)	.07(0) + .19(23) = 4.37
(17, 0)	.07(17) + .19(0) = 1.19
$(\frac{3808}{293}, \frac{483}{293})$	$.07(\frac{3808}{293}) + .19(\frac{483}{293}) = 1.22$

The minimum cost is $1.19.

29. If x Apple IIe's and y IBM PC'c are purchased: Maximize storage: S = 128x + 256y.

Subject to $\begin{cases} x \geq 0, y \geq 0 \\ 900x + 1200y \leq 36{,}000 \\ 5x + 3y \leq 165 \end{cases}$

Corner Points	Objective Function
(0, 0)	128(0) + 256(0) = 0
(0, 30)	128(0) + 256(30) = 7,680
(33, 0)	128(33) + 256(0) = 4,224
$(\frac{300}{11}, \frac{105}{11})$	$128(\frac{300}{11}) + 256(\frac{105}{11}) \approx 5{,}911$

The manager should purchase 30 IBM PC's (for a maximum storage capacity of 7680k.).

9.6 MULTIPLE INTEGRALS, PAGES 301 - 302

1. $\int_0^2 (x^2 y + y^2) dy = x^2 \left(\frac{y^2}{2}\right) + \frac{y^3}{3} \Big|_{y=0}^{y=2} = \left(x^2(2) + \frac{8}{3}\right) - \left(x^2(0) + 0\right) = 2x^2 + \frac{8}{3}$

Chapter 9

3. $\int_1^4 \left(xy^2 - x\right)dx = y^2\left(\frac{x^2}{2}\right) - \frac{x^2}{2} \Big|_{x=1}^{x=4} = \left(y^2 - 1\right)\left(\frac{x^2}{2}\Big|_{x=1}^{x=4}\right) = \left(y^2 - 1\right)\left(8 - \frac{1}{2}\right) = \frac{15}{2}\left(y^2 - 1\right)$

5. Treating x as a constant, let $u = 2y + x^2$. Then $du = 2dy$:

$\int_0^4 x\sqrt{x^2 + 2y}\, dy = x \int_{u=x^2}^{u=x^2+8} u^{\frac{1}{2}}\left(\frac{du}{2}\right) = x\left[\frac{1}{3}u^{\frac{3}{2}}\Big|_{u=x^2}^{u=x^2+8}\right] = \frac{x}{3}\left[\left(x^2+8\right)^{\frac{3}{2}} - x^3\right]$

7. Treating y as a constant, let $u = x + 2y$. Then $du = dx$:

$\int_1^2 3e^{x+2y}dx = 3\int_{u=1+2y}^{2+2y} e^u du = 3e^u \Big|_{1+2y}^{2+2y} = 3\left[e^{2+2y} - e^{1+2y}\right] = 3e^{2y+2} - 3e^{2y+1}$

9. Let $u = x^2 + 3y$. Treating y as a constant, $du = 2xdx$:

$\int_1^5 xe^{x^2+3y}dx = \int_{u=1+3y}^{25+3y} xe^u \frac{du}{2x} = \frac{1}{2}\int_{1+3y}^{25+3y} e^u du = \frac{1}{2}\left(e^{25+3y} - e^{1+3y}\right)$

$= \frac{e^{3y+25}}{2} - \frac{e^{3y+1}}{2}$

11. $\int_1^{e^x} \frac{x}{y} dy = x\ln|y| \Big|_1^{e^x} = x\ln|e^x| - x\ln|1| = x(x) - x(0) = x^2$

13. $\int_1^4\int_0^3 dydx = \int_1^4 \left[y\Big|_0^3\right]dx = \int_1^4 3dx = 3x\Big|_1^4 = 3(4-1) = 9$

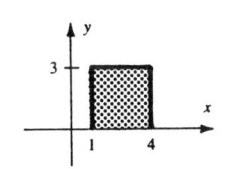

15. $\int_{-3}^2\int_2^4 dxdy = \int_{-3}^2 \left(x\Big|_2^4\right)dy = \int_{-3}^2 (4-2)\,dy = \int_{-3}^2 2dy = 2\left(y\Big|_{-3}^2\right) = 2(2--3) = 10$

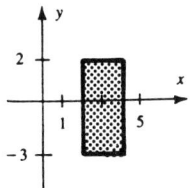

Chapter 9 220

17. $\int_1^3 \int_0^y (x + 3y) \, dx\,dy = \int_1^3 \left(\frac{x^2}{2} + 3xy \Big|_{x=0}^{x=y}\right) dy = \int_1^3 \left(\frac{y^2}{2} + 3y^2\right) dy = \int_1^3 \frac{7}{2} y^2 \, dy = \frac{7}{6} y^3 \Big|_1^3 = \frac{91}{3}$

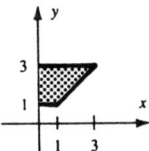

19. $\int_0^2 \int_0^{y^2} xy \, dx\,dy = \int_0^2 \frac{x^2 y}{2} \Big|_{x=0}^{x=y^2} dy = \int_0^2 \frac{1}{2} y^5 \, dy = \frac{y^6}{12} \Big|_0^2 = \frac{16}{3}$

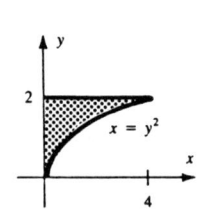

21. $\int_0^5 \int_0^x (2x + y) \, dy\,dx = \int_0^5 \left(2xy + \frac{y^2}{2} \Big|_{y=0}^{y=x}\right) dx = \int_0^5 \left(2x^2 + \frac{x^2}{2}\right) dx = \int_0^5 \frac{5}{2} x^2 \, dx = \frac{5x^3}{6} \Big|_0^5 = \frac{625}{6}$

23. $\int_0^1 \int_{x^3}^{x^2} x^2 y \, dy\,dx = \int_0^1 x^2 \left(\frac{y^2}{2} \Big|_{x^3}^{x^2}\right) dx = \int_0^1 x^2 \left(\frac{x^4}{2} - \frac{x^6}{2}\right) dx = \frac{1}{2} \int_0^1 \left(x^6 - x^8\right) dx$

$= \frac{1}{2} \left(\frac{x^7}{7} - \frac{x^9}{9} \Big|_0^1\right) = \frac{1}{2} \left(\frac{2}{63}\right) = \frac{1}{63}$

25. $\int_1^4 \int_0^3 dy\,dx = \int_0^3 \int_1^4 dx\,dy = \int_0^3 \left(x \Big|_1^4\right) dy = \int_0^3 3 \, dy = 3y \Big|_0^3 = 9$

27. $\int_1^3 \int_0^y (x + 3y) \, dx\,dy = \int_0^1 \int_1^3 (x + 3y) \, dy\,dx + \int_1^3 \int_x^3 (x + 3y) \, dx\,dy$

Chapter 9   221

$$= \int_0^1 \left(xy + \frac{3y^2}{2}\Big|_{y=1}^{y=3}\right) dx + \int_0^1 \left(xy + \frac{3y^2}{2}\Big|_{y=x}^{y=3}\right) dx$$

$$= \int_0^1 \left[\left(3x + \frac{27}{2}\right) - \left(x + \frac{3}{2}\right)\right] dx + \int_0^1 \left[\left(3x + \frac{27}{2}\right) - \left(x^2 + \frac{3x^2}{2}\right)\right] dx$$

$$= \int_0^1 (2x + 12)\, dx + \int_0^1 \left(-\frac{5}{2}x^2 + 3x + \frac{27}{2}\right) dx = x^2 + 12x\Big|_0^1 + \left(-\frac{5x^3}{6} + \frac{3x^2}{2} + \frac{27x}{2}\Big|_0^1\right)$$

$$= 13 + \frac{52}{3} = \frac{91}{3}$$

29. $\displaystyle\int_0^5\int_0^x (2x+y)\, dy\, dx = \int_0^5\int_y^5 (2x+y)\, dx\, dy = \int_0^5 x^2 + xy\Big|_y^5\, dy = \int_0^5 [(25+5y) - (y^2 + y^2)]\, dy$

$$= \int_0^5 (25 + 5y - 2y^2)\, dy = 25y + \frac{5y^2}{2} - \frac{2y^3}{3}\Big|_0^5 = \frac{625}{6}$$

31. $\displaystyle\int_0^3\int_0^2 (x^2+y)\, dx\, dy = \int_0^3 \frac{x^3}{3} + xy\Big|_{x=0}^{x=2}\, dy = \int_0^3 \left(\frac{8}{3} + 2y\right) dy = \frac{8}{3}y + y^2\Big|_0^3 = 17$

33. $\displaystyle\int_1^3\int_0^{1-x} \frac{1}{x}\, dy\, dx = \int_1^3 \frac{1}{x}\left(y\Big|_0^{1-x}\right) dx = \int_1^3 \frac{1}{x}(1-x)\, dx = \int_1^3 \left(\frac{1}{x} - 1\right) dx = \ln|x| - x\Big|_1^3$

$$= (\ln 3 - 3) - (\ln 1 - 1) = \ln 3 - 2$$

35. $\displaystyle\int_1^2\int_0^{y^2} y^3 e^{xy}\, dx\, dy = \int_1^2 y^3\left(\frac{1}{y}e^{xy}\Big|_{x=0}^{x=y^2}\right) dy = \int_1^2 y^2\left(e^{y^3} - 1\right) dy = \int_1^2 y^2 e^{y^3}\, dy - \int_1^2 y^2\, dy$

$$= \frac{1}{3}e^{y^3}\Big|_1^2 - \frac{1}{3}y^3\Big|_1^2 = \frac{1}{3}\left(e^8 - e\right) - \left(\frac{8}{3} - \frac{1}{3}\right) = \frac{e^8}{3} - \frac{e}{3} - \frac{7}{3}$$

37. Let $u = x^2 + y^2$. Treating y as a constant (for the first integration only) $du = 2x\, dx$:

$$\int_2^4\int_1^3 xy\sqrt{x^2+y^2}\, dx\, dy = \int_2^4\left[\int_{u=1+y^2}^{u=9+y^2} xyu^{\frac{1}{2}}\frac{du}{2x}\right] dy = \int_2^4 \frac{y}{2}\int_{1+y^2}^{9+y^2} u^{\frac{1}{2}}\, du = \int_2^4 \frac{y}{2}\left[\frac{2}{3}u^{\frac{3}{2}}\Big|_{1+y^2}^{9+y^2}\right] dy$$

Chapter 9   222

$$= \int_2^4 \left[\frac{y}{3}(9+y^2)^{\frac{3}{2}} - \frac{y}{3}(1+y^2)^{\frac{3}{2}} \right] dy = \frac{1}{3}\int_2^4 y(9+y^2)^{\frac{3}{2}} dy - \frac{1}{3}\int_2^4 y(1+y^2)^{\frac{3}{2}} dy$$

$$= \frac{1}{3}\left[\frac{(9+y^2)^{\frac{5}{2}}}{5} \Big|_2^4 \right] - \frac{1}{3}\left[\frac{(1+y^2)^{\frac{5}{2}}}{5} \Big|_2^4 \right] = \frac{1}{3}\left[625 - \frac{13^{\frac{5}{2}}}{5} \right] - \frac{1}{3}\left[\frac{17^{\frac{5}{2}}}{5} - \frac{5^{\frac{5}{2}}}{5} \right]$$

$$= \frac{1}{15}\left(625 - 13^{\frac{5}{2}} - 17^{\frac{5}{2}} + 5^{\frac{5}{2}} \right) \approx 92$$

39. A. $\displaystyle\int_0^1 \int_{x^2}^{\sqrt{x}} (x+y)\,dy\,dx = \int_0^1 \left[xy + \frac{y^2}{2} \Big|_{x^2}^{\sqrt{x}} \right] dx = \int_0^1 \left(x\sqrt{x} + \frac{x}{2} - x^3 - \frac{x^4}{2} \right) dx$

$$= \frac{2}{5}x^{\frac{5}{2}} + \frac{x^2}{4} - \frac{x^4}{4} - \frac{x^5}{10} \Big|_0^1 = \frac{3}{10}$$

B. $\displaystyle\int_0^1 \int_{y^2}^{\sqrt{y}} (x+y)\,dx\,dy = \int_0^1 \left[\frac{x^2}{2} + xy \Big|_{y^2}^{\sqrt{y}} \right] dy = \int_0^1 \left(\frac{y}{2} + y\sqrt{y} - \frac{y^4}{2} - y^3 \right) dy = \frac{3}{10}$

41. A. $\displaystyle\int_0^5 \int_0^{5-x} 2xy\,dy\,dx = \int_0^5 xy^2 \Big|_0^{5-x} dx = \int_0^5 x(5-x)^2 dx = \int_0^5 (25x - 10x^2 + x^3) dx$

$$= \frac{25x^2}{2} - \frac{10x^3}{3} + \frac{x^4}{4} \Big|_0^5 = \frac{625}{12}$$

B. $\displaystyle\int_0^5 \int_0^{5-y} 2xy\,dx\,dy = \int_0^5 x^2 y \Big|_0^{5-y} dy = \int_0^5 (5-y)^2 y\,dy = \frac{25y^2}{2} - \frac{10y^3}{3} + \frac{y^4}{4} \Big|_0^5 = \frac{625}{12}$

43. $\displaystyle\int_{-3}^2 \int_1^5 5\,dx\,dy = 5\int_{-3}^2 x \Big|_1^5 dy = 5\int_{-3}^2 4\,dy = 20y \Big|_{-3}^2 = 100$

45. $\displaystyle\int_0^1 \int_0^1 xy\sqrt{x^2+y^2}\,dx\,dy$. Let $u = x^2 + y^2$. Treating y as a constant, $du = 2x\,dx$:

$$= \int_0^1 \int_{u=y^2}^{u=y^2+1} \left(xyu^{\frac{1}{2}} \frac{du}{2x} \right) dy = \int_0^1 \frac{y}{2} \int_{y^2}^{y^2+1} u^{\frac{1}{2}} du\,dy = \int_0^1 \frac{y}{2}\left(\frac{2}{3} u^{\frac{3}{2}} \Big|_{y^2}^{y^2+1} \right) dy$$

Chapter 9

$$= \int_0^1 \left[\frac{y}{3}(y^2+1)^{\frac{3}{2}} - \frac{y}{3}(y^3)\right] dy = \frac{1}{3}\int_0^1 y(y^2+1)^{\frac{3}{2}} dy - \frac{1}{3}\int_0^1 y^4 dy$$

$$= \frac{1}{3}\left[\frac{(y^2+1)^{\frac{5}{2}}}{5}\bigg|_0^1\right] - \frac{1}{3}\left[\frac{y^5}{5}\bigg|_0^1\right] = \frac{1}{3}\left(\frac{2^{\frac{5}{2}}}{5} - \frac{1}{5}\right) - \frac{1}{3}\left(\frac{1}{5}\right) = \frac{1}{15}\left(2^{\frac{5}{2}} - 2\right)$$

47. $\displaystyle\int_0^1\int_{x^2}^{\sqrt{x}} 5xy\,dy\,dx = \int_0^1 \frac{5xy^2}{2}\bigg|_{y=x^2}^{y=\sqrt{x}} dx = \int_0^1 \frac{5}{2}(x^2 - x^5)\,dx = \frac{5}{2}\left(\frac{x^3}{3} - \frac{x^6}{6}\bigg|_0^1\right) = \frac{5}{2}\left(\frac{1}{6}\right) = \frac{5}{12}$

49. $\displaystyle\frac{1}{(2-0)(3-0)}\int_0^3\int_0^2 (x-2y)\,dx\,dy = \frac{1}{6}\int_0^3 \frac{x^2}{2} - 2xy\bigg|_0^2 dy = \frac{1}{6}\int_0^3 (2 - 4y)\,dy$

$= \dfrac{1}{6}\left[2y - 2y^2\bigg|_0^3\right] = \dfrac{1}{6}[6 - 18] = -2$

51. $\displaystyle\frac{1}{(5-0)(2-0)}\int_0^2\int_0^5 L^{0.75} K^{0.25}\,dL\,dK = \frac{1}{10}\int_0^2 \frac{L^{1.75}}{1.75} K^{0.25}\bigg|_{L=0}^{L=5} dK = \frac{1}{17.5}\int_0^2 5^{1.75} K^{0.25}\,dK$

$\displaystyle = \frac{5^{1.75}}{17.5}\left[\frac{K^{1.25}}{1.25}\bigg|_0^2\right] = \frac{5^{1.75}}{17.5}\left[\frac{2^{1.25}}{1.25}\right] = \frac{5^{\frac{7}{4}} 2^{\frac{5}{4}}}{175/8} = \frac{8}{175}\sqrt[4]{5^7}\sqrt[4]{2^5} = \frac{8}{175}\left(5\sqrt[4]{5^3}\right)\left(2\sqrt[4]{2}\right)$

$\displaystyle = \frac{80}{175}\left(\sqrt[4]{250}\right) = \frac{16}{35}\sqrt[4]{250}$

9.7 CORRELATION AND LEAST SQUARES APPLICATIONS, PAGES 309 - 310

1. Using Table 9.1, since $.7 = r > .632$ there is a significant linear correlation.
3. Using Table 9.1, since $.4 = r < .463$ there is not a significant linear correlation.
5. Using Table 9.1, since $.732 = |r| > .514$ there is a significant linear correlation.
7. Using Table 9.1, since $.3416 = |r| < .361$ there is not a significant linear correlation.
9. Using Table 9.1, since $.521 = r < .561$ there is not a significant linear correlation.
11. Using Table 9.1, since $.214 = |r| < .396$ there is not a significant linear correlation.
13. $n = 4$, $\Sigma x = 10$, $\Sigma y = 27$, $\Sigma x^2 = 30$, $\Sigma y^2 = 259$, $\Sigma xy = 87$.

Chapter 9

$$r = \frac{(4)(87) - (10)(27)}{\sqrt{4(30) - 10^2}\sqrt{4(259) - 27^2}} = \frac{78}{\sqrt{20}\sqrt{307}} \approx .995$$

Using Table 9.1 there is a linear correlation at a 5% level of significance.

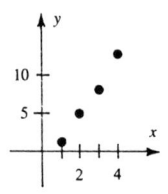

15. $n = 5$, $\Sigma x = 10$, $\Sigma y = 82$, $\Sigma x^2 = 30$, $\Sigma y^2 = 1486$, $\Sigma xy = 127$.

$$r = \frac{(5)(127) - (10)(82)}{\sqrt{5(30) - 10^2}\sqrt{5(1486) - 82^2}} \approx -.9847.$$

Using Table 9.1 there is a linear correlation at a 1% level of significance.

17. $n = 6$, $\Sigma x = 200$, $\Sigma y = 361$, $\Sigma x^2 = 8400$, $\Sigma y^2 = 22293$, $\Sigma xy = 13000$.

$$r = \frac{6(13000) - (200)(361)}{\sqrt{6(8400) - 200^2}\sqrt{6(22293) - 361^2}} \approx .970.$$

Using Table 9.1, there is linear correlation at a 1% level of significance.

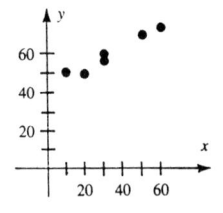

Chapter 9   225

19. See Problem 13. $m = \dfrac{4(87) - (10)(27)}{4(30) - 10^2} = 3.9$, $b = \dfrac{27 - 3.9(10)}{4} = -3$

$y = mx + b \Rightarrow y = 3.9x - 3$.

21. See Problem 17. $m = \dfrac{6(13000) - 200(361)}{6(8400) - 200^2} \approx .5577$, $b = \dfrac{361 - (.5577)(200)}{6} \approx 41.58$

$y = mx + b \Rightarrow y = .56x + 41.58$.

23. $n = 5$, $\Sigma x = 20$, $\Sigma y = 96$, $\Sigma x^2 = 108$, $\Sigma y^2 = 2070$, $\Sigma xy = 308$.

$m = \dfrac{5(308) - 20(96)}{5(108) - 20^2} \approx -2.7143$, $b = \dfrac{96 + 2.7143(20)}{5} \approx 30.0571$

$y = mx + b \Rightarrow y = -2.7143x + 30.0571$

25. $n = 8$, $\Sigma x = 27$, $\Sigma y = 115$, $\Sigma x^2 = 109$, $\Sigma y^2 = 1805$, $\Sigma xy = 432$

$r = \dfrac{8(432) - 27(115)}{\sqrt{8(109) - 27^2}\,\sqrt{8(1805) - 115^2}} \approx .842$

There is a significant correlation at 1%.

27. See Problem 25. $m = \dfrac{8(432) - 27(115)}{8(109) - 27} \approx 2.4545$, $b = \dfrac{115 - (2.4545)(27)}{8} \approx 6.091$,

$y = 2.4545x + 6.091$

29. $n = 5$, $\Sigma x = 84.4$, $\Sigma y = 34675$, $\Sigma x^2 = 1439.22$, $\Sigma y^2 = 307822943$, $\Sigma xy = 564242.5$

$r = \dfrac{5(564242.5) - (84.4)(34675)}{\sqrt{5(1439.22) - 84.4^2}\,\sqrt{5(307822943) - 34675^2}} \approx -.673$

31. $n = 5$, $\Sigma x = 42.9$, $\Sigma y = 45.2$, $\Sigma x^2 = 415.8034$, $\Sigma y^2 = 409.1$, $\Sigma xy = 384.334$

$r = \dfrac{5(384.334) - (42.9)(45.2)}{\sqrt{5(415.8034) - 42.9^2}\,\sqrt{5(409.1) - 45.2^2}} \approx -.719$

33. $n = 5$, $\Sigma x = 34675$, $\Sigma y = 42.9$, $\Sigma x^2 = 307822943$, $\Sigma y^2 = 415.8034$, $\Sigma xy = 348573.03$

$r = \dfrac{5(348573.03) - (34675)(42.9)}{\sqrt{5(307822943) - 34675^2}\,\sqrt{5(415.8034) - 42.9^2}} \approx .901$

Chapter 9

35. $m = \dfrac{6(433309.15) - (42407)(53.94)}{6(359523617) - 42407^2} \approx .00087$, $b = \dfrac{53.94 - (.00087)(42407)}{6} \approx 2.84$

$y = .00087x + 2.84$

9.8 CHAPTER REVIEW, PAGES 311 - 312

1. $P(10, 50) = 10^{.4} \, 50^{.2} \approx 5.49$

3. Since all variables are of first degree, the surface is a plane. The intercepts are:
$x = 0, y = 0 \Rightarrow z = 8$
$x = 0, z = 0 \Rightarrow y = 8$
$y = 0, z = 0 \Rightarrow x = 4$

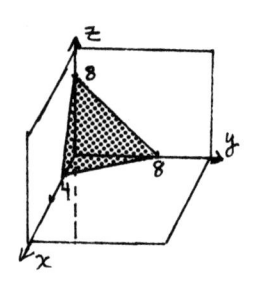

5. $f_x = 10(x - 3y)^9 \cdot \dfrac{\partial}{\partial x}(x - 3y) = 10(x - 3y)^9$

7. $f_x = 16x^3 - 9x^2y^2 + 4x$; $f_{xx} = 48x^2 - 18xy^2 + 4$

9. $\displaystyle\int_0^9 \int_{\sqrt{x}}^{\sqrt{x}+2} y\,dy\,dx = \int_0^9 \dfrac{y^2}{2}\bigg|_{\sqrt{x}}^{\sqrt{x}+2} dx = \int_0^9 \left[\dfrac{(\sqrt{x}+2)^2}{2} - \dfrac{x}{2}\right] dx = \dfrac{1}{2}\int_0^9 (x + 4x^{\frac{1}{2}} + 4 - x)dx$

$= \dfrac{1}{2}\int_0^9 (4x^{\frac{1}{2}} + 4)dx = \dfrac{1}{2}\left[\dfrac{8}{3}x^{\frac{3}{2}} + 4x\right]_0^9 = \dfrac{1}{2}[(72 + 36) - 0] = 54$

11. $f_x = 4x - 3y$, $f_y = -3x + 2y$. To find critical points, solve $\begin{cases} 4x - 3y = 0 \\ -3x + 2y = 0. \end{cases}$ The solution is $(0, 0)$. Using the second derivative test, $A = f_{xx} = 4$; $B = f_{xy} = -3$; $C = f_{yy} = 2$, $D = B^2 - AC = 1$. Since $D > 0$, $(0, 0)$ is a saddle point.

13. The corner points for the feasibility region are $(0, 0)$, $(0, 200)$, $(150, 100)$, $(170, 80)$, and $(210, 0)$.

Corner Points	Objective Function
(0, 0)	$5(0) + 4(0) = 0$
(0, 200)	$5(0) + 4(200) = 800$
(150, 100)	$5(150) + 4(100) = 1150$

Chapter 9

(170, 80)   5(170) + 4(80) = 1170
(210, 0)   5(210) + 4(0) = 1050
Maximum is 1170, occurs at (170, 80).

15. $\int_0^5 \int_{x^2}^{5x} (x+2y)\,dy\,dx = \int_0^5 xy + y^2 \Big|_{x^2}^{5x}\,dx = \int_0^5 (30x^2 - x^3 - x^4)\,dx = 10x^3 - \frac{x^4}{4} - \frac{x^5}{5} \Big|_0^5 = \frac{1875}{4}$

17. $n=8$, $\Sigma x = 323$, $\Sigma y = 725$, $\Sigma x^2 = 14899$, $\Sigma y^2 = 65907$, $\Sigma xy = 29677$.

$$r = \frac{8(29677) - (323)(725)}{\sqrt{8(14899) - 323^2}\sqrt{8(65907) - 725^2}} \approx .658.$$

Not significant at either level. (See Table 9.1.)

19. Revenue = $R(x, y) = (50 - .05x + .001y)x + (130 + .01x - .04y)y$
    $= 50x - .05x^2 + .011xy + 130y - .04y^2$
    $R(5, 8) = 50(5) - .05(5^2) + .011(5)(8) + 130(8) - .04(8^2) = 1286.63$
    $C(5, 8) = 90 + 20(5) + 90(8) = 910$

9.9 STUDENT'S TEST REVIEW AND ADDITIONAL PRACTICE

OBJECTIVES

The material of this chapter is reviewed in the following list of objectives. After each objective there are some practice questions. Answers to these problems immediately follow. Detailed solutions are given for every third problem. For a sample test select the first question of each set and check your answers. Additional practice is given by the other questions in each set. If you are having trouble with a particular type of problem, or if you want additional practice, look back at the indicated section in the test.

[9.1] Objective 1: *Evaluate functions of several variables.* Evaluate the functions in Problems 1-4 for the point (5, -12).

1. $f(x, y) = 3x^2 - 2xy + y^2$ 2. $g(x, y) = e^{\frac{x}{y}}$

3. $k(l, w) = lw$ 4. $P(L, K) = L^{.5}K^{.3}$

Objective 2: *Plot points in three dimensions.*
5. a. (5, 0, 0)   b. (3, -5, 5)   6. a. (0, 1, 0)   b. (-1, 2, 1)
7. a. (0, 0, 10)   b. (5, 10, -10)   8. a. (2, 5, 7)   b. (-2, -5, -4)

Objective 3: *Graph surfaces in space.*

9. $x + y + 2z = 10$ 10. $z = x^2 + y^2$

11. $x^2 + y^2 - z^2 = 0$ 12. $x^2 + y^2 = 9$

Chapter 9

[9.2] Objective 4: *Find partial derivatives.*

13. Find f_x for $f(x, y) = (2x - 5y)^{12}$.

14. Find $g_y(1, 3)$ for $g(x, y) = e^{5xy}$.

15. If $z = T(x, y)$ find $\dfrac{\partial z}{\partial y}$ for $T(x, y) = \dfrac{1000}{\sqrt{x^2 + y^2}}$.

16. If $f(x, y, \lambda) = 2xy + \lambda(2x + 3y - 100)$ find f_λ.

Objective 5: *Find higher order partial derivates.*

Let $z = f(x, y) = 3x^5 - 5x^4y^3 + 2x^2 - 150$ for Problems 17-20.

17. $f_{xx}(x, y)$
18. $\dfrac{\partial^2 z}{\partial y \partial x}$

19. $f_{yx}(1, -1)$
20. $\dfrac{\partial^2 z}{\partial y^2}$

[9.3] Objective 6: *Find the relative maxima, relative minima, and saddle points of a function of two variables.*

21. $f(x, y) = 2x^2 - 3xy + y^2$
22. $f(x, y) = xy - 4x + 3y + 120$

23. $g(x, y) = \dfrac{y}{x}$
24. $g(x, y) = e^{xy}$

[9.4] Objective 7: *Find relative maximums or minimums using the method of Lagrange multipliers.*

25. Find the relative maximum of $f(x, y) = 12 - x^2 - y^2$ subject to $x + y = 10$.

26. Find the relative maximum of $g(x, y) = x^2 + 4y^2$ subject to $x + 2y = 12$.

27. Find two numbers whose sum is 250 and whose product is a maximum.

28. Find the smallest value for a product of two numbers if their difference must be 10.

[9.5] Objective 8: *Find the optimum value for a given objective function subject to a given set of linear constraints.*

29. Maximize $f(x, y) = 5x + 7y$
subject to:
$$\begin{cases} x \geq 0, y \geq 0 \\ 4x + y \geq 7 \\ x + 2y \leq 12 \end{cases}$$

30. Minimize $f(x, y) = 8x + 5y$
subject to:
$$\begin{cases} x \geq 0, y \geq 0 \\ 2x + y \geq 8 \\ x - y \leq 2 \end{cases}$$

Chapter 9

31. Maximize $f(x, y) = 5x + 4y$
subject to:
$$\begin{cases} 2x + y \le 420 \\ 2x + 2y \le 500 \\ 2x + 3y \le 600 \\ x \ge 0, y \ge 0 \end{cases}$$

32. Maximize $f(x, y) = x + y$
subject to:
$$\begin{cases} x \ge 0, y \ge 0 \\ 12x + 150y \le 1200 \\ 6x + 200y \le 1200 \\ 16x + 50y \le 800 \end{cases}$$

[9.6] Objective 9: *Evaluate integrals which are functions of two variables.*

33. $\int_{1}^{3} x^2 y^3 \, dx$

34. $\int_{-1}^{4} 5e^{x+2y} \, dy$

35. $\int_{1}^{e^{\pi}} \frac{x^2}{y} \, dy$

36. $\int_{0}^{1} x^2 y^2 \sqrt{x^3 + 8} \, dx$

Objective 10: *Evaluate iterated integrals.*

37. $\int_{0}^{9} \int_{\sqrt{x}}^{\sqrt{x}+2} y \, dy \, dx$

38. $\int_{0}^{\ln 3} \int_{0}^{x} e^{4y} \, dy \, dx$

39. $\int_{1}^{2} \int_{0}^{\frac{1}{y}} y^3 \, dx \, dy$

40. $\int_{0}^{3} \int_{0}^{y\sqrt{9-y^2}} dx \, dy$

Objective 11: *Evaluate integrals by reversing the order of integration.*

41. $\int_{0}^{8} \int_{\frac{2}{y}}^{\sqrt{2y}} dx \, dy$

42. $\int_{0}^{2} \int_{x}^{4} dy \, dx$

43. $\int_{0}^{4} \int_{2y}^{8} dx \, dy$

44. $\int_{1}^{e^2} \int_{\ln y}^{2} dx \, dy$

Objective 12: *Find a volume between a given surface and a given region by evaluating a double integral.*

45. $\iint_{R} x^2 y^3 \, dx \, dy$ $0 \le x \le 2$
                             $0 \le y \le 1$

46. $\iint_{R} y^{-1} \, dy \, dx$ $1 \le x \le 2$
                           $1 \le y \le e^3$

Chapter 9

47. $\iint\limits_R (x + 2y)\,dA$, R bounded by $y = x^2$ and $y = 5x$

48. $\iint\limits_R xy^2\,dA$, R bounded by $x = y^2$ and $y = x^2$

[9.7] Objective 13: *Given the number of ordered pairs of data, the correlation coefficient, and the significance level, determine whether there is a significant linear correlation.*

49. $n = 20, r = .6, \alpha = 5\%$ 50. $n = 10, r = .6, \alpha = 1\%$

51. $n = 25, r = .5, \alpha = 1\%$ 52. $n = 20, r = .55, \alpha = 5\%$

Objective 14: *Draw a scatter diagram and find the correlation coefficient for a set of data. Also determine whether there is a linear correlation at either the 5% or 1% level.*

53.
x	50	60	70	80	90	100
y	32	91	45	11	64	95

Draw a scatter diagram.

54. The following table compares age with blook pressure.

Age	20	25	30	40	50	35	68	55
Blood pressure	85	91	84	93	100	86	94	92

Draw a scatter diagram.

55. Find the linear correlation coefficient for the data in Problem 53 and determine whether the variables are significantly correlated at either the 1% or 5% level.

56. Find the linear correlation coefficient for the data in Problem 54 and determine whether the variables are significantly correlated at either the 1% or 5% level.

Objective 15: *Find the regression line for a set of data.*

57. Find the regression line for the data in Problem 53.

58. Find the regression line for the data in Problem 54.

[9.1-9.7] Objective 16: *Solve applied problems based on the preceding objectives. For specific examples of the types of applications look at the list of applications in this chapter on page 314.*

59. A company produces two types of skateboards, standard and competition models. The weekly demand and cost equations are given where $p is the price of the standard skateboard and $q is the price of the competition model:

$p = 50 - .05x + .001y$

$q = 130 + .01x - .04y$

for x the weekly demand (in hundreds) for standard skateboards and y the weekly

Chapter 9

demand (in hundreds) for competition skateboards. If the cost function is
$$C(x, y) = 90 + 20x + 90y$$
find the revenue function, R, and evaluate C(5, 8) and R(5, 8).

60. Find $C_x(5, 8)$ and $C_y(5, 8)$ for the cost function given in Problem 59 and interpret each of these.

61. Karlin Enterprises employs between 100 and 500 employees and has a capital investment of between 3 and 5 million dollars. A research company has determined that Karlin's productivity (units per employee per day) is approximated by the formula
$$z = P(x, y) = 4xy - 3x^2 - y^3$$
where x is the size of the labor force (in hundreds) and y is the capital investment (in millions of dollars). Find the marginal productivity of labor when x = 2 and y = 4 and interpret your answer.

62. Find the marginal productivity of capital in Problem 61 when x = 5 and y = 3 and interpret your answer.

63. Find the maximum productivity for Karlin Enterprises (see Problem 61) in terms of labor force and capital investment.

64. Bradbury Bros. Realty plans to open several new branch offices, which will employ either four or six people, and has $1,275,000 in capital for this expansion. A four-person branch requires an initial cash outlay of $175,000 and a six-person branch, $200,000. Bradbury Bros. has also decided not to hire more than 32 people and will not open more than 10 branches. How many of each type of branch should be opened to maximize cash inflow, if it is expected to be $50,000 for four-person branches and $65,000 for six-person branches?

ANSWERS TO STUDENT'S TEST REVIEW AND PRACTICE QUESTIONS

1. 339   2. $e^{-\frac{5}{12}} \approx .659$ 3. k(5, -12) = (5)(-12) = -60   4. Not a real number.

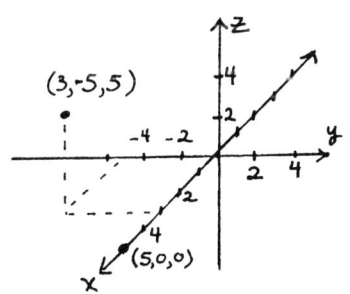

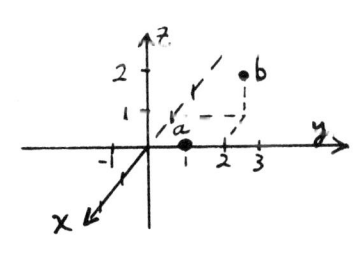

5.   6.

Chapter 9   232

7.

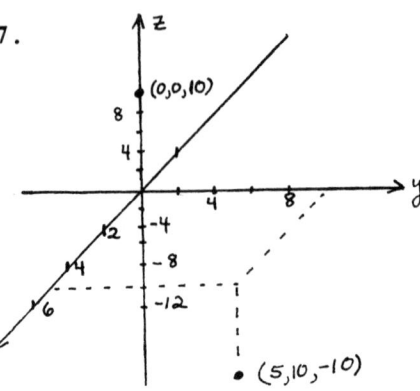

8.

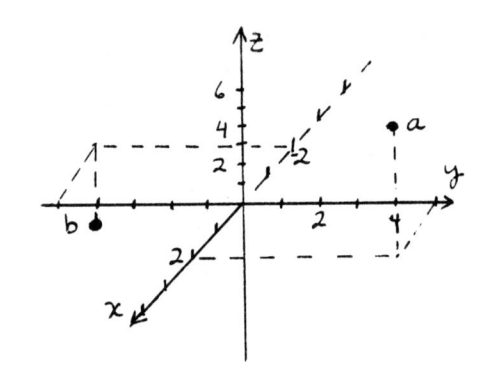

9. The surface is a plane. Set y and z equal to zero to get x = 10, the x-intercept. Set x and y equal to zero to get z = 5, the z-intercept. Set x and z equal to zero to get y = 10, the y–intercept.

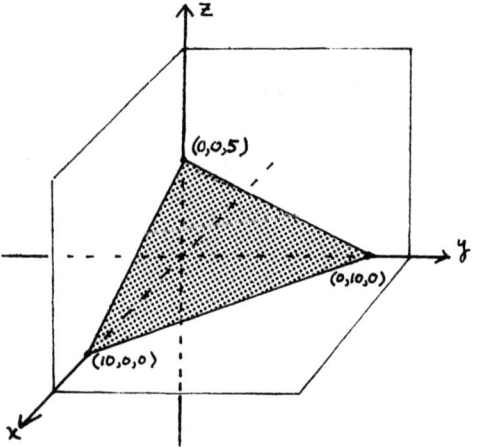

10.

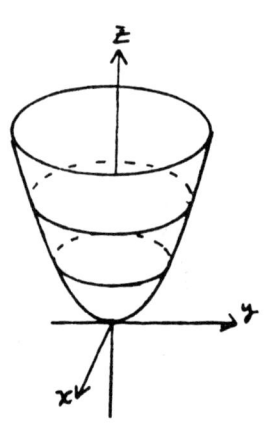

11.

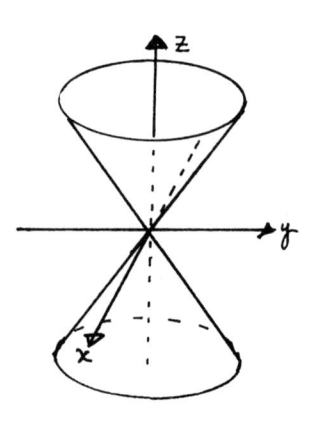

Chapter 9

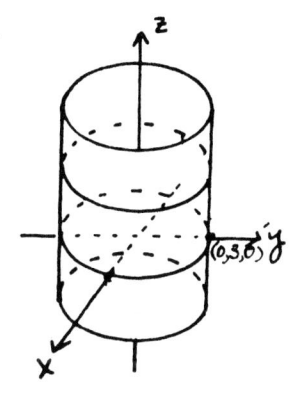

12. [figure: cylinder along z-axis with point (0,3,0) marked]

13. $24(2x - 5y)^{11}$

14. $5e^{15}$

15. Treat x as a constant: $\dfrac{\partial z}{\partial y} = 1000(-\dfrac{1}{2})(x^2 + y^2)^{-\frac{3}{2}} \cdot \dfrac{\partial}{\partial y}(x^2 + y^2) = -500(x^2 + y^2)^{-\frac{3}{2}}(2y)$

$= -1000y(x^2 + y^2)^{-\frac{3}{2}}$

16. $2x + 3y - 100$ **17.** $60x^3 - 60x^2y^3 + 4$

18. $\dfrac{\partial z}{\partial y} = -15x^4y^2$; $\dfrac{\partial^2 z}{\partial y \partial x} = \dfrac{\partial^2 z}{\partial x \partial y} = \dfrac{\partial z}{\partial x}(-15x^4y^2) = -60x^3y^2$ **19.** -60 **20.** $-30x^4y$

21. $f_x = 4x - 3y$, $f_y = -3x + 2y$. Solve $\begin{cases} 4x - 3y = 0 \\ -3x + 2y = 0 \end{cases}$ The only solution is $x = 0$ and $y = 0$.

$f_{xx} = 4$, $f_{xy} = -3$, $f_{yy} = 2$. $D = (-3)^2 - (4)(2) = 1 > 0$.

Since $D > 0$, there is a saddle point at $(0, 0, 0)$.

22. Saddle point at $(-3, 4, 132)$.   **23.** No relative maxima or minima.

24. $g_x = ye^{xy}$, $g_y = xe^{xy}$. Solve $\begin{cases} ye^{xy} = 0 \\ xe^{xy} = 0 \end{cases}$ The only solution is $x = 0$ and $y = 0$.

$g_{xx} = y^2e^{xy}$, $g_{xy} = y(xe^{xy}) + e^{xy}$, $g_{yy} = x^2e^{xy}$. $D = 1^2 - (0)(0) > 0$.

Saddle point at $(0, 0, 1)$.

25. $f(5, 5) = -38$ **26.** $g(6, 3) = 180$

27. Letting x and y be the numbers, we want to maximize $f(x, y) = xy$ subject to $x + y = 250$.

$F(x, y, \lambda) = xy + \lambda(x + y - 250)$. Solve $\begin{cases} F_x = y + \lambda = 0 \\ F_y = x + \lambda = 0 \\ F_\lambda = x + y - 250 = 0 \end{cases}$

Chapter 9   234

The solution is x = 125, y = 125   (λ = -125).

28. The minimum product is -25 (when the two numbers are 5 and -5)

29. f(7, 2.5) = 52.5

30. The feasibility region is unbounded with two corners: (0, 8) and $(\frac{10}{3}, \frac{4}{3})$

$f(0, 8) = 8(0) + 5(8) = 40$; $f(\frac{10}{3}, \frac{4}{3}) = 8(\frac{10}{3}) + 5(\frac{4}{3}) = 33\frac{1}{3}$

The minimum is $f(\frac{10}{3}, \frac{4}{3}) = 33\frac{1}{3}$.

31. f(170, 80) = 1170   **32.** f(50, 0) = 50

33. $y^3 \int_1^3 x^2 dx = y^3 \left(\frac{x^3}{3}\Big|_1^3\right) = y^3 \frac{26}{3} = \frac{26y^3}{3}$ **34.** $\frac{5e^x}{2}(e^8 - e^{-2})$ **35.** πx^2

36. $y^2 \int_0^1 x^2\sqrt{x^3 + 8}\, dx = y^2 \int_0^1 (x^3 + 8)^{\frac{1}{2}} x^2 dx$. Let $u = x^3 + 8$, then $du = 3x^2 dx$:

$\Rightarrow y^2 \int_{x=0}^{x=1} u^{\frac{1}{2}} \frac{du}{3} = \frac{y^2}{3} \left[\frac{2}{3} u^{\frac{3}{2}} \Big|_{x=0}^{x=1}\right] = \frac{2y^2}{9}\left[(x^3+8)^{\frac{3}{2}}\Big|_0^1\right] = \frac{2y^2}{9}\left(27 - 8^{\frac{3}{2}}\right)$

$= 6y^2 - \frac{32}{9}\sqrt{2}$

37. 54   **38.** 2 - ln3

39. $\int_1^2 \int_0^{\frac{1}{y}} y^3 dxdy = \int_1^2 y^3 x \Big|_{x=0}^{x=\frac{1}{y}} dy = \int_1^2 y^2 dy = \frac{y^3}{3}\Big|_1^2 = \frac{7}{3}$ **40.** 9

41. $\int_0^4 \int_{\frac{x}{2}}^{2x} dxdy = \frac{16}{3}$

42. The region of integration is bounded below by the parabola $y = x^2$, above by the line y = 4 and by the y-axis. To reverse the order, x must go from x = 0 to the parabola $x = \sqrt{y}$:

$\int_0^2 \int_{x^2}^4 dydx = \int_0^4 \int_0^{\sqrt{y}} dxdy = \frac{16}{3}$

43. $\int_0^8 \int_0^{\frac{x}{2}} dy\,dx = 16$ 44. $\int_0^2 \int_1^{e^x} dy\,dx = e^2 - 3$

45. $\int_0^1 \int_0^2 x^2 y^3 dx\,dy = \int_0^1 \frac{x^3}{3} y^3 \Big|_0^2 dy = \int_0^1 \frac{8}{3} y^3 dy = \frac{2y^4}{3}\Big|_0^1 = \frac{2}{3}$ 46. 3   47. $\frac{1875}{4}$

48. $\int_0^1 \int_{x^2}^{\sqrt{x}} xy^2 dy\,dx = \int_0^1 (\frac{x^{\frac{5}{2}}}{3} - \frac{x^7}{3}) dx = \frac{3}{56}$ 49. Yes   50. No

51. Using Table 9.1, for n = 25 and α = 1%, we see that r must be at least .505. Since r = .5 there is not a linear correlation at 1% significance.

52. Yes

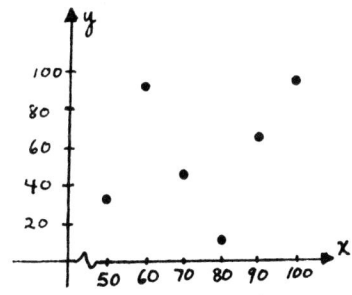

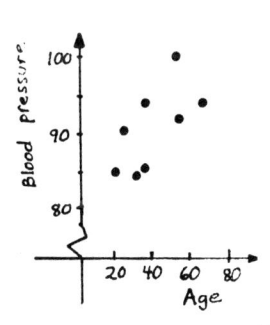

53.

54.

55. r ≈ .3214, not significant at 1% or 5%.   56. r ≈ .6583, not significant at 1% or 5%.

57. n = 6, Σx = 450, Σy = 338, Σx² = 35,500, Σy² = 24,572, Σxy = 26,350

$m = \frac{6(26,350) - (450)(338)}{6(35,500) - (450)^2} = \frac{4}{7}$

$b = \frac{338 - \frac{4}{7}(450)}{6} = \frac{283}{21}$. So the line is

$y = \frac{4}{7}x + \frac{283}{21}$.

58. y = .218x + 81.821

59. R(x, y) = 50x + 130y + .011xy - .05x² - .04y²; C(5, 8) = $190; R(5, 8) = $1,286.63

60. C_x = 20 = C_x(5, 8), C_y = 90 = C_y(5, 8). At a fixed weekly demand for competition skateboards, the change in cost per unit change in x is 20. At a fixed weekly demand for

Chapter 9

standard skateboards, the change in cost per unit change in y is 90.

61. At a capital investment level of 4 million dollars, the rate of increase of productivity per unit change in labor is 4 (units per employee per day) when the size of the labor force is 200.

62. At a labor force of 500, the rate of increase of productivity per unit change in capital investment is -7 (units per employee per day) when the capital investment is 3 million dollars.

63. $P_x = 4y - 6x$, $P_y = 4x - 3y^2$. Solve $\begin{cases} 4y - 6x = 0 \\ 4x - 3y^2 = 0 \end{cases}$ The only solutions are $x = 0$ and $y = 0$ (which we reject), and $x = \frac{16}{27}$ and $y = \frac{8}{9}$. $P_{xx} = -6$, $P_{yy} = -6y$, $P_{xy} = 4$.

$D = 4^2 - (-6)(-6(\frac{8}{9})) < 0$. Since $P_{xx} < 0$ we have a maximum at $(\frac{16}{27}, \frac{8}{9})$. The maximum is $P(\frac{16}{27}, \frac{8}{9}) = \frac{256}{729}$.

64. Let x and y be the number of four-person and six-person branches, respectively.
Maximize $f(x, y) = 50,000x + 65,000y$

Subject to: $\begin{cases} x \geq 0, \ y \geq 0 \\ 175,000x + 200,000y \leq 1,275,000 \\ x + y \leq 10 \\ 4x + 6y \leq 32 \end{cases}$

Maximum is $f(5, 2) = \$380,000$; in other words they should open 5 four-person branches and 2 six-person branches.

MODELING APPLICATION 7: SAMPLE ESSAY
The Cobb-Douglas Production Function *

>Suppose Karlin, Inc. manufactures only one product, which is sold at the price p_0. The firm employs a labor force L, which must be paid an average wage p_1. The firm also requires capital K in terms of tools, buildings, and so forth. The cost of using one unit of capital is p_2. Karlin, Inc. wishes to maximize its profit P. Determine the data which needs to be collected and construct a model which will accomplish Karlin's goal of maximizing profit.

---

* This extended application is adapted from "The Cobb-Douglas Production Function" by Robert Geitz. It is part of the UPMAP series of applications of calculus. © UMAP 1981.

Chapter 9

In economics it is common to use <u>production functions</u> that relate the output of an economic system to various inputs. The most famous and most widely used production function is the Cobb-Douglas function

$$P(L, K) = bL^{\alpha}K^{\beta}$$

relating production to labor and capital. All such functions, of course, represent idealized views of the world; in concrete situations many factors besides labor and capital influence total production.

The necessary steps in constructing any model are to first state the assumptions, next translate these assumptions into equations, derive a production function, and finally to check this function against tabulated data.

<u>Assumptions</u>

Economic data is frequently expressed in relative terms with reference to an arbitrary standard, and it is in this form that we will use it. The reason for this is that the numbers needed to express many economic variables in absolute terms are so large that they are very cumbersome to work with. As an example of how an index is computed, suppose we have the data shown in Table 1.

Table 1 Production Levels

Year	Production	Index(1984=100)
1984	$2,500,000	100
1985	$2,000,000	80
1986	$4,000,000	160
1987	$5,000,000	200

We choose some year to be a standard, and express the production of each other year as a percentage of the standard year's production. For example, if 1984 as the standard, the index for 1985 is:

$$\frac{2{,}000{,}000}{2{,}500{,}000} \times 100 = 80$$

The next step in identifying the assumptions is to identify the variables:

Let P = total production of an economic system (by an economic system we might mean computer industry, lumber, chemicals, or transportation)
Let L = amount of labor involved in this production (this is measured in total number of man-hours worked)
Let K = amount of capital invested in this production (this is measured in terms of the monetary worth of all machinery, equipment, tools, and buildings involved in production)

Cobb and Douglas were concerned with modeling the total economy of the United States. They measured production in terms of the monetary value of all goods produced in the U.S. There are standard indexes of all of these quantities, and the variables P, L, and K are the values of these indexes. For purposes of our model, we will use the information used by Cobb and Douglas as shown in Table 2.

Chapter 9

Table 2 Index values for Production, Labor, and Capital (1899 = 100)

Year	P	L	K	Year	P	L	K	Year	P	L	K
1899	100	100	100	1907	151	140	176	1915	189	156	216
1900	101	105	107	1908	126	123	185	1916	225	183	298
1901	112	110	114	1909	155	143	198	1917	227	198	335
1902	122	117	122	1910	159	147	208	1918	223	201	366
1903	124	122	131	1911	153	148	216	1919	218	196	387
1904	122	121	138	1912	177	155	226	1920	231	194	407
1905	143	125	149	1913	184	156	236	1921	179	146	417
1906	152	134	163	1914	169	152	244	1922	240	161	431

The following assumptions were made by Cobb and Douglas:
1. If either labor or capital vanishes, then so will production.
2. The marginal productivity of labor is proportional to the amount of production per unit of labor.
3. The marginal productivity of capital is proportional to the amount of production per unit of capital.

Translation into mathematical notation

1. $\lim_{L \to 0} P(L, K) = 0$ and $\lim_{K \to 0} P(L, K) = 0$

2. $\dfrac{\partial P}{\partial L} = \alpha \dfrac{P}{L}$ for a constant of porportionality α

3. $\dfrac{\partial P}{\partial L} = \beta \dfrac{P}{L}$ for a constant of porportionality β

Derive a production function (build the model)

The next task is to find a function P that satisfies the equations in (1) and (2) above. Cobb-Douglas found the following production function:

$$P(L, K) = bL^\alpha K^\beta$$

where b is a constant independent of both L and K, $\alpha > 0$ and $\beta > 0$.

Economists use the expression returns to scale. To see that is meant by this expression, replace L by mL and K by mK where m is any fixed positive number:

$$P(mL, mK) = b(mL)^\alpha (mK)^\beta = m^{\alpha + \beta} P(L, k)$$

If $\alpha + \beta = 1$, then P is changed in exactly the same porportion as the inputs L and K.

This situation is described as having <u>constant</u> returns to scale, and is shown in Figure 1a.

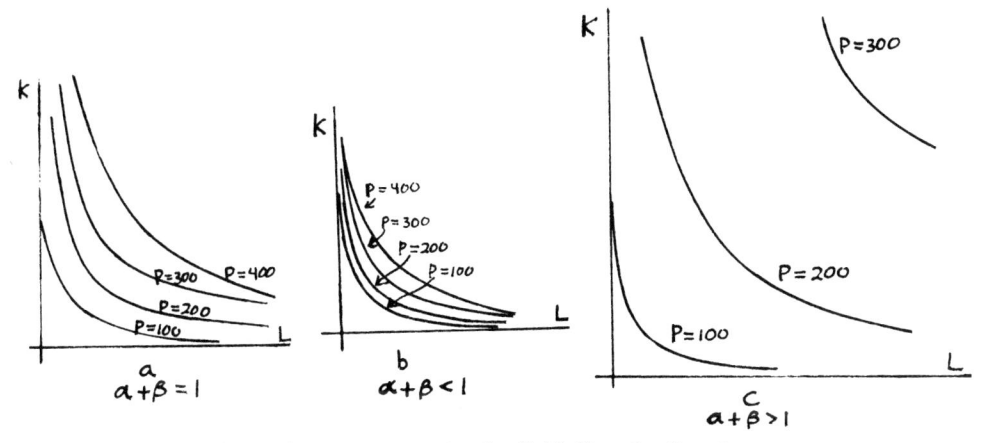

Figure 1 Level curves for the Cobb-Douglas function

If $\alpha + \beta > 1$, then P is changed in greater porportion than the inputs; for example doubling L and K will more than double P. Here we have <u>increasing</u> returns to scale as shown in Figure 1b.

If $\alpha + \beta < 1$, then P is changed in a lesser proportion than the inputs. In this case we have <u>decreasing</u> returns to scale as shown in Figure 1c.

In a more advanced calculus course a result called Euler's theorem is proved which, when applied to the Cobb-Douglas function, yields

$$(\alpha + \beta)P = L\frac{\partial P}{\partial L} + K\frac{\partial P}{\partial K}$$

or

$$P = \frac{1}{\alpha + \beta}L\frac{\partial P}{\partial L} + \frac{1}{\alpha + \beta}K\frac{\partial P}{\partial K}$$

This divides the total production into two parts. The part "due" to labor:

$$\frac{1}{\alpha + \beta}L\frac{\partial P}{\partial L} = \frac{\alpha}{\alpha + \beta}P$$

and the part "due" to capital:

$$\frac{1}{\alpha + \beta}K\frac{\partial P}{\partial K} = \frac{\alpha}{\alpha + \beta}P$$

This says that labor contributes $\dfrac{\alpha}{(\alpha + \beta)}$ of the total production while capital contributes $\dfrac{\beta}{(\alpha + \beta)}$ of the production. This, of course, is merely a felicitous choice of words; there is no production without joint use of both labor and capital. How much should each laborer earn, and what should it cost to use a unit of capital? Since L workers produce goods worth

Chapter 9

$$\frac{1}{\alpha+\beta} L \frac{\partial P}{\partial L}$$

we see that the average amount produced per worker is

$$\frac{1}{\alpha+\beta} \frac{\partial P}{\partial L}$$

Similarly, the use of K units of capital results in the production of goods worth

$$\frac{1}{\alpha+\beta} K \frac{\partial P}{\partial L}$$

the average unit of capital produces

$$\frac{1}{\alpha+\beta} \frac{\partial P}{\partial K}$$

This has led some to propose

$$\frac{1}{\alpha+\beta} \frac{\partial P}{\partial K} \quad \text{and} \quad \frac{1}{\alpha+\beta} \frac{\partial K}{\partial L}$$

as "fair" payments for the use of units of labor and capital.

Check the function against tabulated data

In their study of the United States economy, Cobb and Douglas assumed constant returns to scale ($\alpha + \beta = 1$). Under this assumption the production function is

$$P(L, K) = bL^{\alpha}K^{1-\alpha}$$

We wish to find the values of b and α that make this function most closely match the data in Table 2. The first step is to convert this equation into the equation of a line. Consider what happens when we take the logarithm of each side of this equation.

$$\ln P = \ln b + \alpha \ln L + (1 - \alpha)\ln K$$
$$= \ln b + \alpha \ln(\frac{L}{K}) + \ln K$$

Subtract lnK from both sides:

$$\ln P - \ln K = \ln b + \alpha \ln(\frac{L}{K})$$
$$\ln(\frac{P}{K}) = \alpha \ln(\frac{L}{K}) + \ln b$$
$$y = \alpha x + \ln b \qquad \text{where } y = \ln(\frac{P}{K}) \text{ and } x = \ln(\frac{L}{K})$$

This is the equation of a line with slope α and y-intercept lnb. Now we will use the least square line (Section 9.7) on the n data points. If you carry out this process (a lengthy process; it helps if you do it on a calculator or a computer) you will find

$$\alpha \approx .75 \text{ and } \ln b \approx .01$$

Chapter 9

This says that the Cobb-Douglas production function for the U.S. economy is

$$P(L, K) = 1.01 L^{.75} K^{.25}$$

To see how accurately this function describes the given situation, see the graphs of this function and the actual production for the period 1899-1922 shown in Figure 2.

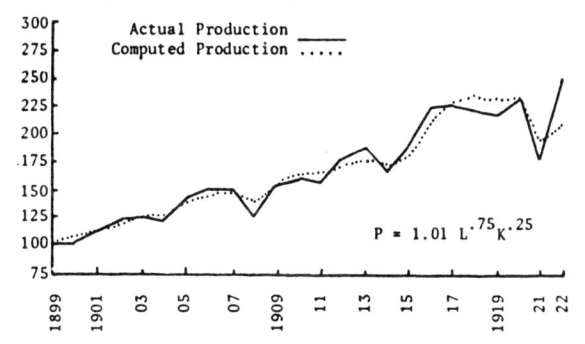

Figure 2 Theoretical and actual curves of production

For Further Study

Since the fundamental paper of Cobb and Douglas, the production function $P(L, K) = bL^{\alpha}K^{\beta}$ has been used in a wide variety of situations, from large scale theories of economic growth to analyses of the decisions facing individual firms. In this application for further study, Karlin, Inc. wishes to maximize its profit (Π) which is easily seen to be

$$\Pi = p_0 P - p_1 L - p_2 K$$

From the Cobb Douglas model, we can suppose that production is given by the function

$$P(L, K) = bL^{\alpha}K^{\beta}C$$

Thus,

$$\Pi = p_0 b L^{\alpha} K^{\beta} - p_1 L - p_2 K$$

The critical points of Π are found by setting $\dfrac{\partial \Pi}{\partial K} = 0$ and $\dfrac{\partial \Pi}{\partial L} = 0$:

$$\frac{\partial \Pi}{\partial L} = p_0 b \alpha L^{\alpha - 1} K^{\beta} - p_1 = 0$$

$$\frac{\partial \Pi}{\partial K} = p_0 b \beta L^{\alpha} K^{\beta - 1} - p_2 = 0$$

By solving these equations simultaneously there is only one critical point (L_0, K_0). Now, apply the second derivative test.

$$A = \frac{\partial^2 \Pi}{\partial L^2} = p_0 b \alpha (\alpha - 1) L^{\alpha - 2} K^{\beta}$$

Chapter 9

$$B = \frac{\partial^2 \Pi}{\partial K \partial L} = p_0 b \alpha \beta L^{\alpha - 1} K^{\beta - 1}$$

$$C = \frac{\partial^2 \Pi}{\partial K^2} = p_0 b \beta (\beta - 1) L^{\alpha} K^{\beta - 2}$$

$$D = B^2 - AC = p_0^2 \alpha^2 \beta^2 L^{2\alpha - 2} K^{2\beta - 2} - p_0^2 b^2 \alpha \beta (\alpha - 1)(\beta - 1) L^{2\alpha - 2} K^{2\beta - 2}$$

$$= p_0^2 b^2 \alpha \beta L^{2\alpha - 2} K^{2\beta - 2} (\alpha + \beta - 1)$$

Since α, β, L and K are all positive, D has the same sign as $\alpha + \beta - 1$. Therefore, a necessary condition for the profit to have a relative maximum is that $\alpha + \beta \leq 1$; the returns to scale must be decreasing or constant.

The second derivative test gives no information if $D = 0$ ($\alpha + \beta = 1$). Putting aside this case for the moment, note that if $\alpha + \beta < 1$ then $D < 0$. Since we know $\alpha > 0$ and $\beta > 0$ and since $\alpha + \beta < 1$, then $\alpha - 1 < 0$. Thus, $A < 0$ so that Π must have a relative maximum at the critical point. This being the only critical point, we may conclude that in the case of decreasing returns to scale, Π has its absolute maximum at the critical point.

Now we return to the case of constant returns to scale. There are only two possibilities: either the maximum profit is zero, obtained when either K or L is zero, or there is no maximum profit. In either case we have shown that we can maximize profit in a nontrivial way only for the case of decreasing returns to scale.

CHAPTER 10
DIFFERENTIAL EQUATIONS

10.1 FIRST ORDER DIFFERENTIAL EQUATIONS, PAGE 319

1. $dy = x^2 dx \implies \int dy = \int x^2 dx \implies y = \dfrac{x^3}{3} + C$

3. $dy = (8x - 10)dx \implies \int dy = \int (8x - 10)dx \implies y = 4x^2 - 10x + C$

5. $dy = (4x^3 - 3x^2 - 5)\,dx \implies \int dy = \int (4x^3 - 3x^2 - 5)\,dx \implies y = x^4 - x^3 - 5x + C$

7. $dy = \dfrac{1}{5}e^x dx \implies \int dy = \int \dfrac{1}{5}e^x dx \implies y = \dfrac{1}{5}e^x + C$

9. $dy = \dfrac{1}{12}(5x+1)^{\frac{1}{2}} dx \implies \int dy = \dfrac{1}{12}\int (5x+1)^{\frac{1}{2}} dx \implies y = \dfrac{1}{12}\left[\dfrac{2}{15}(5x+1)^{\frac{3}{2}}\right] + C$

   $\implies y = \dfrac{1}{90}(5x+1)^{\frac{3}{2}} + C$

11. $y\,dy = x\,dx \implies \int y\,dy = \int x\,dx \implies \dfrac{y^2}{2} = \dfrac{x^2}{2} + C_1 \implies y^2 = x^2 + C \qquad (C = 2C_1)$

13. $y^2 dy = (x^3 - 3)\,dx \implies \int y^2 dy = \int (x^3 - 3)\,dx \implies \dfrac{y^3}{3} = \dfrac{x^4}{4} - 3x + C_1 \implies y^3 = \dfrac{3x^4}{4} - 9x + C$

    $(C = 3C_1)$

15. $\dfrac{dy}{y} = 2x\,dx \implies \int \dfrac{dy}{y} = \int 2x\,dx \implies \ln|y| = x^2 + C_1$

    We could isolate y: $|y| = e^{x^2 + C_1} = e^{x^2}e^{C_1} \implies |y| = C_2 e^{x^2} \qquad (C_2 = e^{C_1})$

    $\implies \pm C_2 e^{x^2} \implies y = Ce^{x^2} \qquad (C = \pm C_2)$

17. Using Theorem 3, where $K = .02$, $P = Me^{.02t}$. Without Theorem 3:

    $\dfrac{dP}{P} = .02\,dt \implies \ln|P| = .02t + C_1 \implies 50\ln|P| = t + C$ ($C = 50C_1$). If we want to solve for P, we would get $P = Me^{.02t}$.

19. Using Theorem 3, where $K = .001$, $N = Me^{.001t}$. Without Theorem 3:

    $\dfrac{dN}{N} = .001\,dt \implies \ln|N| = .001t + C_1 \implies 1000\ln|N| = t + C \qquad (C = 1000\,C_1).$

Chapter 10   244

21. $y^2 dy = x^2 dx$ => $\int y^2 dy = \int x^2 dx$ => $\frac{y^3}{3} = \frac{x^3}{3} + C_1$ => $y^3 = x^3 + C$ ($C = 3C_1$).

Substitute $y = 5$, $x = 0$ and solve for C: $5^3 = 0^3 + C$ => $C = 125$. The particular solution is $y^3 = x^3 + 125$.

23. See Problem 17. $P = Me^{.02t}$. Substitute $P = e^3$, $t = 0$ and solve for the constant M: $e^3 = Me^0$ => $M = e^3$. The particular solution is $P = e^3 e^{.02t}$. An equivalent form is $P = e^{3 + .02t}$ => $\ln P = 3 + .02t$.

25. Multiply both sides by $\frac{dx}{xy}$: $\frac{dy}{y} - \frac{dx}{\sqrt{x}} = 0$ => $\int \frac{dy}{y} - \int x^{-\frac{1}{2}} dx = C_1$ => $\ln |y| - 2x^{\frac{1}{2}} = C$ => $\ln |y| = 2\sqrt{x} + C$. Substitute $y = 1$, $x = 1$ and solve for C: $\ln 1 = 2\sqrt{1} + C$ => $0 = 2 + C$ => $C = -2$. The particular solution is $\ln |y| = 2\sqrt{x} - 2$.

27. $\frac{dy}{y} = \frac{x dx}{1 + x^2}$ => $\ln |y| = \frac{1}{2} \ln |1 + x^2| + C$

Since $1 + x^2$ is always positive, $\ln |y| = \frac{1}{2} \ln (1 + x^2) + C$. Substitute $x = -1$, $y = 2$ and solve for C: $\ln 2 = \frac{1}{2} \ln 2 + C$ => $\frac{1}{2} \ln 2 = C$. The particular solution is $\ln |y| = \frac{1}{2} \ln (1 + x^2) + \frac{1}{2} \ln 2$.

29. If N is the number of crimes and t represents the number of years after 1986, $\frac{dN}{dt} = .03t$.

By Theorem 3, $N = Me^{.03t}$. The number of crimes was 3500 when the year was 1986 => $N = 3500$ when $t = 0$. Substitute this into $N = Me^{.03t}$ and solve for the constant M: $3500 = Me^0$ => $M = 3500$. The particular solution is $N = 3500e^{.03t}$. The number of crimes, N, in 1996 ($N = 10$) will be $N = 3500e^{.03(10)} = 3500e^{.3} \approx 4725$.

31. If N is the number of bacteria and t is the number of hours, $\frac{dN}{dt} = .10t$. By Theorem 3, $N = Me^{.10t}$. Initially there are 12 (million), so substitute $N = 12$ when $t = 0$ and solve for the constant M: $12 = Me^0 = M$. So the particular solution is $N = 12e^{.10t}$. Next, let $N = 1000$ and solve for t: $1000 = 12e^{.10t}$ => $\frac{1000}{12} = e^{.10t}$ => $.10t = \ln (\frac{1000}{12})$ => $t = 10 \ln (\frac{1000}{12}) \approx 44.2$ hours.

Chapter 10

10.2 APPLICATION-GROWTH MODELS, PAGE 326 - 327

1. This is uninhibited growth with a growth rate of 12%, $P_0 = 25,000$ and $t = 5$ (see Example 1). We have $P = 25,000e^{.12(5)} \approx 45,553$.

3. See Example 2. Since 28% is present: $-.0001205479t = \ln(\frac{P}{P_0}) = \ln(.28)$

$t \approx 10,560$ years.

5. Using $P'(x) = 6x^2 - 10 \Rightarrow P(x) = 2x^3 - 10x + C$. Since $P(3) = 1,000$:

$1000 = P(3) = 2(3)^3 - 10(3) + C \Rightarrow C = 976$. Thus, $P(x) = 2x^3 - 10x + 976$.

Now $P(5) = 2(5)^3 - 10(5) + 976 = 1176$. The increase is $P(5) - P(3) = 1176 - 1000 = 176$ items per month.

7. Using the uninhibited growth model, $P = P_0 e^{.0831 t}$ where t is the number of years after 1980 and $P_0 = 2.626$ (billion dollars). In 1990, $t = 10$: $P = 2.626e^{.0831(10)} \approx \$6.028$ billion.

9. Use $P = P_0 e^{rt}$ where t is the number of years after 1970 and $P_0 = 203,302,031$:

$P = 203,302,031 e^{rt}$. Since $P = 226,545,805$ when $t = 10$: $\frac{226,545,805}{203,302,031} = e^{10r}$ or

$r = \dfrac{\ln \dfrac{226,545,805}{203,302,031}}{10} \approx .0108254$. So in 1990, $t = 20$:

$P = 203,302,031 e^{.0108254(20)} \approx 252,447,066$

11. Use $P = P_0 e^{rt}$ where t is the number of years after 1980 and $P_0 = 14,609,000$:

$P = 14,609,000 e^{rt}$. Since $P = 15,575,000$ when $t = 4$: $\frac{15,575,000}{14,609,000} = e^{r(4)}$ or

$r = \dfrac{\ln \dfrac{15,575,000}{14,609,000}}{4} \approx .0160073$. So in 1990, $t = 10$: $P = 14,609,000 e^{.0160073(10)}$

$\approx 17,145,076$

13. Use $P = P_0 e^{rt}$ where t is the number of years after 1984 and r is 10.5%:

$P = 2,487,000 e^{.105 t} \Rightarrow P(6) = 2,487,000 e^{.105(6)} \approx 4,669,618$

15. $\dfrac{dy}{.001y - .7} = dt \Rightarrow \int \dfrac{dy}{.001y - .7} = \int dt \Rightarrow 1000 \ln |.001y - .7| = t + C$

Chapter 10   246

From Problem 14 if there are 200 initially, $1000 \ln |.001(200) - .7| = 0 + C$

$\Rightarrow C \approx -693.147$. Thus, $1000 \ln |.001y - .7| = t - 693.147$

In one week, t = 168 (hours): $1000 \ln |.001y - .7| = 168 - 693.147 \Rightarrow$

$\ln |.001y - .7| = -.525147 \Rightarrow |.001y - .7| = e^{-.525147} \approx .59147$

If we assume growth rate is positive, $.001y - .7 = .59147 \Rightarrow y = 1{,}291$

17. Using the limited growth model, $P(t) = .5(1 - e^{-rt})$.

$.2 = P(10) = .5 [1 - e^{-r(10)}] \Rightarrow r = \dfrac{\ln(\frac{.2}{.5})}{-10} \approx .05108256$

We have $P(t) = .5(1 - e^{-.05108256\,t})$. We want to find t such that P = 30%:

$.30 = .5(1 - e^{-.05108256\,t}) \Rightarrow \dfrac{.30}{.5} = 1 - e^{-.05108256\,t} \Rightarrow e^{-.05108256\,t} = .4$

$\Rightarrow t = \dfrac{\ln(.4)}{-.05108256} \approx 18$ times per week.

19. $\dfrac{dP}{L-P} = K dt \Rightarrow \displaystyle\int \dfrac{dP}{L-P} = \displaystyle\int K dt \Rightarrow -\ln|L - P| = Kt + C$

Since L = 1 and L - P ≥ 0, we have -ln (1 - P) = Kt + C. When t = 0, P = 0:

-ln (1 - 0) = K(0) + C $\Rightarrow$ C = 0. This gives us -ln (1 - P) = Kt $\Rightarrow$ ln (1 - P) = -Kt

$\Rightarrow 1 - P = e^{-Kt} \Rightarrow P = 1 - e^{-Kt}$.

10.3 SPECIAL SECOND ORDER DIFFERENTIAL EQUATIONS, PAGES 331 - 332

1. $\dfrac{dy}{dx} = \displaystyle\int -24x\,dx = -12x^2 + C_1$

$y = \displaystyle\int (-12x^2 + C_1)dx = -4x^3 + C_1 x + C_2$

Using y = 5 when x = 0: $5 = -4(0)^3 + C_1(0) + C_2 \Rightarrow C_2 = 5$

We have $y = -4x^3 + C_1 x + 5$. Using y = 29 when x = 2:

$29 = -4(2)^3 + C_1(2) + 5 \Rightarrow C_1 = 28$. The particular solution is $y = -4x^3 + 28x + 5$.

3. $\dfrac{dy}{dx} = \displaystyle\int (10 - 6x)dx = 10x - 3x^2 + C_1$

$y = \displaystyle\int (10x - 3x^2 + C_1)dx = 5x^2 - x^3 + C_1 x + C_2$

Chapter 10   247

Using y = 8 when x = 0: $8 = 5(0)^2 - 0^3 + C_1(0) + C_2$ => $C_2 = 8$

We have $y = 5x^2 - x^3 + C_1x + 8$. Using y = 52 when x = 4:

$52 = 5(4)^2 - 4^3 + C_1(4) + 8$ => $C_1 = 7$. The particular solution is $y = 5x^2 - x^3 + 7x + 8$

5. $\dfrac{dy}{dx} = \int(2x^2 - 3x + 5)dx = \dfrac{2x^3}{3} - \dfrac{3x^2}{2} + 5x + C_1$

$y = \int(\dfrac{2x^3}{3} - \dfrac{3x^2}{2} + 5x + C_1)dx = \dfrac{x^4}{6} - \dfrac{x^3}{2} + \dfrac{5x^2}{2} + C_1x + C_2$

Using y = 3 when x = 0: $3 = \dfrac{0^4}{6} - \dfrac{0^3}{2} + \dfrac{5(0)^2}{2} + C_1(0) + C_2$

We have $y = \dfrac{x^4}{6} - \dfrac{x^3}{2} + \dfrac{5x^2}{2} + C_1x + 3$. Using y = 12 when x = 2:

$12 = \dfrac{2^4}{6} - \dfrac{2^3}{2} + \dfrac{5(2)^2}{2} + C_1(2) + 3$ => $C_1 = \dfrac{1}{6}$. The particular solution is

$y = \dfrac{x^4}{6} - \dfrac{x^3}{2} + \dfrac{5x^2}{2} + \dfrac{x}{6} + 3$.

7. $y'' = 4 - 3x$ => $y' = \int(4 - 3x)dx = 4x - \dfrac{3x^2}{2} + C_1$

$y = \int(4x - \dfrac{3x^2}{2} + C_1)dx = 2x^2 - \dfrac{x^3}{2} + C_1x + C_2$

Using y = 4500 when x = 0: $4500 = 2(0)^2 - \dfrac{0^3}{2} + C_1(0) + C_2$.

We have $y = 2x^2 - \dfrac{x^3}{2} + C_1x + 4500$. Using y = 50 when x = 10:

$50 = 20(10)^2 - \dfrac{10^3}{2} + C_1(10) + 4500$ => $C_1 = -415$. The particular solution is

$y = 2x^2 - \dfrac{x^3}{2} - 415x + 4500$.

9. See Example 2. Let y' = p so that $y'' = \dfrac{dp}{dx}$: $xy'' = 2y'$ => $x\dfrac{dp}{dx} = 2p$. Separate the

variables: $\dfrac{dp}{p} = 2\dfrac{dx}{x}$ => $\ln|p| = 2\ln|x| + k$.

Let $k = \ln C_1$: $\ln|p| = 2\ln|x| + \ln C_1 = \ln x^2 + \ln C_1$ => $\ln|p| = \ln C_1 x^2$ => $p = \pm C_1 x^2$

Substituting $p = \dfrac{dy}{dx}$: $\dfrac{dy}{dx} = \pm C_1 x^2$

Chapter 10   248

$$y = \int \pm C_1 x^2 dx = \pm \frac{C_1 x^3}{3} + C_2$$

Using $y = -4$ when $x = 0$: $-4 = \pm \frac{C_1(0)^3}{3} + C_2$. We have $y = \pm \frac{C_1 x^3}{3} - 4$. Using $y = 8$

when $x = 3$: $8 = \pm \frac{C_1(3)^3}{3} - 4 \Rightarrow C_1 = \pm \frac{4}{3}$. The particular solution is $y = \pm \frac{4}{9} x^3 - 4$.

11. Subtract y from both sides to get the form $ay'' + by' + cy = 0$: $2y'' - y' - y = 0$. Using Theorem 4, the auxiliary equation is:

$$2x^2 - m - 1 = 0 \Rightarrow (2m + 1)(m - 1) = 0 \Rightarrow m = -\frac{1}{2} \text{ or } 1.$$

The general solution is $y = Ae^{-\frac{1}{2}x} + Be^x$.

Using $y = 0$ when $x = 0$: $0 = Ae^0 + Be^0 = A + B$. Using $y = 1$ when $x = 4$:

$1 = Ae^{-2} + Be^4$. We need to solve the system

$$\begin{cases} A + B = 0 \\ e^{-2}A + e^4 B = 1 \end{cases} \text{The solution is } A = \frac{e^2}{1 - e^6} \text{ and } B = \frac{e^2}{e^6 - 1}$$

The particular solution is $y = \left(\frac{e^2}{1 - e^6}\right) e^{-\frac{x}{2}} + \left(\frac{e^2}{e^6 - 1}\right) e^x$.

13. $\frac{dy}{dx} = \int e^x dx = e^x + A$, where A is a constant.

$y = \int (e^x + A) dx = e^x + Ax + B$, where B is a constant.

15. $\frac{dy}{dx} = \int (1 - e^{-x}) dx = x + e^{-x} + A$, where A is a constant.

$y = \int (x + e^{-x} + A) dx = \frac{x^2}{2} - e^{-x} + Ax + B$, where B is a constant.

17. $y'' = 5x^2 - 2 \Rightarrow y' = \int (5x^2 - 2) dx = \frac{5x^3}{3} - 2x + A$

$y = \int (\frac{5x^3}{3} - 2x + A) dx = \frac{5x^4}{12} - x^2 + Ax + B$, where A and B are constants.

19. $y'' = 2 - x \Rightarrow y' = \int (2 - x) dx = 2x - \frac{x^2}{2} + A$

$y = \int (2x - \frac{x^2}{2} + A) dx = x^2 - \frac{x^3}{6} + Ax + B$, where A and B are constants.

Chapter 10

21. Using Theorem 4, the auxiliary equation is $6m^2 + 23m - 4m = 0$. The solutions are $m_1 = \frac{1}{6}$ and $m_2 = -4$. The general solution is $y = Ae^{m_1 x} + Be^{m_2 x} = Ae^{\frac{1}{6}x} + Be^{-4x}$.

23. Using Theorem 4, the auxiliary equation is $m^2 - 9 = 0$. The solutions are $m_1 = 3$ and $m_2 = -3$. The general solution is $y = Ae^{m_1 x} + Be^{m_2 x} = Ae^{3x} + Be^{-3x}$.

25. Since the variable y is missing, let $p = \frac{dy}{dx}$ so that $\frac{dp}{dx} = \frac{d^2 y}{dx^2}$.

$xy'' + 2y' = 0 \Rightarrow x\frac{dp}{dx} + 2p = 0$. Separating variables, $\frac{dp}{p} = -\frac{2dx}{x} \Rightarrow \ln|p|$

$= -2\ln|x| + k$. Let $k = \ln C_1$ and use properties of logarithms: $\ln|p| = \ln x^{-2} + \ln C_1$

$= \ln C_1 x^{-2} \Rightarrow p = \pm C_1 x^{-2}$

$\frac{dy}{dx} = \pm C_1 x^{-2} \Rightarrow y = \int \pm C_1 x^{-2} dx = \pm C_1 x^{-1} + B$

Letting $A = \pm C_1$ we have $y = Ax^{-1} + B$.

27. Let $p = \frac{dy}{dx}$ so the differential equation becomes

$2\frac{dp}{dx} + (x^3)\frac{dp}{dx} - 3x^2 p = 0 \Rightarrow (2 + x^3)\frac{dp}{dx} = 3x^2 p \Rightarrow \frac{dp}{p} = \frac{3x^2}{2 + x^3}dx$. Integrating both

sides we get $\ln|p| = \ln|2 + x^3| + C_1$. Letting $C_1 = \ln C_2$, $\ln|p| = \ln|2 + x^3| + \ln C_2$

$= \ln C_2 |2 + x^3|$. Therefore, $p = \pm C_2 (2 + x^3)$. Substituting $\frac{dy}{dx}$ for p,

$\frac{dy}{dx} = \pm C_2 (2 + x^3) \Rightarrow y = \int \pm C_2 (2 + x^3) dx = \pm C_2 (2x + \frac{x^4}{4}) + C_3$. Letting $A = \pm C_2$

and $B = C_3$, we have $y = A(2x + \frac{x^4}{4}) + B$.

29. Using Theorem 4, write the equation as $D'' - MD = 0$. The auxiliary equation is $m^2 - M = 0$. The solutions are $m_1 = \sqrt{M}$ and $m_2 = -\sqrt{M}$. The general solution is:

$D = Ae^{m_1 t} + Be^{m_2 t} = Ae^{\sqrt{M}\, t} + Be^{-\sqrt{M}\, t}$.

Chapter 10

10.4 CHAPTER REVIEW, PAGES 332 - 333

1. $dy = (12x^3 - 6x^2 + 5)dx$

    $\int dy = \int (12x^3 - 6x^2 + 5)dx$

    $y = 3x^4 - 2x^3 + 5x + C$

3. $9\dfrac{dy}{dx} = 14x - 18$

    $dy = (\dfrac{14}{9}x - 2) dx$

    $y = \dfrac{7}{9}x^2 - 2x + C$

5. $8y\,dy = (x + 3)\,dx$

    $\int 8y\,dy = \int (x + 3)\,dx$

    $4y^2 = \dfrac{x^2}{2} + 3x + C_1$

    $8y^2 = x^2 + 6x + C_2$

7. $\dfrac{dy}{y} = 10x\,dx$

    $\int \dfrac{dy}{y} = \int 10x\,dx$

    $\ln |y| = 5x^2 + C_1$

    $|y| = e^{5x^2 + C_1} = e^{C_1} e^{5x^2} = C_2 e^{5x^2}$ $(e^{C_1} = C_2)$

    $|y| = C_2 e^{5x^2}$ => $y = C_3 e^{5x^2}$ $(C_3 = \pm C_2)$

9. $y^2\,dy = x^4\,dx$

    $\int y^2\,dy = \int x^4\,dx$

    $\dfrac{y^3}{3} = \dfrac{x^5}{5} + C.$ Substituting $y = -1$ and $x = 0$,

    $-\dfrac{1}{3} = 0 + C.$ So $\dfrac{y^3}{3} = \dfrac{x^5}{5} - \dfrac{1}{3}$ => $5y^3 = 3x^5 - 5$.

11. $\dfrac{(4 - y^2)\,dy}{y} = x\,dx$

    $\int (\dfrac{4}{y} - y)\,dy = \int x\,dx$

    $4\ln |y| - \dfrac{y^2}{2} = \dfrac{x^2}{2} + C.$ Substituting $y = 1$ and $x = 1$,

    $4 \ln 1 - \dfrac{1}{2} = \dfrac{1}{2} + C$

    $0 - \dfrac{1}{2} = \dfrac{1}{2} + C$

    $-1 = C.$ So $4 \ln |y| - \dfrac{y^2}{2} = \dfrac{x^2}{2} - 1$

13. $y'' = x^3 - 8$

    $y' = \dfrac{x^4}{4} - 8x + C_1$

    $y = \dfrac{x^5}{20} - 4x^2 + C_1 x + C_2$

$$8 \ln |y| = x^2 + y^2 - 2$$
$$\ln y^8 = x^2 + y^2 - 2$$
$$y^8 = e^{x^2 + y^2 - 2}$$

15. $8y'' - 17y' + 2y = 0$. Auxiliary equation is $8m^2 - 17m + 2 = 0 \Rightarrow (m-2)(8m-1) = 0$
$\Rightarrow m = \dfrac{1}{8}$ or 2. The solution to the differential equation is $y = Ae^{\frac{x}{8}} + Be^{2x}$.

17. The auxiliary equation is $m^2 - m - 6 = 0 \Rightarrow (m-3)(m+2) = 0 \Rightarrow m = 3$ or -2. The solution to the differential equation is $y = Ae^{3x} + Be^{-2x}$. First substitute $x = 0$ and $y = 0$,
then $x = 1$ and $y = 5$: $\begin{cases} 0 = A + B \\ 5 = Ae^3 + Be^{-2} \end{cases}$

Solving this system for A and B gives $A = \dfrac{5e^2}{e^5 - 1}$, $B = \dfrac{-5e^2}{e^5 - 1}$. The solution of the

differential equation is $y = \dfrac{5e^2}{e^5 - 1} e^{\frac{x}{8}} - \dfrac{5e^2}{e^5 - 1} e^{2x}$.

19. $\dfrac{dP}{1-P} = K dt \Rightarrow \int \dfrac{dP}{1-P} = \int K dt \Rightarrow -\ln|1-P| = Kt + C_1 \Rightarrow \ln|1-P| = -Kt - C_1$

$|1 - P| = e^{-Kt - C_1} \cdot C_2 e^{-Kt} \quad (C_2 = e^{-C_1})$

$1 - P = C_3 e^{-Kt} \quad (C_3 = \pm C_2)$

$P = 1 - C_3 e^{-Kt}$. Initially 0% are aware of the new proof, which means $P = 0$ when $t = 0$:

$0 = 1 - C_3 \Rightarrow C_3 = 1$. So $P = 1 - e^{-Kt}$. We are given $P = 25\%$ when $t = 6$:

$.25 = 1 - e^{-6K} \Rightarrow .75 = e^{-6K} \Rightarrow K = \dfrac{\ln .75}{-6} \approx .047947 \Rightarrow P = 1 - e^{-.047947t}$.

If $P = 75\%$, $.75 = 1 - e^{-.047947t} \Rightarrow .25 = e^{-.047947t} \Rightarrow \dfrac{\ln .25}{-.047947} = t$

$\Rightarrow t \approx 29$ (months)

10.5 STUDENT'S TEST REVIEW AND ADDITIONAL PRACTICE
OBJECTIVES

The material of this chapter is reviewed in the following list of objectives. After each objective there are some practice questions. Answers to these problems immediately follow. For a sample test select the first question of each set and check your answers. Additional practice is given by the other questions in each set. If you are having trouble

Chapter 10

with a particular type of problem, or if you want additional practice, look back at the indicated section in the test.

[10.1] Objective 1: *Solve first order differential equations using Differential Equation Theorem 1.*

1. $\dfrac{dy}{dx} = 8x^3 - 2x^2 + 1$
2. $\dfrac{dy}{dx} = \sqrt{x} + 5$

3. $3y' = 2x - 5$
4. $6y' - 5 = 11x$

Objective 2: *Solve first order differential equations using Differential Equation Theorem 2.*

5. $4yy' = x + 5$
6. $yy' = e^{2x+1} - x$

7. $\dfrac{dy}{dx} = 5xy$
8. $\dfrac{dy}{dx} = 3x^3y^2 - 2xy^2 - y^2$

[10.2] Objective 3: *Solve first order differential equations for a particular solution.*

9. $\dfrac{dy}{dx} = x^3 y^{-2}$ for $y = -5$ when $x = 0$

10. $y'x^2 = y$ for $y = 1$ when $x = 2$

11. $2xy = y'$ for $y = e^2$ when $x = 1$

12. $\dfrac{dy}{dx} = \dfrac{xy}{5-y^2}$ for $y = 1$ when $x = 0$

[10.3] Objective 4: *Solve second order differential equations where both the variables y and y' are missing.*

13. $\dfrac{d^2y}{dx^2} = 15 - 3x^2$ where $y = 3$ if $x = 0$ and $y = 9$ if $x = 2$

14. $y'' = 2x^2 + 5x - 3$ where $y = -1$ if $x = 0$ and $y = 5$ if $x = 1$

15. $2x^3 - y'' = 4$

16. $x^{-2} y'' = 6x$

Objective 5: *Solve second order differential equations where either the variable x or the variable y is missing.*

Chapter 10

17. $y'' - 16y = 0$

18. $\dfrac{8d^2y}{dx^2} = \dfrac{17dy}{dx} - 2y$

19. $y'' = y$ where $y = 0$ when $x = 0$ and $y = \dfrac{2e}{1 - e^2}$ when $x = 1$

20. $y'' = y' + 6y$ where $y = 0$ when $x = 0$ and $y = 5$ when $x = 1$

[10.1-10.3] Objective 6: *Solve applied problems based on the preceding objectives. For specific examples of the types of applications look at the list of applications in this chapter on page 314.*

21. Suppose that Melville's Scooter Sales are increasing at a constant rate of 5% per year. If the present sales are 1250 scooters per year, how many scooters could they expect to sell in 3 years?

22. In 1987 a new mathematical theorem was proved, and a diffusion of information model is

$$\dfrac{dP}{dt} = k(1 - P)$$

P is the percentage of mathematicians who are aware of the new proof after t months. Find the constant k if it takes 6months for one-fourth of the mathematicians to hear of the theorem. How long will it take for 90% of the mathematicians to hear of the new result?

23. Pertec stock has enjoyed a growth rate according to the model

$$\dfrac{dV}{dt} = k(100 - V)$$

where V is the value of the stock (per share) after t months. If the stocks issue price was $10 and if the value was $40 after 18 months, find the value of k. What is the limiting value for this stock?

24. If inflation in Italy over the past 5 years has averaged 20.1%, how long will it take 10,000 lire to be worth what 100 lire is worth today?

ANSWERS TO STUDENT'S TEST REVIEW AND PRACTICE QUESTIONS

1. $y = 2x^4 - \dfrac{2}{3}x^3 + x + C$ 2. $y = \dfrac{2}{3}x^{\tfrac{3}{2}} + 5x + C$ 3. $y = \dfrac{1}{3}x^2 - \dfrac{5}{3}x + C$

4. $y = \dfrac{11}{12}x^2 + \dfrac{5}{6}x + C$ 5. $y^2 = \dfrac{1}{4}x^2 + \dfrac{5}{2}x + C$ 6. $y^2 = e^{2x+1} - x^2 + C$

Chapter 10

7. $y = Ae^{\frac{5x^2}{2}}$
8. $\dfrac{1}{y} = -\dfrac{3}{4}x^4 + x^2 + x + C$
9. $y^3 = \dfrac{3}{4}x^4 - 125$
10. $y = \pm e^{\frac{1}{2}} \cdot e^{-\frac{1}{x}}$

11. $y = \pm e^{x^2 + 1}$
12. $10 \ln |y| - y^2 = x^2 - 1$
13. $y = -\dfrac{1}{4}x^4 + \dfrac{15}{2}x^2 - 10x + 3$

14. $y = \dfrac{1}{6}x^4 + \dfrac{5}{6}x^3 - \dfrac{3}{2}x^2 + \dfrac{13}{2}x - 1$
15. $y = \dfrac{1}{10}x^5 - 2x^2 + Ax + B$

16. $y = \dfrac{3}{10}x^5 + Ax + B$
17. $y = Ae^{4x} + Be^{-4x}$
18. $y = Ae^{\frac{x}{8}} + Be^{2x}$

19. $y = \dfrac{-2e^2}{(1-e^2)^2}e^x + \dfrac{2e^2}{(1-e^2)^2}e^{-x}$

20. $y = \dfrac{5e^2}{e^5 - 1}e^{3x} + \dfrac{5e^2}{1 - e^5}e^{-2x}$ or $\dfrac{5}{e^5 - 1}(e^{3x+2} - e^{2-2x})$

21. 1452
22. $k \approx -.0479701$; 48 months
23. $k \approx -.02252584$; $100
24. About 22.9 years

MODELING APPLICATION 8: SAMPLE ESSAY

The Battle of Trafalgar*

    The Franch Revolution, beginning in 1789, paved the way to power for Napoleon Bonaparte, who overran much of Europe. Britain's security depended on its control of the seas. In 1805, combined French and Spanish fleets were met off Cape Trafalgar by the British fleet under Admiral Nelson. In a prebattle memorandum, Nelson adopted a plan which can be mathematically verified. Nelson decided to break the enemy line in two (see Figure 1), concentrating 32 of his ships on the 23 ships at the rear of the enemy's line, and using his remaining 8 to fight a delaying action against the remaining 23 enemy. At Trafalgar, Nelson lost his life but won the battle, a victory which was the threshold of British dominance of the seas in the 19th century. Show this mathematical verification.

---

* This extended application is adapted from "Differential Equations and the Battle of Trafalgar" by David H. Nash from the College Mathematics Journal, March 1985, pp. 98-102.

Chapter 10

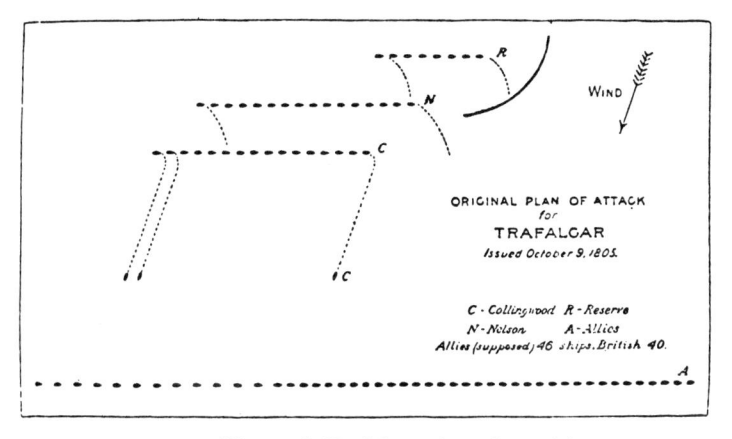

Figure 1 Trafalgar plan of attack*

Early this century, the British engineer Frederick Lanchester developed what he called the N-square law of fighting strength. This law states that after an engagement of M units of the red army against N units of the blue army, where $M > N$,

(the number of red units left)$^2 = M^2 - N^2$

the number of blue units left $= 0$

Assuming Lanchester's formula for the battle of Trafalgar, Nelson's 32 ships against 23 would give him not just a 9 (i.e. 32 - 32 = 9) ship advantage, but a

$\sqrt{32^2 - 23^2} = 22$ (22.5) ship advantage. Against the other 8 British ships, the remaining 23 enemy would have a $\sqrt{32^2 - 23^2} = 22$ (21.56) ship advantage. Put another way,

    Nelson's expection: destroy 23 enemy ships with 32 of his own and have 22 (22.25) of the 32 left.
    Enemy's expection: destroy the other 8 British ships with its remaining 23 and emerge with 22(21.56)
    Nelson's remaining 22.25 ships would give him a slight advantage over the enemy's remaining 21.56 ships.

We now look at some of the mathematics behind the N-square law. Let $r(t)$ and $b(t)$, respectively, denote the number of red army and blue army units at time t. Assume that $r(t)$ and $b(t)$ are differentiable, and let

$$\begin{cases} r'(t) = -Bb(t) \\ b'(t) = -Rr(t) \end{cases}$$

---

* From <u>The Life of Nelson</u>, Vol. II, by A. T. Mahan (London: Samson Low, Marston & Co., 1897, p. 345).

where B and R are positive constants. These equations state that the loss rate for a side is proportional to the number of its enemy. Lanchester identified the constants B and R as measures of the fighting effectiveness of the blue units and the red units, respectively. We will make, as Lanchester did, the conservative assumption that both sides have the same "fighting effectiveness" constants; that is B = R, and solve the system.

First, solve each equation independently to obtain

$$\begin{cases} r + b = Cae^{-Bt} \\ r - b = Cbe^{Bt} \end{cases}$$

With work that is beyond the scope of this book, the solution to this system of differential equations is

$$\begin{cases} r(t) = .5\,[r_0(e^{Bt} + e^{-Bt}) - b_0(e^{Bt} - e^{-Bt})] \\ b(t) = -.5\,[r_0(e^{Bt} - e^{-Bt}) - b_0(e^{Bt} + e^{Bt})] \end{cases}$$

Assume $r_0 > 0$ and $b_0 > 0$, then $b(t) = 0$ can be solved for t only if $r_0 > b_0$. The result is that $b(t) = 0$ when t assumes the value

$$t_1 = \left(\frac{1}{2B}\right) \ln\left[\frac{r_0 + b_0}{r_0 - b_0}\right]$$

Multiplying corresponding sides of the original system together

$$[r(t)]^2 - [b(t)]^2 = r_0^2 - b_0^2,$$

for all t. In particular, when $r_0 > b_0$ and $t = t_1$, then

$$[r(t_1)]2 = r_0^2 - b_0^2$$

It can be shown that the reds win (in the sense that they reduce the enemy to 0 units in a finite time) if and only if

$$Rr_0^2 > Bb_0^2$$

In this problem, there is a stalement in the sense that neither side reduces the other to exactly 0 units in a finite time. It appears that Lanchester's "fighting effectiveness" constant multiplied by the square of the number of fighting units of a side at any time is a useful measure for determining who wins. Lanchester called it "fighting strength". It is easy to show, for example, that equal fighting strengths imply that the ratio $\frac{r(t)}{b(t)}$ is

Chapter 10

constant over time, and that the fractional attrition rates, $\dfrac{r'(t)}{r(t)}$ and $\dfrac{b'(t)}{b(t)}$, are also equal. One way for a side to improve its new fighting strength position over an enemy is to break the battle into independent subbattles.

Define the (combined) fighting strength of two disjoint detachments on one side, each of which is independently fighting one of two disjoint detachments of the enemy, to be the sum of the fighting strengths of its two individual detachments. This is a reasonable definition because it can be shown that if winners of battles between detachments redeploy (if they are opponents), and fight each other in a second round, a side with a (combined) fighting strength advantage will maintain an advantage and win in the obvious sense of annihilating both enemy detachments in a finite length of time, whereas equal (combined) fighting strengths implies the stalemate that neither side ever completely annihilates the other. In fact, the preceding follows easily from the N–square law, which implies that in each subbattle a difference between detachment fighting strengths is constant; hence, so are sums of differences. Therefore, opponents enter a second round, if one is needed, with the same net fighting strengths over the enemy as held initially.

Lanchester's additivity assumption ignores the complication that independent battles between detachments may end at different times. In practice, it could make sense for units remaining after a completed subbatle to go to the immediate aid of an allied detachment still fighting its own subbattle. Lanchester assumed each side would wait for all subbattles between detachments to terminate before using any leftover units in a second round. For the eqation which is solved for t_p theoretically, a detachment of 23 enemy ships would detroy 8 British in less than half the time it would take 32 of Nelson's to detroy 23 enemy. Nelson anticipated such a disadvantage but, correctly, as it turned out, did not expect the enemy to redeploy efficiently to press its advantage. Thus, the Lanchester assumption was not a bad one for Trafalgar.

Although it is almost certain Nelson did not explicitly know these mathematical details, he evidently chose a smart plan. Incidentally, Nelson's stated reason for not using the line-ahead formation was that it would be impossible to manage effectively under the weather and other circumstances. Furthermore, an initial attack at the middle of the enemy line was ordered partly as an attempt to capture the opposing cammander-in-chief there. Nelson's memorandum also shows that he ideally wanted to use his entire fleet of 40 ships against the rear 26 of the enemy, and supposed the remaining 20 enemy ships would not be in a position to engage him at all until he had gained a victory over the 26. So if Nelson had used the N-square law, he might have computed his effective ship advantage as

$$\sqrt{40^2 - 26^2 - 20^2} = 22.89$$

rather than Lanchester's .69. The option Lanchester analyzed concentrated the maximum number of Nelson's ships, 8, for delaying advantage. Of course, all of the preceding applies continuous mathematics to an essentially discrete situation. Real battles and leaders do not necessarily conform to clear-cut mathematics. But sometimes mathematics can help in strategic analyses and explicitly or implicitly give one side a strategic edge.

APPENDIX A
REVIEW OF ALGEBRA

A.1 ALGEBRA PRETEST, PAGE 336
PART I
1. Even though the 5 is also a factor of $5x^2y$, the best answer is numerical coefficient. Answer is E.
3. $(-2)(-2)(-2) = (4)(-2) = -8$. Answer is A.
5. $a^2 - 2ab + b^2 = (-3)^2 - 2(-3)(5) + 5^2 = 9 + 30 + 25 = 64$. Answer is C.
7. $3x - 5x = (3 - 5)x = -2x$. Answer is D.
9. $(-3y^2)^3 = (-3)^3(y^2)^3 = -27y^6$. Answer is D.
11. $3x + 12 = 6$
    $3x = -6$ Subtracted 12 from both sides.
    $x = -2$ Divided both sides by 3.
    Answer is C.
13. $2x - 16 = 3x - 9$
    $-16 = x - 9$ Subtracted 2x from both sides.
    $-7 = x$ Added 9 to both sides.
    Answer is B.
15. $3(x + 2) - (x - 4y) + (x + y) = 3x + 6 - x + 4y + x + y = 3x + 5y + 6$. Answer is D.
17. False. The left side of the equation is 2xy. The right side of the equation if 4xy.
19. False. This is "almost" true. Since $0^2 = 0$, x^2 could be zero. A <u>true</u> statement would be, "x^2 is non-negative".
21. False. Do not "cancel" terms. Had the "+" been "·" a true statement would result:
$$\frac{A \cdot C}{B \cdot C} = \frac{A}{B}$$
23. False. $0/5$ is defined and equals 0. The expression $5/0$ is undefined.
25. True. A third expression equal to both sides could be $\frac{A + C}{B}$.
27. True. One way to see this is simply choose values for x and y and observe that the square of a number is the same as the square of the number's opposite. Another way to see the validity is to expand each side: $(x - y)^2 - x^2$ $2xy + y^2 = y^2 - 2xy + x^2 = (y - x)^2$.
29. True. $\frac{1}{x}y\frac{1}{z} = \frac{1}{x} \cdot \frac{y}{1} \cdot \frac{1}{z} = \frac{1 \cdot y \cdot 1}{x \cdot 1 \cdot z} = \frac{y}{xz}$

PART II
1. Because $2\pi - 7$ is negative, $|2\pi - 7| = 7 - 2\pi$. Answer is B.
3. The distance between any two points x and y is |x - y|, which must be non-negative. The

Appendix A   260

distance between 5 and $\sqrt{10}$ is <u>exactly</u> $|5 - \sqrt{10}| = 5 - \sqrt{10}$. An approximation is listed for B, but the <u>exact</u> answer is C.

5. $(x + y)^2 = (x + y)(x + y) = x^2 + 2xy + y^2$. Answer is C.

7. Factor each: $10x^3 = 2 \cdot 5 \cdot x^3$, $5x^2 = 5 \cdot x^2$, $25x = 5^2 \cdot x$. All three have 5x in common, and no more. Answer is A.

9. $15x^3 - 15xy^2 = 15x(x^2 - y^2) = 15x(x + y)(x - y)$. Answer is B.

11. Remember to reverse the direction of an inequality when multiplying or dividing both sides by a negative number:

    $-2x \leq 8$

    $-\dfrac{2x}{-2} \geq \dfrac{8}{-2}$

    $x \geq -4$

    This solution includes all <u>real</u> numbers, not just integers as listed in B, greater than or equal to 4. Answer is D.

13. $x^2 - 7x + 12 = 0$

    $(x - 3)(x - 4) = 0$

    $x - 3 = 0$ or $x - 4 = 0$

    $x = 3$ or $x = 4$.   Answer is C.

15. Critical values are $x = -1$ and $x = 2$:

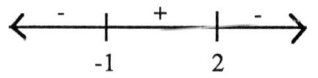

    Choose a number left of -1, say $x = -2$, and substitute into $(x + 1)(2 - x) < 0$:
    $(-2 + 1)(2 - -2) < 0$ => $(-1)(4) < 0$, which is true.

    Choose a number between -1 and 2, say $x = 0$, and substitute into $(x + 1)(2 - x) < 0$:
    $(0 + 1)(2 - 0) < 0$ => $(1)(2) < 0$, which is false.

    Choose a number right of 2, say $x = 3$, and substitute into $(x + 1)(2 - x) < 0$:
    $(3 + 1)(2 - 3) < 0$ => $(4)(-1) < 0$, which is true.

    The solution is all numbers left of -1; all numbers right of 2: $x < -1$ or $x > 2$. Answer is A.

17. $\dfrac{x^2 - 5x + 4}{x + 3} \cdot \dfrac{x^2 + 2x - 3}{x - 4} = \dfrac{(x - 1)(x - 4)}{x + 3} \cdot \dfrac{(x + 3)(x - 1)}{x - 4} = \dfrac{(x - 1)(x - 4)(x + 3)(x - 1)}{(x + 3)(x - 4)}$

    $= (x - 1)(x - 1) = (x - 1)^2$. Answer is D.

19. If we add the equations: $x + y = 4$

    $\underline{x - y = 2}$

    $2x + 0 = 6$ => $2x = 6$ => $x = 3$.

    Answer is C.

Appendix A

A.2 REAL NUMBERS, PAGES 341 - 343

1. **a.** -9 is not a natural or whole number, but is an integer. Answer is I, Q, R. **b.** $\sqrt{30}$ is not a rational number. Answer is Q', R.
3. **a.** 5 is a natural number. Answer is N, W, I, Q, R. **b.** $\sqrt{4} = 2$ which is a natural number. Answer is N, W, I, Q, R.
5. **a.** Since all repeating decimals are rational numbers, $.\overline{4}$ is a rational number. Answer is Q, R. **b.** Interpreting .5252 $\cdots$ as a repeating decimal, $.\overline{52}$, it is a rational number. Answer is Q, R.
7. **a.** Since $.381 = \frac{381}{1000}$, it is a rational number. Answer is Q, R. **b.** Since π is irrational, $\frac{\pi}{6}$ is also irrational. Answer is Q', R.
9. **a.** $\frac{16}{0}$ is undefined. **b.** $\frac{0}{16} = 0$ which is not a natural number but is a whole number. Answer is W, I, Q, R.
11. Since all prime numbers are natural numbers P is shown inside N.

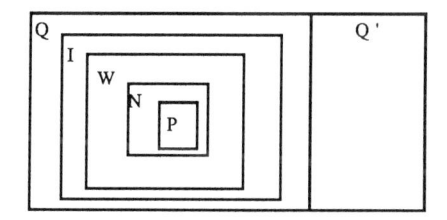

13. All even numbers are whole numbers but not natural numbers; (0 is even but not natural) so E is shown inside W but not entirely inside N.

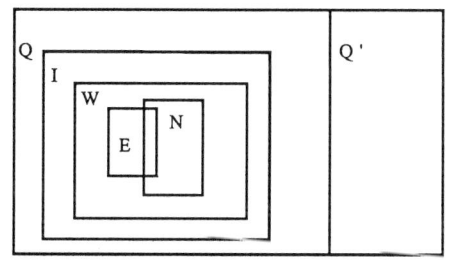

15. False   17. False   19. True   21. True. $|5| = 5$ 23. True   25. False. A counterexample is $9 + 8 = 17$.   27. True   29. True   31. True   33. False
35. False. A counterexample is 9, which is an odd integer but not prime.   37. False
39. True   41. 9   43. 19   45. $-\pi$ 47. Since $2 - \pi$ is negative, $|2 - \pi| = \pi - 2$.
49. Since $\pi - 10$ is negative, $|\pi - 10| = 10 - \pi$.   51. Since $\sqrt{20} - 4$ is positive, $|\sqrt{20} - 4| = \sqrt{20} - 4$.   53. Since $\sqrt{50} - 8$ is negative, $|\sqrt{50} - 8| = 8 - \sqrt{50}$.   55. Since $2\pi - 7$ is negative, $|2\pi - 7| = 7 - 2\pi$.   57. If $x \leq -3$ then $x + 3$ is non-positive. Therefore,

Appendix A

$|x + 3| = -(x + 3)$.   **59.** If $y \geq 5$ then $y - 5$ is non-negative. Therefore, $|y - 5| = y - 5$.
61. If $t < -10$ then $4 + 3t$ is negative. Therefore, $|4 + 3t| = -(4 + 3t)$.   **63.** $|(-5) - (-15)| = 10$
65. $|(-2) - (9)| = 11$ **67.** $|\pi - 2| = \pi - 2$ **69.** $|\sqrt{3} - 2| = 2 - \sqrt{3}$

A.3 ALGEBRAIC EXPRESSIONS, PAGE 348

1. $3 + 2 \cdot 5 = 3 + 10 = 13$ **3.** $5 + 3 \cdot 2 = 5 + 6 = 11$ **5.** $(-6)^2 = (-6)(-6) = 36$
7. $-5^2 = -5 \cdot 5 = -25$ **9.** $-7^2 = -7 \cdot 7 = -49$ **11.** $8 - 5(2 - 7) = 8 - 5(-5) = 8 - -25 = 33$
13. $\dfrac{(-2)6 + (-4)}{-4} + \dfrac{(-3)(-4)}{6} = \dfrac{-12 + -4}{-4} + \dfrac{12}{6} = \dfrac{-16}{-4} + \dfrac{12}{6} = 4 + 2 = 6$
15. $x^2 = (-3)^2 = 9$ **17.** $y^2 = (2)^2 = 4$ **19.** $z^2 = (-1)^2 = 1$
21. $x + y - z = -3 + 2 - (-1) = 0$ **23.** $(xy)^2 + xy^2 + x^2y = ((-3)(2))^2 + (-3)(2)^2 + (-3)^2(2)$
$= (-6)^2 + (-3)(4) + (9)(2) = 36 + (-12) + 18 = 42$
25. $x^2 - w^2(y^2 + z^2) = (-3)^2 - (-4)^2((2)^2 + (-1)^2) = 9 - 16(4 + 1) = 9 - 16(5) = 9 - 80 = -71$
27. $\dfrac{x - z}{y} = \dfrac{(-3) - (-1)}{2} = \dfrac{-3 + 1}{2} = \dfrac{-2}{2} = -1$

Note: For Problems 29 - 41 the middle step is shown for you. According to the directions, you should do this step <u>mentally</u>.

29. a. $(x - 2)(x + 6) = x^2 - 2x + 6x - 12 = x^2 + 4x - 12$
    b. $(x + 5)(x - 4) = x^2 + 5x - 4x - 20 = x^2 + x - 20$
31. a. $(x - 5)(x - 3) = x^2 - 5x - 3x + 15 = x^2 - 8x + 15$
    b. $(x + 3)(x - 4) = x^2 + 3x - 4x - 12 = x^2 - x - 12$
33. a. $(2y + 1)(y - 1) = 2y^2 + y - 2y - 1 = 2y^2 - y - 1$
    b. $(2y - 3)(y - 1) = 2y^2 - 3y - 2y + 3 = 2y^2 - 5y + 3$
35. a. $(2y + 3)(3y - 2) = 6y^2 + 9y - 4y - 6 = 6y^2 + 5y - 6$
    b. $(2y + 3)(3y + 2) = 6y^2 + 9y + 4y + y = 6y^2 + 13y + 6$
37. a. $(x + y)(x - y) = x^2 + xy - xy - y^2 = x^2 - y^2$
    b. $(a + b)(a - b) = a^2 + ab - ab - b^2 = a^2 - b^2$
39. a. $(x + 2)^2 = (x + 2)(x + 2) = x^2 + 2x + 2x + 4 = x^2 + 4x + 4$
    b. $(x - 2)^2 = (x - 2)(x - 2) = x^2 - 2x - 2x + 4 = x^2 - 4x + 4$
41. a. $(a + b)^2 = (a + b)(a + b) = a^2 + ab + ab + b^2 = a^2 + 2ab + b^2$
    b. $(a - b)^2 = (a - b)(a - b) = a^2 - ab - ab + b^2 = a^2 - 2ab + b^2$
43. $(x - y - z) + (2x + y - 3z) = x + 2x - y + y - z - 3z = 3x - 4z$
45. $(x + 3y - 2z) + (3x - 5y + 3z) = x + 3x + 3y - 5y - 2z + 3z = 4x - 2y + z$
47. $(2x - y) - (2x + y) = 2x - y - 2x - y = -2y$
49. $(6x - 4y) - (4x - 6y) = 6x - 4y - 4x + 6y = 2x + 2y$
51. $(x + y - 5) - (2x - 3y + 4) = x + y - 5 - 2x + 3y - 4 = x - 2x + y + 3y - 5 - 4 = -x + 4y - 9$
53. $(6x^2 - 3x + 2) - (2x^2 + 5x + 3) = 6x^2 - 3x + 2 - 2x^2 - 5x - 3 = 6x^2 - 2x^2 - 3x - 5x + 2 - 3$
$= 4x^2 - 8x - 1$

Appendix A 263

55. $(2x^2 - 3x - 5) - (5x^2 - 6x + 4) = 2x^2 - 3x - 5 - 5x^2 + 6x - 4 = 2x^2 - 5x^2 - 3x + 6x - 5 - 4$
    $= -3x^2 + 3x - 9$
57. $(x^2 - x) + (x - 3) - (x - x^2) = x^2 - x + x - 3 - x + x^2 = x^2 + x^2 - x + x - x - 3 = 2x^2 - x - 3$
59. $(x^2 - 7) - (3x + 4) - (x - 2x^2) = x^2 - 7 - 3x - 4 - x + 2x^2 = x^2 + 2x^2 - 3x - x - 7 - 4$
    $= 3x^2 - 4x - 11$
61. $(x + 1)^3 = (x + 1)(x + 1)(x + 1) = (x + 1)(x^2 + 2x + 1) = x^3 + 2x^2 + x + x^2 + 2x + 1$
    $= x^3 + 3x^2 + 3x + 1$
63. $(3x^2 - 5x + 2) + (x^3 - 4x^2 + x - 4) = x^3 + 3x^2 - 4x^2 - 5x + x + 2 - 4 = x^3 - x^2 - 4x - 2$
65. $(3x^2 - 5x + 2) - (5x + 1) = 3x^2 - 5x + 2 - 5x - 1 = 3x^2 - 10x + 1$
67. $(5x + 1)(3x^2 - 5x + 2) = 15x^3 - 25x^2 + 10x + 3x^2 - 5x + 2 = 15x^3 - 22x^2 + 5x + 2$

A.4 FACTORING, PAGES 352 - 353

1. $20xy - 12x = 4x(5y - 3)$ 3. $6x - 2 = 2(3x - 1)$ 5. $xy + xz^2 + 3x = x(y + z^2 + 3)$
7. $a^2 + b^2$ does not factor further   9. $m^2 - 2mn + n^2 = (m - n)(m - n) = (m - n)^2$
11. Both terms have the quantity $(4x - 1)$. Factoring this out gives $(4x - 1)(x + 3)$.
13. $2x^2 + 7x - 15 = (2x\ ?\ )(x\ ?\ )$

    Trying pairs of numbers whose product is -15:

    $(2x - 15)(x + 1) = 2x^2 - 13x - 15$. Middle term does not check.

    $(2x - 1)(x + 15) = 2x^2 + 29x - 15$. Middle term does not check.

    $(2x - 5)(x + 3) = 2x^2 + x - 15$. Middle term does not check.

    $\boxed{(2x - 3)(x + 5)} = 2x^2 + 7x - 15$. Checks.

15. $3x^2 - 5x - 2 = (3x\ ?\ )(x\ ?\ )$. Trying pairs of numbers whose product is -2:

    $(3x - 1)(x + 2) = 3x^2 + 5x - 2$. Middle term does not check.

    $(3x - 2)(x + 1) = 3x^2 + x - 2$. Middle term does not check.

    $\boxed{(3x + 1)(x - 2)} = 3x^2 - 5x - 2$. Checks.

17. First factor out 2. $2x^2 - 10x - 48 = 2(x^2 - 5x - 24) = 2(x\ ?\ )(x\ ?\ )$. Trying pairs of numbers whose product is -24:

    $2(x - 2)(x + 12) = 2(x^2 + 10x - 24)$. Middle term does not check.

    $2(x - 4)(x + 6) = 2(x^2 + 2x - 24)$. Middle term does not check.

    $\boxed{2(x - 8)(x + 3)} = 2(x^2 - 5x - 24)$. Checks.

19. First factor out x^2: $12x^4 + 11x^3 - 15x^2 = x^2(12x^2 + 11x - 15)$. Try $x^2(3x\ ?\ )(4x\ ?\ )$.
    We will try pairs of numbers whose product is -15:

Appendix A   264

$x^2(3x - 1)(4x + 15) = x^2(12x^2 + 41x - 15)$. Does not check.

$x^2(3x - 5)(4x + 3) = x^2(12x^2 - 11x - 15)$. Does not check.

$\boxed{x^2(3x + 5)(4x - 3)} = x^2(12x^2 + 11x - 15)$. Checks.

21. Try $4x^4 - 17x^2 + 4 = (2x^2\ ?\ )(2x^2\ ?\ )$. We will try pairs of numbers whose produce is 4:

$(2x^2 - 1)(2x^2 - 4) = 4x^2 - 10x^2 + 4$. Does not check.

$(2x^{32} - 2)(2x^2 - 2) = 4x^2 - 8x^2 + 4$. Does not check.

There are other possibilities, but let's abandon $(2x^2\ ?\ )(2x^2\ ?\ )$ and work with $(4x^2\ ?\ )(x^2\ ?\ )$. We will again try pairs of numbers whose product is 4:

$(4x^2 - 2)(x^2 - 2) = 4x^4 - 10x^2 + 4$. Does not check.

$(4x^2 - 1)(x^2 - 4) = 4x^2 - 17x^2 + 4$. Checks.

We are not done. Each factor is the difference of two squares:

$(4x^2 - 1)(x^2 - 4) = \boxed{(2x + 1)(2x - 1)(x + 2)(x - 2)}$

23. This is the difference of two squares:

$(x - y)^2 - 1 = ((x - y) + 1)((x - y) - 1) = (x - y + 1)(x - y - 1)$

25. This is the difference of two squares:

$(5a - 2)^2 - 9 = ((5a - 2) + 3)((5a - 2) - 3) = (5a + 1)(5a - 5) = (5a + 1)(5(a - 1))$

$= 5(a - 1)(5a + 1)$

27. Factor out $\dfrac{1}{25}$, then recognize the difference of squares:

$\dfrac{4}{25}x^2 - (x + 2)^2 = \dfrac{1}{25}(4x^2 - 25(x + 2)^2) = \dfrac{1}{25}(2x + 5(x + 2))(2x - 5(x + 2))$

$= \dfrac{1}{25}(7x + 10)(-3x - 10) = -\dfrac{1}{25}(7x + 10)(3x + 10)$

29. This is the difference of squares: $(a + b)^2 - (x + y)^2$

$= ((a + b) + (x + y))((a + b) - (x + y)) = (a + b + x + y)(a + b - x - y)$

31. Try $(2x\ ?\ )(x\ ?\ )$. We'll try pairs of numbers whose product is -6:

$(2x - 1)(x + 6) = 2x^2 + 11x - 6$. Does not check.

$\boxed{(2x - 3)(x + 2)} = 2x^2 + x - 6$. Checks.

33. Work with $(6x\ ?\ )(x\ ?\ )$. Trying pairs of numbers whose product is -8 gives:

$(6x - 4)(x + 2) = 6x^2 + 8x - 8$. Does not check.

Appendix A	265

$(6x + 1)(x - 8) = 6x^2 - 47x - 8$. Does not check.

$\boxed{(6x - 1)(x + 8)} = 6x^2 + 47x - 8$. Checks.

35. Work with $(6x \ ?\ )(x \ ?\ )$. Trying pairs of numbers whose product is 8, we get:

$(6x + 4)(x + 2) = 6x^2 + 16x + 8$. Does not check.

$(6x - 1)(x - 8) = 6x^2 - 49x + 8$. Does not check.

$\boxed{(6x + 1)(x + 8)} = 6x^2 + 49x + 8$. Checks.

37. If we work with $(2x \ ?\ )(2x \ ?\ )$ and try pairs of numbers whose product is -12, we will eventually see no combination checks. Working with $(4x \ ?\ )(x \ ?\ )$, we try pairs of numbers whose product is -12:
$(4x - 1)(x + 12) = 4x^2 + 47x - 12$. Does not check.

$\boxed{(4x - 3)(x + 4)} = 4x^2 + 13x - 12$. Checks.

39. If we work with $(3x \ ?\ )(3x \ ?\ )$ and try pairs of numbers whose product is 12, we will eventually see that no combination checks. Working with $(9x \ ?\ )(x \ ?\ )$, we try pairs of numbers whose product is 12:
$(9x - 1)(x - 12) = 9x^2 - 109x + 12$. Does not check.

$\boxed{(9x - 2)(x - 6)} = 9x^2 - 56x + 12$. Checks.

41. $10x^2 - 9 - x^4 = -(x^4 - 10x^2 + 9)$
$= -(x^2 - 1)(x^2 - 9)$
$= -(x + 1)(x - 1)(x + 3)(x - 3)$

43. $(x^2 - \frac{1}{4})(x^2 - \frac{1}{9}) = \frac{1}{4}(4x^2 - 1)\frac{1}{9}(9x^2 - 1)$
$= \frac{1}{36}(2x + 1)(2x - 1)(3x + 1)(3x - 1)$

45. This is the difference of two squares:
$(x^2 - 3x - 6)^2 - 2^2 = [(x^2 - 3x - 6) + 2][(x^2 - 3x - 6) - 2]$
$= [x^2 - 3x - 4][x^2 - 3x - 8]$
$= [(x - 4)(x + 1)][x^2 - 3x - 8]$
$= (x - 4)(x + 1)(x^2 - 3x - 8)$

47. $2(x + y)^2 - 5(x + y)(a + b) - 3(a + b)^2 = [2(x + y) \ ?\ ][(x + y) \ ?\ ]$

We want to try two quantities whose product is $-3(a + b)^2$.

$= [2(x + y) + 3(a + b)][(x + y) - (a + b)] = 2(x + y)^2 + (x + y)(a + b) - 3(a + b)^2$

which does not check.

Appendix A

$$[2(x + y) + (a + b)][(x + y) - 3(a + b)] = 2(x + y)^2 - 5(x + y)(a + b) - 3(a + b)^2$$

which checks. Removing parentheses, we have $[2x + 2y + a + b][x + y - 3a - 3b]$.

A.5 LINEAR EQUATIONS AND INEQUALITIES, PAGE 356

1. $3x = 5x - 4$ Subtract 5x from both sides.
    $-2x = -4$ Divide both sides by -2.
    $x = 2$

3. $7x + 10 = 5x$ Subtract 7x from both sides.
    $10 = -2x$ Divide both sides by -2.
    $-5 = x$

5. $3x + 22 = 1 - 4x$ Add 4x to both sides.
    $7x + 22 = 1$ Subtract 22 from both sides.
    $7x = -21$ Divide both sides by 7.
    $x = -3$

7. $2x - 13 = 7x + 2$ Subtract 2x from both sides.
    $-13 = 5x + 2$ Subtract 2 from both sides.
    $-15 = 5x$ Divide both sides by 5.
    $-3 = x$

9. $7x + 18 = -2x$ Subtract 7x from both sides.
    $18 = -9x$ Divide both sides by -9.
    $-2 = x$

11. $8x - 3 = 15 - x$ Add x to both sides.
    $9x - 3 = 15$ Add 3 to both sides.
    $9x = 18$ Divide both sides by 9.
    $x = 2$

13. $2(x - 1) = 1 + 3(x - 2)$ Remove parentheses and simplify.
    $2x - 2 = 1 + 3x - 6 = 3x - 5$ Subtract 2x from both sides.
    $-2 = x - 5$ Add 5 to both sides.
    $3 = x$

15. $5 - 2x = 1 + 3(x - 2)$ Remove parentheses and simplify.
    $5 - 2x = 1 + 3x - 6 = 3x - 5$ Add 2x to both sides.
    $5 = 5x - 5$ Add 5 to both sides.
    $10 = 5x$ Divide both sides by 5.
    $2 = x$

17. $3x - 2 < 7$ Add 2 to both sides.
    $3x \leq 9$ Divide both sides by 3.
    $x \leq 3$

19. $4 - 5x > 29$ Subtract 4 from both sides.

Appendix A

 $-5x > 25$ — Divide both sides by -5 (reverse inequality).
 $x < -5$

21. $9x + 7 \geq 5x - 9$ — Subtract 5x from both sides.
 $4x + 7 \geq -9$ — Subtract 7 from both sides.
 $4x \geq -16$ — Divide both sides by 4.
 $x \geq -4$

23. $2(3 - 4x) \leq 30$ — Divide both sides by 2.
 $3 - 4x < 15$ — Subtract 3 from both sides.
 $-4x < 12$ — Divide both sides by -4 (reverse inequality).
 $x > -3$

25. $7 < x + 2 < 11$ — Subtract 2 from all three sides.
 $5 < x < 9$

27. $-2 < x - 1 < 3$ — Add 1 to both sides.
 $-1 < x < 4$

29. $9 < 1 - 2x < 15$ — Subtract 1 from both sides.
 $8 < -2x < 14$ — Divide both sides by -2 (reverse inequality).
 $-4 > x > -7$ — This can be rewritten as:
 $-7 < x < -4$

31. $2(x - 3) - 5x = 3(1 - 2x)$ — Remove parentheses.
 $2x - 6 - 5x = 3 - 6x$ — Combine like terms.
 $-3x - 6 = 3 - 6x$ — Add 6x to both sides.
 $3x - 6 = 3$ — Add 6 to both sides.
 $3x = 9$ — Divide both sides by 3.
 $x = 3$

33. $5(x - 1) + 3(2 - 4x) > 8$ — Remove parentheses.
 $5x - 5 + 6 - 12x > 8$ — Combine like terms.
 $-7x + 1 > 8$ — Subtract 1 from both sides.
 $-7x > 7$ — Divide both sides by -7 (reverse inequality).
 $x < -1$

35. $4(x - 2) + 1 \leq 3(x + 1)$ — Remove parentheses.
 $4x - 8 + 1 \leq 3x + 3$ — Combine similar terms.
 $4x - 7 \leq 3x + 3$ — Subtract 3x from both sides.
 $x - 7 \leq 3$ — Add 7 to both sides.
 $x \leq 10$

37. $2(4 - 3x) > 4(3 - x)$ — Remove parentheses.
 $8 - 6x > 12 - 4x$ — Add 6x to both sides.
 $8 > 12 + 2x$ — Subtract 12 from both sides.
 $-4 > 2x$ — Divide both sides by 2.

Appendix A

$\qquad$ -2 > x

39. $2(3 - 7x) \geq -4 - (5 - x)$ Remove parentheses.
$6 - 14x \geq -4 - 5 + x$ Combine similar terms.
$6 - 14x \geq -9 + x$ Add 14x to both sides.
$6 \geq -9 + 15x$ Add 9 to both sides.
$15 \geq 15x$ Divide both sides by 15.
$1 \geq x$

41. $3(1 - x) - 5(x - 2) = 5$ Remove parentheses.
$3 - 3x - 5x + 10 = 5$ Combine similar terms.
$13 - 8x = 5$ Subtract 13 from both sides.
$-8x = -8$ Divide both sides by -8.
$x = 1$

43. $6(2x + 5) = 4(3x + 1)$ Remove parentheses.
$12x + 30 = 12x + 4$ Subtract 12x from both sides.
$30 = 4$ This is a contradiction.
No solution.

45. $3(2x - 5) = 5(4x - 3)$ Remove parentheses.
$6x - 15 = 20x - 15$ Subtract 6x from both sides.
$-15 = 14x - 15$ Add 15 to both sides.
$0 = 14x$ Divide both sides by 14.
$0 = x$

47. $5(1 - 2x) = 3(x - 4) - (13x - 17)$ Remove parentheses.
$5 - 10x = 3x - 12 - 13x + 17$ Combine like terms.
$5 - 10x = -10x + 5$ This is an identity.
Any real number is a solution.

A.6 QUADRATIC EQUATIONS AND INEQUALITIES, PAGE 362

1. $x^2 + 2x - 15 = 0$ **3.** $x^2 + 7x - 18 = 0$
$(x + 5)(x - 3) = 0$ $(x + 9)(x - 2) = 0$
$x + 5 = 0$ or $x - 3 = 0$ $x + 9 = 0$ or $x - 2 = 0$
$x = -5$ or $x = 3$ $x = -9$ or $x = 2$

5. $10x^2 - 3x - 4 = 0$ **7.** $x^2 + 5x - 6 = 0$
$(2x + 1)(5x - 4) = 0$ $(x - 1)(x + 6) = 0$
$2x + 1 = 0$ or $5x - 4 = 0$ $x - 1 = 0$ or $x + 6 = 0$
$x = \dfrac{-1}{2}$ or $x = \dfrac{4}{5}$ $x = 1$ or $x = -6$

Appendix A

9. $x^2 - 10x + 25 = 0$
   $(x - 5)(x - 5) = 0$
   $x - 5 = 0$
   $x = 5$

11. $12x^2 + 5x - 2 = 0$
    $(3x + 2)(4x - 1) = 0$
    $3x + 2 = 0$ or $4x - 1 = 0$
    $x = -\frac{2}{3}$ or $x = \frac{1}{4}$

13. $5x^2 - 4x + 1 = 0$ $5x^2 - 4x + 1$ does not factor.
    $x = \frac{-b \pm \sqrt{b^2 - 4ac}}{21} = \frac{4 \pm \sqrt{16 - 20}}{10} = \frac{4 \pm \sqrt{-4}}{10}$ No real solution, since $\sqrt{-4}$ is not real.

15. $4x^2 - 5 = 0$
    $x^2 = \frac{5}{4}$
    $x = \pm\sqrt{\frac{5}{4}} = \pm\frac{\sqrt{5}}{2}$

17. $3x^2 = 7x$
    $3x^2 - 7x = 0$
    $x(3x - 7) = 0$
    $x = 0$ or $x = \frac{7}{3}$

19. $3x^2 = 5x + 2$
    $3x^2 - 5x - 2 = 0$
    $(3x + 1)(x - 2) = 0$
    $3x + 1 = 0$ or $x - 2 = 0$
    $x = -\frac{1}{3}$ or $x = 2$

21. Critical points are $x = 6$ and $x = 2$.
    Pick a number to the left of 2, say $x = 0$: $(0 - 6)(0 - 2) \geq 0$ true.
    Pick a number between 2 and 6, say $x = 4$: $(4 - 6)(4 - 2) \geq 0$ false.
    Pick a number to the right of 6, say $x = 7$: $(7 - 6)(7 - 2) \geq 0$ true.
    The answer is $x \leq 2$ or $x \geq 6$.

23. Critical points are $x = 0$ and $x = -3$.
    Pick a number to the left of -3, say $x = -4$: $(-4)(-4 + 3) < 0$ false.
    Pick a number between -3 and 0, say $x = -1$: $(-1)(-1 + 3) < 0$ true.
    Pick a number to the right of 0, say $x = 1$: $(1)(1 + 3) < 0$ false.
    The answer is $-3 < x < 0$.

25. Critical points are $x = -2$ and $x = 8$.
    Pick a number to the left of -2, say $x = -3$: $(-3 + 2)(8 - (-3)) \leq 0$ true.
    Pick a number between -2 and 8, say $x = 0$: $(0 + 2)(8 - 0) \leq 0$ false.
    Pick a number to the right of 8, say $x = 9$: $(9 + 2)(8 - 9) \leq 0$ true.
    The answer is $x \leq -2$ or $x \geq 8$.

27. Critical points are $1/3$ and 4.
    Pick a number to the left of $1/3$, say $x = 0$: $(1 - 3(0))(0 - 4) < 0$ true.
    Pick a number between $1/3$ and 4, say $x = 1$: $(1 - 3(1))(1 - 4) < 0$ false.
    Pick a number to the right of 4, say $x = 5$: $(1 - 3(5))(5 - 4) < 0$ true.
    The answer is $x < 1/3$ or $x > 4$.

Appendix A

29. $x^2 - 9 \geq 0$

   $(x + 3)(x - 3) \geq 0$. Critical points are $x = -3$ and $x = 3$.
   Pick a number to the left of -3, say $x = -4$: $(-4 + 3)(-4 - 3) \geq 0$ true.
   Pick a number between -3 and 3, say $x = 0$: $(0 + 3)(0 - 3) \geq 0$ false.
   Pick a number to the right of 3, say $x = 4$: $(4 + 3)(4 - 3) \geq 0$ true.
   The answer is $x \leq -3$ or $x \geq 3$.

31. Since there are no solutions to $x^2 + 9 = 0$, there are no critical points.
   Pick any number, say $x = 1$: $(1)^2 + 9 \leq 0$ false. There is no solution.

33. $x^2 - x - 6 > 0$

   $(x - 3)(x + 2) > 0$. Critical points are $x = 3$ and $x = -2$.
   Pick a number to the left of -2, say $x = -3$: $(-3 - 3)(-3 + 2) > 0$ true.
   Pick a number between -2 and 3, say $x = 0$: $(0 - 3)(0 + 2) > 0$ false.
   Pick a number to the right of 3, say $x = 4$: $(4 - 3)(4 + 2) > 0$ true.
   The answer is $x < -2$ or $x > 3$.

35. $5x - 6 \geq x^2$

   $0 \geq x^2 - 5x + 6$
   $0 \geq (x - 2)(x - 3)$. Critical points are $x = 2$ and $x = 3$.
   Pick a number to the left of 2, say $x = 0$: $0 \geq (0 - 2)(0 - 3)$ false.
   Pick a number between 2 and 3, say $x = 2.5$: $0 \geq (2.5 - 2)(2.5 - 3)$ true.
   Pick a number to the right of 3, say $x = 4$: $0 \geq (4 - 2)(4 - 3)$ false.
   The answer is $2 \leq x \leq 3$.

37. $5 - 4x \geq x^2$

   $0 \geq x^2 + 4x - 5$
   $0 \geq (x + 5)(x - 1)$. Critical points are $x = -5$ and $x = 1$.
   Pick a number to the left of -5, say $x = -6$: $0 \geq (-6 + 5)(-6 - 1)$ false.
   Pick a number between -5 and 1, say $x = 0$: $0 \geq (0 + 5)(0 - 1)$ true.
   Pick a number to the right of 1, say $x = 2$: $0 \geq (2 + 5)(2 - 1)$ false.
   The answer is $-5 \leq x \leq 1$.

39. Since $x^2 - 2x - 2$ does not factor, use the quadratic formula to get critical points:

   $x = \dfrac{-b \pm \sqrt{b^2 - 4ac}}{2a} = \dfrac{2 \pm \sqrt{4 + 8}}{2} = 1 \pm \sqrt{3}$.

   Pick a number to the left of $1 - \sqrt{3}$, say $x = -2$: $(-2)^2 - 2(-2) - 2 < 0$ false.
   Pick a number between $1 - \sqrt{3}$ and $1 + \sqrt{3}$, say $x = 0$: $0^2 - 2(0) - 2 < 0$ true.
   Pick a number to the right of $1 + \sqrt{3}$, say $x = 3$: $3^2 - 2(3) - 2 < 0$ false.
   The answer is $1 - \sqrt{3} < x < 1 + \sqrt{3}$.

Appendix A

A.7 RATIONAL EXPRESSIONS, PAGE 365

1. $\dfrac{2}{x+y} + 3 = \dfrac{2}{x+y} + \dfrac{3(x+y)}{x+y} = \dfrac{2+3x+3y}{x+y}$

3. $\dfrac{x^2 - y^2}{2x+2y} = \dfrac{(x+y)(x-y)}{2(x+y)} = \dfrac{x-y}{2}$

5. $\dfrac{3x^2 - 4x - 4}{x^2 - 4} = \dfrac{(3x+2)(x-2)}{(x+2)(x-2)} = \dfrac{3x+2}{x+2}$

7. $\dfrac{3}{x+y} + \dfrac{5}{2x+2y} = \dfrac{3}{x+y} \cdot \dfrac{2}{2} + \dfrac{5}{2(x+y)} = \dfrac{6}{2(x+y)} + \dfrac{5}{2(x+y)} = \dfrac{11}{2(x+y)}$

9. Any non-zero number raised to the power of 0 is 1.

11. $(x^2 - 36) \cdot \dfrac{3x+1}{x+6} = (x+6)(x-6) \cdot \dfrac{3x+1}{x+6} = (x-6)(3x+1) = 3x^2 - 17x - 6$

13. $\dfrac{x+3}{x} + \dfrac{3-x}{x^2} = \dfrac{x(x+3)}{x^2} + \dfrac{3-x}{x^2} = \dfrac{x^2 + 3x + 3 - x}{x^2} = \dfrac{x^2 + 2x + 3}{x^2}$

15. $\dfrac{x}{x-1} + \dfrac{x-3}{1-x} = \dfrac{x}{x-1} + \dfrac{3-x}{x-1} = \dfrac{3}{x-1}$

17. $\dfrac{2x+3}{x^2} + \dfrac{3-x}{x} = \dfrac{2x+3}{x^2} + \dfrac{x(3-x)}{x^2} = \dfrac{2x+3}{x^2} + \dfrac{3x - x^2}{x^2} = \dfrac{3 + 5x - x^2}{x^2}$

19. $\dfrac{x}{y} + 2 + \dfrac{y}{x} = \dfrac{x^2}{xy} + \dfrac{2xy}{xy} + \dfrac{y^2}{xy} = \dfrac{x^2 + 2xy + y^2}{xy} = \dfrac{(x+y)^2}{xy}$

21. $\dfrac{x}{y} - 2 + \dfrac{y}{x} = \dfrac{x^2}{xy} - \dfrac{2xy}{xy} + \dfrac{y^2}{xy} = \dfrac{x^2 - 2xy + y^2}{xy} = \dfrac{(x-y)^2}{xy}$

23. $\dfrac{1}{3xy^2} + xy + \dfrac{1}{x^3y} = \dfrac{x^2}{3x^3y^2} + \dfrac{3x^4y^3}{3x^3y^2} + \dfrac{3y}{3x^3y^2} = \dfrac{x^2 + 3x^4y^3 + 3y}{3x^3y^2}$

25. $\dfrac{4x-12}{x^2-49} \div \dfrac{18-2x^2}{x^2-4x-21} = \dfrac{4x-12}{x^2-49} \cdot \dfrac{x^2-4x-21}{18-2x^2} = \dfrac{4(x-3)}{(x+7)(x-7)} \cdot \dfrac{(x-7)(x+3)}{-2(x+3)(x-3)}$

$= \dfrac{-2}{(x+7)}$

27. $\dfrac{1}{x^2+1} - \dfrac{x^2}{x^2+1} = \dfrac{1-x^2}{x^2+1}$

29. $\dfrac{4x^2}{x^4 - 2x^3} + \dfrac{8}{4x - x^3} - \dfrac{-4}{x+2} = \dfrac{4x^2}{x^3(x-2)} + \dfrac{8}{-x(x+2)(x-2)} + \dfrac{4}{x+2}$

$= \dfrac{4}{x(x-2)} - \dfrac{8}{x(x+2)(x-2)} + \dfrac{4}{x+2} = \dfrac{4(x+2)}{x(x+2)(x-2)} - \dfrac{8}{x(x+2)(x-2)} + \dfrac{4x(x-2)}{x(x+2)(x-2)}$

$= \dfrac{4(x-1)}{(x+2)(x-2)}$

Appendix A

A.8 SYSTEMS OF EQUATIONS, PAGES 368 - 369

1. Graph both lines on the same axes:
They intersect at (3, 2), so the solution is x = 3 and y = 2.

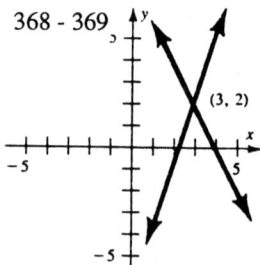

3. Graph both lines on the same axes:
They intersect at (6, -3), so the solution is x = 6 and y = -3.

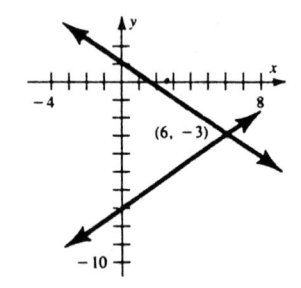

5. Graph both lines on the same axes:
They are the same line, which means the system is dependent.

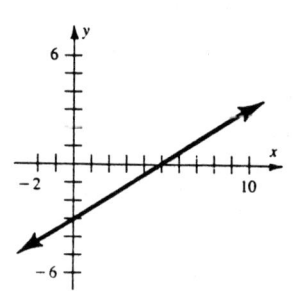

7. Since a = 3b - 7, substitute this for a in the other equation:
3b - 7 = -2b + 8 => 5b = 15 => b = 3. Since a = 3b - 7, a = 3(3) - 7 = 2. The solution is a = 2 and b = 3.

9. Since $m = \frac{2}{3} n - 7$, substitute this for m in the other equation:
$2n + 3(\frac{2}{3} n - 7) = 3$ => 2n + 2n - 21 = 3 => 4n - 21 = 3 => n = 6.
Since $m = \frac{2}{3} n - 7$, $m = \frac{2}{3}(6) - 7 = -3$. The solution is m = -3 and n = 6.

11. Substitute $\frac{2}{3}p - 3$ for q in the first equation:
$2p - 3(\frac{2}{3} p - 3) = 9$ => 2p - 2p + 9 = 9 => 9 = 9, which is always true. The system is dependent.

Appendix A

13. Add the equations: $\begin{cases} x + y = 2 \\ 2x - y = 1 \end{cases}$

    $3x = 3 \implies x = 1.$

    Substitute $x = 1$ into the first equation: $x + y = 2 \implies 1 + y = 2 \implies y = 1$. The solution is $x = 1$ and $y = 1$.

15. Multiply the first equation by 3, and multiply the second equation by 4. Then add the resulting equations:

    $\begin{cases} 3x - 4y = 3 \implies 9x - 12y = 9 \\ 5x + 3y = 5 \implies 20x + 12y = 20 \end{cases}$

    $29x = 29 \implies x = 1.$

    Now substitute $x = 1$ into the first equation: $3x - 4y = 3 \implies 3 - 4y = 3 \implies y = 0$. The solution is $x = 1$ and $y = 0$.

17. Multiply the first equation by 2, and then add the resulting equations:

    $\begin{cases} 7x + y = 5 \implies 14x + 2y = 10 \\ 14x - 2y = -2 \implies 14x - 2y = -2 \end{cases}$

    $28x = 8 \implies x = \dfrac{2}{7}$

    Substitute $x = \dfrac{2}{7}$ into the first equation: $7x + y = 5 \implies 7(\dfrac{2}{7}) + y = 5 \implies y = 3$. The solution is $x = \dfrac{2}{7}$ and $y = 3$.

19. We will use the linear combination method. Multiply the second equation by 2, and then add:

    $\begin{cases} 5x + 4y = 5 \implies 5x + 4y = 5 \\ 15x - 2y = 8 \implies 30x - 4y = 16 \end{cases}$

    $35x = 21 \implies x = \dfrac{3}{5}$

    Substitute $x = \dfrac{3}{5}$ into the first equation: $5(\dfrac{3}{5}) + 4y = 5 \implies 4y = 2 \implies y = \dfrac{1}{2}$. The solution is $x = \dfrac{3}{5}$ and $y = \dfrac{1}{2}$.

21. Since $y = \dfrac{1}{2}x + 5$, substitute this into the first equation:

    $4x - 2y = -28 \implies 4x - 2(\dfrac{1}{2}x + 5) = -28 \implies 4x - x - 10 = -28 \implies 3x = -18 \implies x = -6$

    Now substitute $x = -6$ into $y = \dfrac{1}{2}x + 5$: $y = \dfrac{1}{2}(-6) + 5 = 2$. The solution is $x = -6$ and $y = 2$.

Appendix A

23. Substitute $y = 2x - 1$ into the second equation:
$2x - 1 = -3x - 9 \Rightarrow 5x = -8 \Rightarrow x = -\frac{8}{5}$. Now substitute $x = -\frac{8}{5}$ into the first equation: $y = 2(-\frac{8}{5}) - 1 \Rightarrow y = 2(-\frac{8}{5}) - 1 = -\frac{21}{5}$. The solution is $x = -\frac{8}{5}$ and $y = -\frac{21}{5}$.

25. Multiply the first equation by -3, and multiply the second equation by 2. Then add the resulting equations:

$$\begin{cases} 2x - 3y - 5 = 0 \Rightarrow -6x + 9y + 15 = 0 \\ 3x - 5y + 2 = 0 \Rightarrow 6x - 10y + 4 = 0 \end{cases}$$
$$-y + 19 = 0 \Rightarrow y = 19.$$

Now substitute $y = 19$ into the first equation: $2x - 3(19) - 5 = 0 \Rightarrow x = 31$. The solution is $x = 31$ and $y = 19$.

27. Multiply the first equation by b, and multiply the second equation by -a:

$$\begin{cases} ax + by = 1 \Rightarrow abx + b^2y = b \\ bx + ay = 0 \Rightarrow -abx - a^2y = 0 \end{cases}$$
$$(b^2 - a^2)y = b \Rightarrow y = \frac{b}{b^2 - a^2}$$

Multiply the first equation by a, and the second equation by -b:

$$\begin{cases} ax + by = 1 \Rightarrow a^2x + aby = a \\ bx + ay = 0 \Rightarrow -b^2x - aby = 0 \end{cases}$$
$$(a^2 - b^2)x = a \Rightarrow x = \frac{a}{a^2 - b^2}$$

29. Divide the top equation by -.06, and then add:

$$\begin{cases} .12x + .06y = 228 \Rightarrow -2x - y = -3800 \\ x + y = 2000 \qquad \Rightarrow x + y = 2000 \end{cases}$$
$$-x = -1800 \Rightarrow x = 1800$$

Substitute this into $x + y = 2000 \Rightarrow 1800 + y = 2000 \Rightarrow y = 200$.

Appendix A

A.9 CHAPTER REVIEW, PAGE 370

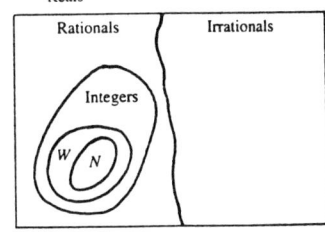

1. Reals / Rationals / Irrationals / Integers / W, N (Venn diagram)

3. $|\sqrt{10} - 5| = 5 - \sqrt{10}$

5. $-(-6)^2 + 5(-6)(3 - 2(-3)) = -36 + 5(-6)(9) = -306$

7. $(x + y)(x + y) = x^2 + 2xy + y^2$

9. Try $(3x\ ?\ )(2x\ ?\ )$. Use pairs of numbers whose product is -5:
   $(3x + 1)(2x - 5) = 6x^2 - 13x - 5$. Does not check.
   $(3x + 5)(2x - 1) = 6x^2 + 7x - 5$. Does not check.
   Abandon $(3x\ ?\ )(2x\ ?\ )$, and try $(6x\ ?\ )(x\ ?\ )$. Use pairs of numbers whose product is -5:
   $(6x - 5)(x + 1) = 6x^2 + x - 5$. Does not check.
   $\boxed{(6x + 1)(x - 5)} = 6x^2 - 29x - 5$. Checks.

11. $4 - 3x = 2(5 - 2x)$ Remove parentheses.
    $4 - 3x = 10 - 4x$ Add 4x.
    $4 + x = 10$ Subtract 4.
    $x = 6$

13. $2(x + 1) + 5 < 3(1 - x)$ Remove parentheses.
    $2x + 2 + 5 < 3 - 3x$ Add 3x and simplify.
    $5x + 7 < 3$ Subtract 7 and divide by 5.
    $x < -\dfrac{4}{5}$

15. $x = 5x^2$
    $x - 5x^2 = 0$
    $x(1 - 5x) = 0$
    $x = 0$ or $1 - 5x = 0$
    $x = 0$ or $x = \dfrac{1}{5}$

17. The critical points are $x = 1$ and $x = 2/3$. Substitute a number to the left of $2/3$, say $x = 0$, into the inequality: $(1 - 0)(3(0) - 2) > 0$, which is false. Substitute a number between $2/3$ and 1, say $x = 5/6$, into the inequality: $(1 - \dfrac{5}{6})(3(\dfrac{5}{6}) - 2) > 0$, which is true. Finally, substitute a number to the right of 1, say $x = 2$, into the inequality: $(1 - 2)(3(2) - 2) > 0$, which is false. The solution is all numbers between $\dfrac{2}{3}$ and $1 \Rightarrow \dfrac{2}{3} < x < 1$.

19. $\dfrac{7x}{30} - \dfrac{y}{12} = \dfrac{14x}{60} - \dfrac{5y}{60} = \dfrac{14x - 5y}{60}$

Appendix A

A.10 STUDENT'S TEST REVIEW AND ADDITIONAL PRACTICE OBJECTIVES

The material of this chapter is reviewed in the following list of objectives. After each objective there are some practice questions. Answers to these problems immediately follow. Detailed solutions are given for every third problem. For a sample test select the first question of each set and check your answers. Additional practice is given by the other questions in each set. If you are having trouble with a particular type of problem, or if you want additional practice, look back at the indicated section in the test.

[A.2] **Objective 1:** *Classify numbers as natural, whole, integer, rational, irrational or real.*

1. Classify the following numbers by listing the set(s) into which they fall:

   $-8, \dfrac{5}{6}, .2\overline{3}, \sqrt{169}, \dfrac{3}{0}$

2. Draw a diagram showing how the integers, rationals, and reals are related.

Objective 2: *Find the absolute value of a number or an expression.*

3. $-|-5|$
4. $|\sqrt{1000} - 35|$ (do not approximate)
5. $|x - 4|$ if $x < -2$

Objective 3: *Find the distance between pairs of points on a number line.*

6. $A(-6)$ and $B(11)$
7. $C(2\pi)$ and $D(-4)$
8. $E(4)$ and $F(\sqrt{20})$

[A.3] **Objective 4:** *Simplify numerical expressions.*

9. a. -8^2 b. $(-8)^2$ c. $-(-8)^2$ d. -1^0

10. $6 + 5 \cdot 2 - 8 + 4 + 2$
11. $\dfrac{6 + 4(-3)}{-2}$
12. $\dfrac{12 - 2(3)}{-4^2}$

Objective 5: *Evaluate algebraic expressions.* Let $x = -1, y = 2,$ and $z = -3$.

13. $x - (yz - 2z)$
14. $xy - yz + 4xz$
15. a. $-x^2$ b. $(-x)^2$
16. a. $(-y)^2$ b. $-y^2$

Objective 6: *Mentally multiply binomials using FOIL.*

17. a. $(x - 2)(x + 7)$ b. $(x + 3)(x - 3)$
18. a. $(x + 3)^2$ b. $(2x - 1)^2$
19. a. $(x + 2y)(x - y)$ b. $(x - 3y)(x + 3y)$
20. a. $(2x - 3)(x + 5)$ b. $(3x - 1)(2x + 3)$

Objective 7: *Simplify algebraic expressions.*

21. $(x - y - z) + (2x + y - 3z)$
22. $(5x^2 + 3x - 5) - (2x^2 + 2x + 3)$
23. $(3 - x^2) - (5 - x) - (3x - 2)$
24. $(x + 2)(2x^2 + 3x + 2)$

[A.4] **Objective 8:** *Factor polynomials.*

25. $x^2 - 5x + 6$
26. $x^2 - 9x + 14$
27. $25 - x^2$
28. $4x^2 - 12x + 8$
29. $9x^4 - 40x^2 + 16$
30. $4x^4 - 13x^3 + 9x^2$

Appendix A

[A.5] Objective 9: *Solve linear equations.*

31. $90 = 5x - 10$
32. $6 - 5x = 5 - 9x$
33. $3 - 4x = 3(5 - 2x)$
34. $5(x - 5) + 2 = 3(x - 3)$

Objective 10: *Solve linear inequalities.*

35. $3x - 5 > 16$
36. $9 - x < -4$
37. $x < 3(2 + x)$
38. $2(x - 2) + 3 \leq 5(x + 1)$

Objective 11: *Solve compound inequalities.*

39. $-8 \leq x + 5 \leq -3$
40. $5 \leq x + 1 < 15$
41. $6 < -x < 10$
42. $-1 < 5 - 2x \leq 1$

[A.6] Objective 12: *Solve quadratic equations.*

43. $(3x - 1)(5x + 2) = 0$
44. $(2x + 1)(x - 4) = 11$
45. $6x^2 + 19x = -15$
46. $x^2 - 6x + 1 = 0$
47. $x^2 - 6x + 7 = 0$
48. $2x^2 + 1 = 2x$

Objective 13: *Solve quadratic inequalities.*

49. $(x - 5)(x - 7) < 0$
50. $x^2 - 4x \geq 0$
51. $6x^2 > 11x + 10$
52. $x^2 - 2x + 3 < 0$

[A.7] Objective 14: *Simplify rational expressions.*

53. $\dfrac{2x}{15} - \dfrac{y}{12}$
54. $\dfrac{5x}{x - 1} + 1$
55. $\dfrac{3x + 2}{9x^2 - 6x + 1} + \dfrac{1}{3x - 1}$
56. $\dfrac{x^2 - 9}{x - 2} + \dfrac{x^2 + x - 6}{x^2 - 4}$

[A.8] Objective 15: *Solve systems of linear equations by graphing.*

57. $\begin{cases} x + y = 4 \\ x - y = 0 \end{cases}$
58. $\begin{cases} y = 2x + 1 \\ y = 10 - x \end{cases}$
59. $\begin{cases} 2x - 3y = -2 \\ 4x - 5y = 0 \end{cases}$
60. $\begin{cases} 5x + 2y = 1 \\ 8x + 3y = 3 \end{cases}$

Objective 16: *Solve systems of linear equations by substitution.*

61. $\begin{cases} y = 2x + 1 \\ x + y = 10 \end{cases}$
62. $\begin{cases} y = x - 8 \\ x + 2y + 10 = 0 \end{cases}$
63. $\begin{cases} x + 3y = 7 \\ 2x + 2y = 7 \end{cases}$
64. $\begin{cases} x + 3y = 1 \\ x - 2y = -9 \end{cases}$

Objective 17: *Solve systems of linear equations by linear combinations.*

Appendix A

65. $\begin{cases} x + y = 7 \\ x - y = 3 \end{cases}$ 66. $\begin{cases} 3x + 2y = 5 \\ 2x + y = 6 \end{cases}$

67. $\begin{cases} x - 3y = -2 \\ 5x - 9y = -5 \end{cases}$ 68. $\begin{cases} 5x + y = 4 \\ 9x - 5y = -3 \end{cases}$

ANSWERS TO STUDENT'S TEST REVIEW AND PRACTICE QUESTIONS

1.

Number	Natural	Whole	Set Integer	Rational	Irrational	Real
-8	No	No	Yes	Yes	No	Yes
$\frac{5}{6}$	No	No	No	Yes	No	Yes
$.\overline{23}$	No	No	No	Yes	No	Yes
$\sqrt{169}$	Yes	Yes	Yes	Yes	Yes	Yes
$\frac{3}{0}$	No	No	No	No	No	No

2. R ⊃ Q ⊃ I   **3.** -5   **4.** $35 - \sqrt{1000}$ **5.** $4 - x$ **6.** 17

7. $2\pi + 4$ **8.** $\sqrt{20} - 4$ **9. a.** -64   **b.** 64   **c.** -64   **d.** -1   **10.** 14   **11.** 3
12. $-\frac{3}{8}$ **13.** -1   **14.** 16   **15. a.** -1   **b.** 1   **16. a.** 4   **b.** -4
17. a. $x^2 + 5x - 14$ **b.** $x^2 - 9$ **18. a.** $x^2 + 6x + 9$ **b.** $4x^2 - 4x + 1$
19. a. $x^2 + xy - 2y^2$ **b.** $x^2 - 9y^2$ **20. a.** $2x^2 + 7x - 15$ **b.** $6x^2 + 7x - 3$
21. $3x - 4z$ **22.** $3x^2 + x - 8$ **23.** $-x^2 - 2x$ **24.** $2x^3 + 7x^2 + 8x + 4$
25. $(x - 2)(x - 3)$ **26.** $(x - 2)(x - 7)$ **27.** $(5 + x)(5 - x)$ **28.** $4(x - 1)(x - 2)$
29. $(3x + 2)(3x - 2)(x + 2)(x - 2)$ **30.** $x^2(4x - 9)(x - 1)$ **31.** $x = 20$ **32.** $x = -\frac{1}{4}$
33. $x = 6$ **34.** $x = 7$ **35.** $x > 7$ **36.** $x > 13$ **37.** $x > -3$ **38.** $x \geq -2$
39. $-13 \leq x \leq -8$ **40.** $4 \leq x < 14$ **41.** $-10 < x < -6$ **42.** $2 \leq x < 3$
43. $x = \frac{1}{3}$ or $-\frac{2}{5}$ **44.** $x = 5$ or $-\frac{3}{2}$ **45.** $x = -\frac{5}{3}$ or $-\frac{3}{2}$ **46.** $x = 3 \pm 2\sqrt{2}$
47. $x = 3 \pm \sqrt{2}$ **48.** No real solution   **49.** $5 < x < 7$ **50.** $x \leq 0$ or $x \geq 4$

Appendix A

51. $x < -\frac{2}{3}$ or $x > \frac{5}{2}$ **52.** No real solution **53.** $\frac{8x - 5y}{60}$ **54.** $\frac{6x - 1}{x - 1}$

55. $\frac{6x + 1}{(3x - 1)^2}$ **56.** $\frac{(x - 3)(x + 2)}{x - 2}$

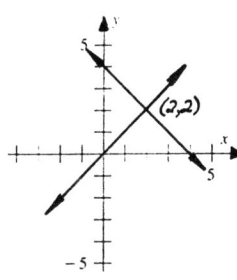

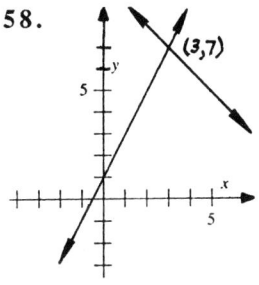

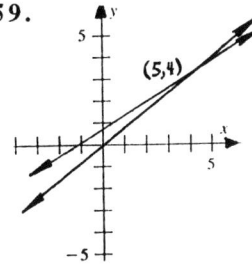

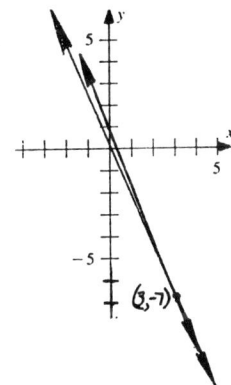

57. **58.** **59.**

60. **61.** (3, 7) **62.** (2, -6) **63.** (1, 2)

64. (-5, 2) **65.** (5, 2) **66.** (7, -8) **67.** $(\frac{1}{2}, \frac{5}{6})$ **68.** $(\frac{1}{2}, \frac{3}{2})$